# STATISTICS
## A FRESH APPROACH

# STATISTICS
## A FRESH APPROACH
### SECOND EDITION

**DONALD H. SANDERS**

M. J. Neeley School of Business
Texas Christian University

**A. FRANKLIN MURPH**

M. J. Neeley School of Business
Texas Christian University

**ROBERT J. ENG**

Babson College

**McGRAW-HILL BOOK COMPANY**

New York  St. Louis  San Francisco  Auckland  Bogotá  Hamburg
Johannesburg  London  Madrid  Mexico  Montreal  New Delhi  Panama  Paris
São Paulo  Singapore  Sydney  Tokyo  Toronto

**STATISTICS: A FRESH APPROACH**

Copyright © 1980, 1976 by McGraw-Hill, Inc. All rights reserved. Printed in the United States of America. No part of this publication may be reproduced, stored in a retrieval system, or transmitted, in any form or by any means, electronic, mechanical, photocopying, recording, or otherwise, without the prior written permission of the publisher.

1234567890DODO7832109

This book was set in Times Roman by Ruttle, Shaw & Wetherill, Inc. The editors were Charles E. Stewart and Edwin Hanson; the designer was Anne Canevari Green; the production supervisor was Dominick Petrellese. The drawings were done by Danmark & Michaels, Inc.
R. R. Donnelley & Sons Company was printer and binder.

Library of Congress Cataloging in Publication Data

Sanders, Donald H
    Statistics.

    Includes index.
    1.  Statistics.  I.  Murph, A. Franklin, joint
author.  II.  Eng, Robert J., joint author.
III.  Title.
QA276.12.S26    1980      519.5      79-14930
ISBN 0-07-054667-3

**TO MY WIFE**
—A.F.M.

**TO MY PARENTS**
—R.J.E.

**TO THOSE WHO OPEN THIS BOOK WITH DISMAY**
—D.H.S.

# CONTENTS

# PART 2
# SAMPLING IN THEORY AND PRACTICE

# PART 3
# COPING WITH CHANGE

# PART 4
# CONCLUDING TOPICS

# PREFACE

> If I had only one day left to live,
> I would live it in my statistics
> class—it would seem so much longer.
>
> —Quote in a university student calendar

It's that time again—time to attempt once more to present the subject of statistics in an interesting (and sometimes humorous) way so that a period spent on the subject will not seem to students to represent the eternity suggested by the above quote.

Actually, our experience has convinced us that most students for which this book was written probably do accept the fact that an educated citizen must have an understanding of basic statistical tools to function in a modern world that is becoming increasingly dependent on quantitative information. But most of them have never placed the solving of mathematical problems at the top of their list of favorite things to do. In fact, many of them probably don't care much for math (may even be terrified of the subject and consider it a foreign language) and have probably heard numerous disturbing rumors about statistics courses.

A motivating force behind the preparation of this text is the distinct possibility that the student misgivings and apathy implicit in the introductory quote are related in some way to the unfortunate fact that many existing statistics books are rigorously written, mathematically profound, precisely detailed— and excruciatingly dull!

Therefore, you will find that the *main difference between this text and many others* is that an attempt has been made here to (1) communicate with students rather than lecture to them, (2) present material in a rather relaxed and informal way without omitting the more important concepts, (3) recapture student attention with occasional quotes, ridiculous names, and unlikely situations of a humorous nature, and (4) utilize an intuitive and/or commonsense approach to develop concepts whenever possible. In short, this book is written for students rather than statisticians, and its intent is to convince them that the study of statistics can be a lively, interesting, and rewarding experience.

More specifically, *the purpose of this book is* to introduce students at an early stage in a college program to many of the important concepts and procedures they are likely to need in order (1) to evaluate such daily inputs as organizational reports, newspaper and magazine articles, and radio and television commentaries, (2) to improve their ability to make better decisions over a wide range of topics, and (3) to improve their ability to measure and cope with changing conditions both at home and on the job. But *the purpose of this book is not* to produce professional statisticians, because it is recognized that most users of this text will be consumers rather than producers of statistical information. Therefore, the emphasis has been placed on explaining statistical procedures and interpreting the resulting conclusions. However, the *mathematical demands are very modest—no college-level math background is required or assumed.* (The treatment of probability and probability distributions, for example, is limited to the bare essentials.)

**TEXT ORGANIZATION AND REVISION FEATURES**

This edition is organized into *four parts*. Each of these parts is introduced by a brief essay that explains the purpose of the part and identifies the chapters included in the part. Each chapter, in turn, is introduced with *new opening pages* containing *learning objectives* and a *chapter outline*. Problems and discussion questions are found at the end of each chapter and in the *self-testing review sections* that are included in most chapters. These problems and questions support the learning objectives of the chapter. A brief summary of the parts of the text, along with some more specific comments about the revisions made in this edition, is presented below.

**Part 1: Descriptive Statistics**

The subject of the *four chapters* in Part 1 is *descriptive* statistics. After introductory materials are introduced in Chapter 1, and after examples of how statistical methods have been improperly used are discussed in Chapter 2, Chapters 3 and 4 deal with measures of central tendency and dispersion. New examples of bias and updated tables and charts have been included in Chapter 2. Greater emphasis is given in Chapter 3 to the real lower and upper class limits in a frequency distribution; a new section on summarizing qualitative data has been added; and additional problems have been included. Alternative formulas and additional problems have been inserted into Chapter 4.

**Part 2: Sampling in Theory and Practice**

*Statistical inference* concepts are considered in the *seven chapters* of Part 2. After the foundations for the material on sampling applications have been presented in Chapters 5 and 6, Chapter 7 shows how sample data may be used in *estimating* population parameters. Chapters 8 through 11 then focus on *hypothesis-testing* procedures. The following changes and additions have been made in this edition:

1  New problems have been added to Chapter 5.

2  A simplified approach has been used to compute $\sigma_{\bar{x}}$ in Chapter 7; material on determining sample size has been moved into the chapter, and a new self-testing review section and end-of-chapter problems have been added.

3  *Chapter 9*—"Testing Hypotheses and Making Decisions: Two-Sample Procedures"—*is new for this edition.*

**4** *Chapter 10 —* "Comparison of Three or More Sample Means: Analysis of Variance" *— is also new for this edition.*

**5** *And a third new chapter is Chapter 11 —* "Comparison of Several Sample Percentages: Chi-Square Analysis."

**Part 3: Coping with Change**

The *three chapters* in Part 3 focus attention on the *measurement and prediction of change.* Chapter 12 looks at how index number procedures may be used to measure changes in economic conditions, while Chapters 13 and 14 show how time-series analysis and regression and correlation techniques may be used in forecasting. Some of the revisions in this part include the updating of examples, charts, and problems in Chapters 12 and 13 and the incorporation of recent changes made in government-produced price indexes in Chapter 12.

**Part 4: Concluding Topics**

Several additional quantitative tools and techniques available to the statistician and decision maker that have not been considered in the preceding 14 chapters have been included in this part. For example, *Chapter 15, a new chapter for this edition, deals with some nonparametric statistical methods.* Although this chapter could logically have been included in Part 2, we have elected to place it in this final part. In addition, the last chapter (Chapter 16) is a brief essay that describes a few of the tools and techniques that cannot be considered in any detail.

**USE OF THIS BOOK**

This book is written for use in an early one- or two-term course in statistics. As noted earlier, no college-level math background is required or assumed. The organization of this book into four parts permits a certain amount of *modular flexibility.* Although the order of presentation is logical and has been used successfully in many other texts, there is no necessary reason why chapters and even parts must be covered in the sequence in which they appear in this volume. Depending on the objectives of the course(s), some (but certainly not all) of the ways in which this book might be used are as follows:

| Part | | Part | | Part | | Part | | Part | |
|---|---|---|---|---|---|---|---|---|---|
| 1 | | 1 | | 1 | first | 1 | one | 1 | |
| 2 | one | 3 | one | 3 | term | 2 | term | Chap. 12 | |
| 3 | term | 2 | term | | | Chap. 15 | | 2 (selected | one |
| 4 | | 4 | | 2 | second | | | chapters) | or |
| | | | | 4 | term | | | Chap. 15 | two |
| | | | | | | | | Chap. 13 | terms |
| | | | | | | | | Chap. 14 | |

The chapters in Part 3 are essentially free-standing and may be covered in sequence or independently. Chapter 15 may be included with the material in Part 2, or it may be used in place of some of the material in Part 2.

One inevitable limitation of a book such as this involves the priorities given to the various statistical topics. Some may feel that there is an appalling lack of coverage of topics that should be included. Certainly, many of the subjects briefly discussed in the last chapter—"Where Do We Go from Here?"—could each have been expanded to lengthy chapters with the possible result that the finished product would be twice its present size. And undoubtedly others will feel that we have devoted entirely too much space to some subject—e.g., sampling applications in Part 2. Since this book is written for beginning students, however, and since we never seem to be able to finish all the material we originally set out to cover in our own courses, we have preferred to err in the direction of keeping the size (and cost) of the book down.

Finally, it is customary at about this point in a preface (although there is always the question whether anyone is reading a preface at this point) for the author(s) to acknowledge their accountability for all errors of omission and commission. Because of the nature of the writing assignments and final editing in this project, however, the undersigned must accept the bulk of the responsibility for the remaining errors. (Of course, the mistake *you* find will be blamed on either Frank Murph or Bob Eng!)

**Donald H. Sanders**

# STATISTICS
## A FRESH APPROACH

# DESCRIPTIVE STATISTICS

The procedures for collecting, classifying, summarizing, and presenting quantitative facts are an important part of the subject of statistics. Earlier introductory statistics books, in fact, dealt almost exclusively with these descriptive procedures.

In this part of the text we will see how various statistical measures may be used (and misused) to describe and summarize the relationship that exists between variables. And although the focus in Part 3 is on the measurement and prediction of change, the three chapters in that section of the book will also deal largely with descriptive statistical procedures.

In contrast to earlier introductory statistics books, however, current texts also give considerable emphasis to the statistical inference procedures used to make decisions on the basis of computed measures taken from samples. (Statistical inference is the subject of the chapters in Part 2 of this book.) But even in statistical inference applications, the computed measures used as the basis for decision making are descriptive in nature. Thus, a knowledge of descriptive statistics is needed by all consumers of quantitative information — i.e., by all educated citizens.

The chapters included in Part 1 are:

1 "Let's Get Started"
2 "Liars, #$%*& Liars, and a Few Statisticians"
3 "Statistical Description: Frequency Distributions and Measures of Central Tendency"
4 "Statistical Description: Measures of Dispersion and Skewness"

# CHAPTER 1

# LET'S GET STARTED

**LEARNING OBJECTIVES**

After studying this chapter and answering the discussion questions, you should be able to:

☞ Explain the meaning of such terms as "statistics," "descriptive statistics," "statistical inference," "sample," "population," "parameter," and "census."

☞ Understand and explain why a knowledge of statistics is needed.

☞ Outline the basic steps in the statistical problem-solving methodology.

In O. Henry's *The Handbook of Hymen*, Mr. Pratt is wooing the wealthy Mrs. Sampson. Unfortunately for Pratt, he has a poet for a rival. To compensate for his romantic disadvantage, Pratt selects a book of quantitative facts to dazzle Mrs. Sampson.

> *"Let us sit on this log at the roadside," says I, "and forget the inhumanity and ribaldry of the poets. It is in the glorious columns of ascertained facts and legalized measures that beauty is to be found. In this very log we sit upon, Mrs. Sampson," says I, "is statistics more wonderful than any poem. The rings show it was sixty years old. At the depth of two thousand feet it would become coal in three thousand years. The deepest coal mine in the world is at Killingworth, near Newcastle. A box four feet long, three feet wide, and two feet eight inches deep will hold one ton of coal. If an artery is cut, compress it above the wound. A man's leg contains thirty bones. The Tower of London was burned in 1841."*
>
> *"Go on, Mr. Pratt," says Mrs. Sampson. "Them ideas is so original and soothing. I think statistics are just as lovely as they can be."*

It is possible (even likely) that you do not at this time share Mrs. Sampson's view of statistics. Oh, you probably accept the fact that an understanding of statistical tools is necessary in the modern world. But you have never placed the solving of mathematical problems at the top of your list of favorite things to do; you have possibly heard disturbing rumors about statistics courses; and you have not been eagerly looking forward to this day when you must crack open a statistics book. If the comments made thus far in this paragraph are applicable to you, you need not be apologetic about your possible misgivings. After all, many statistics books are rigorously written, mathematically profound, precisely detailed — and excruciatingly dull!

It will not be possible in this book, of course, to avoid the use of mathematical formulas to solve statistical problems and demonstrate important statistical theories. However, a knowledge of advanced mathematics is certainly *not* required to grasp the material presented in this book. In fact, you may be relieved to know that a beginning-level high school algebra course should adequately prepare you for virtually all the math required (e.g., performing such tough operations as addition, subtraction, multiplication, division, and finding square roots).

You will find in the pages and chapters that follow that an attempt has been made to (1) communicate with you rather than lecture to you, (2) present material in a rather relaxed and informal way without omitting the more important concepts, (3) recapture your attention with occasional quotes, ridiculous names, and unlikely situations of a humorous nature, and (4) utilize an intuitive and/or commonsense approach to develop concepts whenever possible. In short, this book is written for beginning students rather than statisticians, and its intent is to convince you that the study of statistics is a lively, interesting, and rewarding experience. (If Mr. Pratt could convince Mrs. Sampson, then maybe you too can be converted.)

In the remaining pages of this chapter let us now briefly outline the (1) *purpose and organization of the text,* (2) *need for statistics,* (3) *statistical problem-solving methodology,* and (4) *role of the computer in statistics.*

## PURPOSE AND ORGANIZATION OF THE TEXT

To do is to be—J.-P. Sartre
To be is to do—I. Kant
Do be do be do—F. Sinatra[1]

### Purpose of This Book

*The purpose of this book is* to introduce you to many of the important statistical concepts and procedures you are likely to need in order (1) to evaluate such daily inputs as organizational reports, newspaper and magazine articles, and radio and television commentaries, (2) to improve your ability to make better decisions on such wide-ranging topics as the quality of a particular product, the public servant likely to be thrown out of office, or the salesperson to believe, and (3) to improve your ability to measure and cope with changing conditions both at home and on the job.

*On the other hand, the purpose of this is not* to make a professional statistician out of you, because it is recognized that you are unlikely to be seeking such a career. Therefore, since you will more likely be primarily a *consumer* rather than a *producer* of statistical information, the emphasis of this book will be placed on explaining statistical procedures and interpreting the resulting conclusions. In short, the following dialogue from K. A. C. Manderville's *The Undoing of Lamia Gurdleneck* concludes with an important message that is kept in mind throughout this text:[2]

> *"You haven't told me yet," said Lady Nuttal, "what it is your fiancé does for a living."*
>
> *"He's a statistician," replied Lamia, with an annoying sense of being on the defensive.*
>
> *Lady Nuttal was obviously taken aback. It had not occurred to her that statisticians entered into normal social relationships. The species, she would have surmised, was perpetuated in some collateral manner, like mules.*
>
> *"But Aunt Sara, it's a very interesting profession," said Lamia warmly.*
>
> *"I don't doubt it," said her aunt, who obviously doubted it very much. "To express anything important in mere figures is so plainly impossible that there must be endless scope for well-paid advice on how to do it. But don't you think that life with a statistician would be rather, shall we say, humdrum?"*
>
> *Lamia was silent. She felt reluctant to discuss the surprising depth of emotional possibility which she had discovered below Edward's numerical veneer.*
>
> *"It's not the figures themselves," she said finally, "it's what you do with them that matters."*

---

[1] Have you ever noticed that chapters and sections of chapters in learned books and academic treatises are often preceded by quotations such as these that are selected by the author for some reason? In some cases a quotation is intended to emphasize a point to be presented; in other cases (often in the more erudite sources) there appears to be no discernible reason for the message, and it forever remains a mystery to the reader. In this particular case, the quotations from the above philosophers unfortunately fall into the *latter category!* However, we will from time to time throughout the book attempt to use quotations (from such authorities as Aldous Huxley, Mark Twain, and Winnie-the-Pooh) for the more valid purpose of emphasizing a point.

[2] Frontispiece from Maurice G. Kendall and Alan Stuart, *The Advanced Theory of Statistics,* vol. 2: *Inference and Relationships,* Hafner Publishing Company, Inc., New York, 1967.

**Organization of the Book**

This book is organized into 4 parts and 16 chapters in order to achieve its purpose. Before we go into more organizational detail, however, it is probably appropriate to pause here just long enough to define a few terms. The word "statistics" commonly has two different meanings. Lady Nuttal or Mrs. Sampson would likely define the word in its *first context* as being essentially the same as "figures" or "data"—i.e., they might consider the word "statistics" to be *plural* and synonymous with "numerical facts." In addition to meaning quantitative facts, however, the word "statistics" is also used in a *second way* and in a *singular* sense to refer to a subject of study in the same way that "mathematics" refers to such a subject.

Early statistics books dealt primarily with the procedures for *describing* data by means of classifying, summarizing, and communicating methods and techniques. The emphasis was generally on the *gathering* and *classification* of data, and then on the use of *summary* measures such as averages that would effectively describe the characteristics or basic structure of the subject being studied. The preparation of charts and tables to discover relationships and to interpret and *communicate* the measured values also received considerable attention. Obviously, *statistical description is still an important part of the subject of statistics.* Sales or revenue data, for example, may be classified or grouped by (1) volume, size, or quantity, (2) geographic location, or (3) type of product or service. To be of value, masses of data must often be condensed or sifted—i.e., summarized—so that resulting information will be concise and effective. A general sales manager, for example, may be interested only in the average monthly total sales of particular stores. Although he or she could be given a report that breaks sales down by department, product, and sales clerk, such a report would more likely be of interest to a department manager. Once the data have been classified and summarized, it is then often necessary that the processed information be presented or communicated in a usable form—e.g., through the use of appropriate tables and charts—to the final user.

Let's now briefly summarize some of the points made in the preceding paragraphs and get back to the organization of the text. Although the word "statistics" may refer to numerical data, it will more generally be used in this book to refer to *the body of principles and procedures developed for the collection, classification, summarization, interpretation, and communication of numerical data, and for the use of such data.* The subject of Part 1 of this book (the first four chapters) is *descriptive statistics—a term applied to the procedures of data collection, classification, summarization, and presentation.* A knowledge of descriptive statistical procedures will help you evaluate information presented in reports, articles, and broadcasts, and it will improve your ability to measure and thus cope with changing economic conditions. In Chapter 2, for example, we will see *how statistical methods have been improperly used* by various individuals and groups to confuse or deliberately mislead people. Many of the invalid uses presented involve descriptive procedures. And in Chapter 3 we will consider data collection and classification and the various *measures of central tendency* (or averages) frequently used by administrators and economists. Other important measures used to describe the amount of *dispersion or spread* in the data are discussed in Chapter 4.

In addition to descriptive statistics, however, another aspect of the subject of statistics which must receive a great deal of emphasis in any modern text is inferential statistics. *Statistical inference* involves drawing conclusions about a group under consideration—the *population* or *universe*—on the basis of data obtained from a *sample* selected from the population. It should be noted here that "population" (as used in statistics) is a term that is not limited to a group of people but instead refers to the total of any kind of unit under consideration. Thus, a population could be the parts assembled in a production run, the number of frozen chickens in a shipment, the number of credit accounts of a firm, or even the number of people in an organization. In contrast to a *census,* which is a study of the entire population, a sample, of course, is some selected segment of the population that is studied in order to (1) *make estimates* about some unknown population characteristic[3]—e.g., the percentage of consumers who like a new product—or (2) *make tests* to determine if assumptions or claims about an unknown population characteristic are likely to be acceptable—e.g., the claim of a salesperson that the average life expectancy of his or her product is greater than that of the product you are currently using. Thus, statistical inference goes beyond the mere description of sample data and becomes a tool to aid the decision maker in reducing the level of uncertainty that would have existed without any sample data. Figure 1-1 summarizes a number of the points made in the preceding paragraphs.

Part 2 of the book is entitled "Sampling in Theory and Practice" and includes Chapters 5 through 11. Chapters 5 and 6 present, in a simple and straightforward way, the conceptual foundations for the material on sampling applications. Then, in Chapters 7 through 11, the use of sample data to make estimates about, and test assumptions about, some unknown population characteristic is presented.

Part 3 of the book ("Coping with Change") contains elements of statistical inference, but the material in Chapters 12, 13, and 14 is essentially descriptive. However, the focus of Part 3 is on the *measurement and prediction of change.* In Chapter 12, for example, we look at *index number procedures* that have been developed to *measure* relative changes in economic conditions over time.[4] Then, in Chapter 13, we examine a *time-series forecasting approach* that consists of (1) analyzing past data to detect reasonably dependable patterns and then (2) projecting these patterns into the future to arrive at future expectations. Finally, in Chapter 14 we study a forecasting approach that consists of identifying and analyzing reasonably predictable independent factors having a significant influence on the dependent variable to be predicted.

Part 4 ("Concluding Topics") includes Chapters 15 and 16. These final chapters discuss some of the additional tools available to the analyst and decision maker that have not been considered elsewhere in this text. (You may be relieved to know that in this book we can only scratch the surface; many of our chapters are the subjects of entire texts that are encyclopedic in length.)

Most of the chapters in this book contain *self-testing review sections* following

---

[3] A *population characteristic* such as a population percentage or a population average (arithmetic mean) is called a *parameter*. A *sample characteristic* (arithmetic mean or percentage) is called a *statistic*. We will review these definitions again in Chapter 6.

[4] In Chapter 12, for example, you will be introduced to the Consumer Price Index, which measures the average changes in prices of many consumer goods over time and which may be the most important statistical series published regularly by the federal government. You see references to this measure all the time on television and in the newspapers.

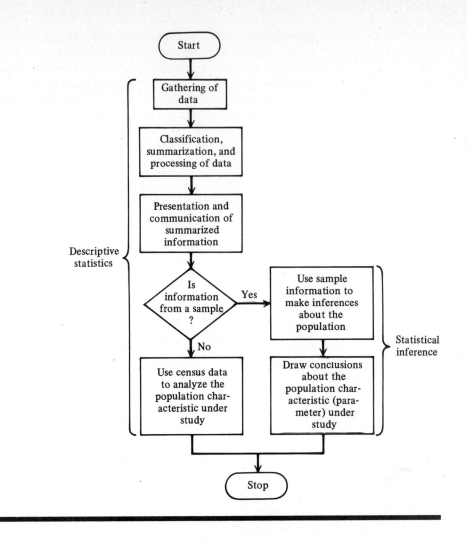

**FIGURE 1-1**

the presentation of important material. You are encouraged to test your grasp of the reviewed subjects by answering the questions and/or problems in these review sections before proceeding to the next topic. Answers to self-testing reviews are given at the end of the chapters. You will also find at the end of most chapters (1) a list of the *important terms and concepts* presented in the chapter, (2) *additional problems*, and (3) *topics for discussion*.

**NEED FOR STATISTICS**

At the beginning of this chapter the following sentence appeared: "Oh, you probably accept the fact that an understanding of statistical tools is necessary in the modern world." Perhaps that statement was premature; perhaps you don't accept the fact at all. It *is* a fact, however, that you need a knowledge of statistics to help you (1) *describe and understand relationships*, (2) *make better decisions*, and (3) *cope with change*.

## Describing Relationships between Variables

Since the amount of quantitative data that is gathered, processed, and presented to citizens and to organizational decision makers for one purpose or another has increased so rapidly, it is necessary that we be able to sift through this almost overwhelming mass in order to *identify and describe* those sometimes obscure *relationships between variables* that are often so important in decision making. The following examples serve to illustrate the *need for statistical analysis in understanding relationships:*

1 A businessperson can, by summarizing masses of revenue and cost data, compare the average return on investment during one time period with related figures from previous periods. (A number of decisions might hinge on the outcome of this comparison of descriptive measures.)

2 A government or public health official may come to conclusions about the relationship between smoking and/or obesity and a variety of diseases by applying statistical techniques to an enormous amount of input data. (These conclusions, in turn, may lead to decisions that affect millions of people.)

3 A marketing researcher may use statistical procedures to describe the relationships between the demand for a product and such characteristics as the income, family size and composition, age, and ethnic background of the consumers of the product. On the basis of these relationships, advertising and distribution efforts may then be directed toward those groups that represent the most profitable market.

4 An educator may use statistical techniques to determine if there is a significant relationship between scholastic aptitude test scores and the grade-point averages of students at her school. If there is such a relationship, she can make predictions about the probable academic success of an applicant on the basis of the test score.

## Aiding in Decision Making

An administrator can use statistics as an aid to making better decisions in the face of uncertainty. Consider the following examples:

1 Suppose that you are the manager in charge of purchasing for a large food-processing plant that packages frozen fried-chicken dinners. You are responsible for purchasing dressed broilers in shipments of 100,000 birds. Standards have been established specifying that the average weight of the birds in a shipment should be 32 ounces. (Birds over that weight tend to be too tough; lighter birds are too scrawny.) The truck of a new supplier rolls up to the unloading dock with a shipment you have agreed to take provided that the weight and quality standards are met. The supplier's salesperson assures you that the shipment will meet your standards. Should you accept the shipment on the basis of this claim? Probably not. Rather, you could use statistical inference techniques to select a random *sample* of, say, 100 birds from the *population* of 100,000 broilers. You could then weigh each bird in the sample, compute the average weight of the 100 birds, perform some other calculations, and then reach a conclusion about the average weight of the population of 100,000 chickens. Given this information,

you could then make a more informed decision on whether or not to accept the shipment.

2   Suppose again that you are the production manager of a plant that produces shotgun shells. It is known that certain variations will exist in the shells produced— e.g., there will likely be some variation in shot patterns produced and in shot velocity—but that these variations can be tolerated so long as they do not exceed specified limits more than 1 percent of the time. By using a statistically designed sampling plan, you would be able to arrive at reliable conclusions or inferences about the quality of a production run. Your conclusions or inferences would be based on tests conducted by firing a relatively small number of shells randomly selected from those produced during the run.

3   Finally, suppose that the manager of Big-Wig Executive Hair Stylists, Hugo Bald, has advertised that 90 percent of the firm's customers are satisfied with the company's services. If Ms. Polly Tician, a political activist, feels that this is an exaggerated statement which might require some legal action, she could use statistical inference techniques to decide whether or not to bring suit against Mr. Bald. (The answer to this question of whether or not to sue Hugo is found in Chapter 8.)

In the first example above, you *could* have weighed all 100,000 birds to determine the average weight. Such an approach, however, would have been *expensive* and *time-consuming*. In the second example, you *could* have determined the quality of the product by testing all the shells produced, but since such *testing is destructive*, there would have been nothing left for you to sell. (In both examples, of course, you received statistical information which permitted you to make better decisions.) And in the third example, Polly can test Hugo's claim before deciding about taking legal action (with the real possibility of a countersuit if it turns out that she is wrong).

**Coping with Change**

To plan is to decide in advance on a future course of action; therefore, plans and decisions are based on expectations about future events and/or relationships. Thus, virtually everyone is required to employ some forecasting process or technique to arrive at a future expectation. Although statistical procedures will obviously not enable us to predict the future with unerring accuracy, there are, as the following examples indicate, useful *statistical aids that may help to measure current change and to improve the forecasting process*:

1   Government statisticians periodically gather price data on about 400 different items from over 50 urban areas to compute a single monthly summary figure—an *index number*—that *measures overall changes in price levels* between the current period and some period in the past. (Similar index numbers are prepared, for example, to measure the changes in physical *quantities* of items produced.) Thus, a union leader (or you) could use information on changes in price levels to determine what has happened to the purchasing power of the dollar prior to entering into wage negotiations.

**2** Suppose that a sales manager has sales figures on a product line extending over a 10-year period. If, after studying these *time-series* figures, the sales manager has reason to believe that an identifiable past pattern will persist, she can build a forecast of future sales by using statistical procedures to project the past pattern into the future. She can also adjust her future sales forecast to take into account *seasonal variations*—e.g., higher sales, perhaps, during December than during February.

**3** Let's assume that a personnel manager has noted that job applicants who score high on a manual dexterity test later tend to perform well in the assembling of a product, while those with lower test scores tend to be less productive. By applying a statistical technique known as *regression analysis* (the subject of Chapter 14), the manager can predict or forecast how productive a new applicant will be on the job on the basis of how well he performs on the test.

If you have not already noticed, a comparison of (1) the above sections on the reasons why you need to have a knowledge of statistics with (2) the purpose and organization of this book stated earlier will show a number of similarities that can be summed up as follows: The purpose of the parts of this book is to help you acquire the statistical understanding you need in order to better describe and understand relationships, make better decisions, and more effectively cope with changing conditions.

## STATISTICAL PROBLEM-SOLVING METHODOLOGY

One *could* approach a problem-solving situation in a way somewhat analogous to the following illustration:

> In the comic dialect of the last century, a satirical almanac offers a solution to astrological signs:
>
> Tew kno exackly whare the sighn iz, multiply the day ov the month bi the sighn, then find a dividend that will go into a divider four times without enny remains, subtrakt this from the sighn, add the furst quoshunt tew the last divider, then multiply the whole ov the man's body bi all the sighns, and the result will be jist what yu are looking after.[5]

Or, one could elect to emulate the following example of Mark Twain in *Sketches Old and New*:

> If it would take a cannon ball 3 1/3 seconds to travel four miles, and 3 3/8 seconds to travel the next four, and 3 5/8 to travel the next four, and if its rate of progress continued to diminish in the same ratio, how long would it take to go fifteen hundred million miles?
>
> —Arithmeticus
> Virginia, Nevada

---

[5] Josh Billings, *Old Probability: Perhaps Rain—Perhaps Not*, G. W. Carleton and Company, Publishers, New York, 1879.

*I don't know.*

*—Mark Twain*

In most statistical problem-solving situations, however, it is appropriate to follow a more scientific approach. Several steps are followed in arriving at rational answers to statistical problems, and when one of these steps is ignored, the final results may be invalid, inaccurate, or needlessly expensive. The *basic steps in statistical problem solving are:*

1 *Identifying the problem or opportunity.* The manager or researcher must first understand and correctly define the problem or opportunity that he or she faces. Quantitative information useful at this time includes data outlining the nature and scope of the problem—e.g., production shortages and sales back orders—facts about the population to be studied, and the impact of the situation on resources such as personnel, materials, money, and time.

2 *Gathering available facts.* Data must be gathered that are accurate, timely, as complete as possible, and *relevant to the problem* being considered. Sources of data may be classified into *internal* and *external* categories. Internal business and economic data are found in accounting, production, and marketing departments, as well as in other areas within the organization. In addition to facts supplied by trade associations, customers, and suppliers, external data may be facts gathered (1) from *business periodicals* such as *Business Week, Sales Management, Wall Street Journal,* and *Chain Store Age;* and (2) from *government* publications such as the *Census of the United States,* the *Census of Business,* the *Survey of Current Business,* the *Statistical Abstract of the United States,* the *Monthly Labor Review,* and the *Federal Reserve Bulletin.* It is generally preferable to gather data from *primary sources*—i.e., from those organizations or agencies that initially gather the data and first publish them—rather than from *secondary sources*—i.e., organizations or agencies that republish the data. This is true because secondary sources may be subject to reproduction errors and may not explain how the facts were gathered or what limitations exist to their use.[6] Sherlock Holmes summarized the importance of data gathering in *The Adventure of the Copper Beeches* with this comment: "Data! Data! Data! I can't make bricks without clay."

3 *Gathering new original data.* In many cases the data needed by the analyst are simply not available from other sources, and so there is no alternative but for the analyst to gather them. There are *advantages to gathering new data,* for the ana-

---

[6] They may also fail to indicate how the variables being measured are defined by the primary source. During the depression month of November 1935, for example, the National Industrial Conference Board estimated the number unemployed at about 9 million; the National Research League estimated 14 million; and the Labor Research Association topped them all with a figure of 17 million. These estimates varied, of course, primarily because of differences in the way unemployment was defined.

lyst who is aware of the problem can participate in defining the variables and in determining the ways that the variables will be measured so that the resulting facts will possess the properties needed for the solution of the problem. There is a variety of *methods for obtaining desired data*. A common data-gathering practice involves the use of *personal interviews*. An interviewer generally asks a respondent the prepared questions that appear on a *schedule* form and then records the answers in the spaces provided on the form. This data-gathering approach permits the interviewer to clarify any terms that are not understood by the respondent, and it results in a high percentage of usable returns; however, it is an expensive approach and is subject to possible errors introduced by the interviewer's manner in asking questions. Interviews may sometimes be conducted over the *telephone*. (This is usually less expensive, but, of course, some households either do not have telephones or have unlisted numbers, and this may bias the results of a survey.) Another common data-gathering practice involves the use of *mail questionnaires*. As a general rule, the questions must be designed so that they can be answered by check marks or with a few words. The use of questionnaires is often less expensive than personal interviews, but the percentage of usable returns is generally lower, and those who do respond may not always be the ones to whom the questionnaire was addressed, and/or they may respond because of a nonrepresentative interest in the survey subject matter.

4  *Classifying and summarizing the data*. Once the data have been collected, the next step is to organize or group the facts for purposes of study. Identifying items with like characteristics and arranging them into groups or classes, we have seen, is called *classifying*. Production data can be classified, for example, by product made, location of production plant, and production process used. Classifying is sometimes accomplished by a shortened, predetermined method of abbreviation known as *coding*. Code numbers may be used, for example, to designate persons (social security number, payroll number), places (zip code, sales district number), and things (part number, catalog number). Once the data have been arranged into ordered classes, it is then possible to reduce the amount of detail to a more usable form by *summarization*. Tables, charts, and numerical descriptive values such as *measures of central tendency* (or averages) and *measures of dispersion* (the extent of the spread or scatter of the data about the central tendency measure) are tools used in summarizing.

5  *Presenting the data*. Summarized information in the form of tables, charts, and key quantitative measures facilitates problem understanding, helps identify relationships, and helps the analyst in presenting and communicating the important points to other interested parties.

6  *Analyzing the data*. The problem solver must *interpret* the results of the preceding steps, use the descriptive measures computed as the basis for making any statistical inferences that may be of value, and employ any statistical aids that may help in seeking out the most attractive possible courses of action. (The appropriateness of the options selected is, of course, determined by the problem solver's skill and the quality of his or her information.) The decision maker must then weigh the options in light of established goals in order to arrive at the one

**FIGURE 1-2**

**Statistical problem-
solving methodology**

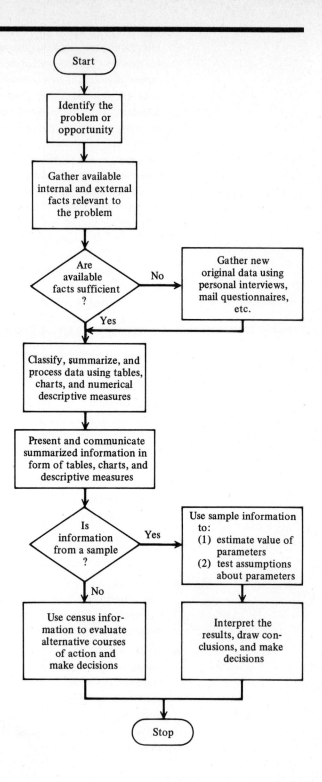

plan or decision that represents the best solution to the problem. Again, the correctness of this choice depends on analytical skill[7] (including the ability to apply appropriate quantitative techniques) and information quality.

We may summarize the above steps (and modify Fig. 1-1) in the graphical presentation shown in Fig. 1-2.

## ROLE OF THE COMPUTER IN STATISTICS

**Statistician: Modern name for member of Pythagorean Brotherhood. Secrets of the Brotherhood are closely guarded, but password is thought to be *sigma*. Acolytes stay all night in temple of their idol feeding it IBM cards. Believe that if idol is not fed it will be slain by its enemy the Budget Cutter and the Brotherhood consigned to a purgatory of desk calculators. Viewed with suspicion by other tribes because of their excessive interest in deviations.**
**—Royall Brandis**

As the above facetious definition suggests, there is definitely a role in statistics for the computer. In fact, a computer may be efficiently used in any processing operation that has one or more of the following characteristics:

1  *Large volume of input.* The greater the volume of data that must be processed to produce needed information, the more economical computer processing becomes relative to other possible methods.

2  *Repetition of projects.* Because of the expense involved in preparing a task for computer processing, it is frequently most economical to use the computer for repetitive projects.

3  *Desired and necessary greater speed in processing.* The greater the need for timely information, the greater will be the value of a computer relative to alternative (and slower) methods.

4  *Desired and necessary greater accuracy.* Computer processing will be quite accurate if the task to be performed has been properly prepared.

5  *Processing complexities that require electronic help.* In some situations in which large numbers of interacting variables are present, there is no alternative to the computer. For example, decision making with such statistical tools as linear programming and simulation[8] generally requires the use of a computer.

There are many statistical techniques that use a large volume of input data, are repetitive, and produce information that needs to be both timely and accurate. In

---

[7] The story has often been told of a scientist who trained a flea to jump when a bell was sounded. Then, the scientist would ring the bell, and after each jump by the flea he would pull off one of the insect's legs. When the poor flea was down to its last leg, the scientist rang the bell, and the flea flopped over weakly. The scientist then removed the last leg, again rang the bell, and observed that the flea did not move at all. His conclusion: When you remove all of a flea's legs, it goes deaf.

[8] These tools are briefly discussed in Chapter 16.

Chapter 13, for example, we will compute an index of seasonal variation. Although our example problem has been simplified to reduce the number of calculations necessary, it still took the author some time to perform the arithmetic with a desk calculator. A computer can do this repetitive task using a much larger data input volume in less than a minute. Similarly, in Chapter 14, regression analysis computations that can become quite tedious if done by hand may be performed in a very short time by a computer. As a matter of fact, in these and other areas of statistical analysis, the computer has made it possible to apply techniques that were previously limited in use because of the magnitude of the calculations required. (John Adams spent 2 years in the 1840s laboriously calculating the position of the planet Neptune. His work could now be duplicated with greater accuracy by a computer in a little over a minute.)

Since the emphasis of this book is on explaining statistical procedures and interpreting the resulting conclusions, we shall not discuss the preparation of computer programs to solve the statistical problems presented.[9] However, programs to solve virtually all the measures and procedures discussed in this book are commonly found in larger computer centers engaged in business data processing. In addition, so-called canned programs to process statistical applications are generally found in the program libraries of time-shared computing services.

## SUMMARY

The purpose of this book is to help you become an intelligent consumer of statistics, and the emphasis will be placed on explaining statistical procedures and interpreting the resulting conclusions. The book is organized into 4 parts and 16 chapters in order to achieve its purpose. Part 1 and much of Part 3 are devoted to a study of descriptive statistics, while Part 2 takes up the subject of statistical inference. A knowledge of both descriptive statistics and statistical inference is needed by the educated citizen in order to (1) describe and understand relationships, (2) make better decisions, and (3) cope with change.

Six basic steps in a statistical problem-solving procedure are listed and discussed in this chapter. In the chapters that follow we will use these steps over and over again, although we will not specifically refer to them by number. It is hoped that we will not *misuse* statistical procedures in the ways shown in the next chapter!

---

[9] One of the authors of this book modestly insists that you consider one or more of his four computer texts (e.g., Donald H. Sanders, *Computers in Business: An Introduction,* 4th ed., McGraw-Hill Book Company, New York, 1979) for more information on computer usage. His coauthors suggest that any other good introductory text would probably also be satisfactory.

**Important Terms and Concepts**

1. Statistics (plural)
2. Statistics (singular)
3. Descriptive statistics
4. Statistical inference
5. Measures of central tendency
6. Population
7. Sample
8. Parameter
9. Statistic (see footnote 3)
10. Census
11. Index number
12. Internal data
13. External data
14. Primary sources
15. Secondary sources
16. Personal interview
17. Mail questionnaire
18. Classifying
19. Summarizing

**Topics for Discussion**

1. Discuss the ways in which the word "statistics" may be used.
2. What is the difference between descriptive statistics and inferential statistics?
3. "Statistical description is still an important part of the subject of statistics." Discuss this statement.
4. How can sample data be used in decision making?
5. "A statistic may be used to estimate a parameter." Discuss this statement.
6. How can a knowledge of statistics help you *(a)* describe relationships between variables? *(b)* make better decisions? and *(c)* cope with change?
7. "Since nobody can know exactly what the future will bring, it is a waste of time to try to develop forecasts." Discuss this statement.
8. Identify and discuss the steps in the statistical problem-solving methodology.
9. What is the difference between primary and secondary sources of data?
10. "Some households either do not have telephones or have unlisted numbers, and this may bias the results of a survey." Discuss this statement.
11. Compare and contrast the personal interview and mail questionnaire approaches to gathering new original data.
12. *(a)* Discuss the characteristics that may permit the efficient use of a computer. *(b)* How can computers be used by statisticians?

# CHAPTER 2
## LIARS, #$%*& LIARS, AND A FEW STATISTICIANS

Consider the following comments that have appeared in print with distressing frequency:

> *"There are three kinds of lies: lies, damned lies, and statistics."* —Benjamin Disraeli

> *"Get the facts first and then you can distort them as much as you please."* —Mark Twain

> *"In earlier times they had no statistics, and so they had to fall back on lies. Hence the huge exaggerations of primitive literature —giants or miracles or wonders! They did it with lies and we do it with statistics; but it is all the same."* —Stephen Leacock

> *"He uses statistics as a drunken man uses lampposts —for support rather than illuminations."* —Andrew Lang

> *"It ain't so much the things we don't know that get us in trouble. It's the things we know that ain't so."* —Artemus Ward

And since the above comments deal primarily with statistics, it is not surprising that someone has also said that "if all the statisticians in the world were laid end to end — it would be a good thing!"

These unfavorable opinions of statistics are, unfortunately, probably held by many people who in earlier times uncritically accepted statistical conclusions only to discover later that they had been misled. For it is true that statistical methods have been improperly used by writers, advertising agencies, statisticians, politicians, used-car salespeople, bureaucrats, physicians—the list is virtually endless—to confuse people or deliberately to mislead them.[1]

In this chapter we shall point out a few of the more common ways statistical procedures have been misapplied. As a consumer of statistical information, you should be alert to the possibility of misleading statistical conclusions.[2] It is no exaggeration to say that a course in statistical analysis is well worth the time spent if it only succeeds in helping educated citizens do a better job of distinguishing between valid and invalid uses of quantitative techniques. In the pages that follow, we will consider a few examples of how statistics have been misapplied, under these headings: (1) *the bias obstacle,* (2) *aggravating averages,* (3) *disregarded dispersions,* (4) *the persuasive artist,* (5) *the post hoc ergo propter hoc trap,* (6) *antics with semantics,* (7) *the trend must go on,* (8) *follow the bouncing base (if you can),* and

---

[1] It is unnecessary to dwell on the motives of those who improperly use statistical procedures. Regardless of whether technical errors are innocently made or whether valid statistical information is deliberately twisted, oversimplified, or selectively distorted, the results are the same— people are misinformed and misled.

[2] For more information on the misuses of statistics, the following sources are recommended: Stephen K. Campbell, *Flaws and Fallacies in Statistical Thinking,* Prentice-Hall, Inc., Englewood Cliffs, N.J., 1974; Darrell Huff, *How to Lie with Statistics,* W. W. Norton & Company, Inc., New York, 1954; and W. Allen Wallis and Harry V. Roberts, *Statistics: A New Approach,* The Free Press, Glencoe, Ill., 1956, pp. 64–89.

(9) *spurious (and curious) accuracy*. The final section of the chapter is a brief discussion on *avoiding pitfalls*.

**THE BIAS OBSTACLE**

We have seen that an early step in statistical problem solving is gathering relevant data. Unfortunately, the use of poorly worded and/or biased questions during this early step may lead to worthless conclusions. The wording of questionnaires sent out by members of Congress to their constituents, for example, is often haphazard and often fails to produce focused answers. Even worse, questions are sometimes slanted to elicit answers that reinforce the congressperson's own biases. For example, in considering the military spending issue, New York Representative Frederick Richmond asked constituents if they favored "elimination of waste in the defense budget," and 95 percent naturally replied "yes." Sam Stratton, another New York Representative and a strong Pentagon champion, put the question of military spending to his constituents in this way:[3]

> *This year's defense budget represents the smallest portion of our national budget devoted to defense since Pearl Harbor. Any substantial cuts . . . will mean the U.S.A. is no longer number one in military strength. Which one of the following do you believe?*
>
> *"A. We must maintain our number-one status."* [*Nearly 63 percent voted for that.*]
>
> *"B. I don't mind if we do become number two behind Russia."* [*Only 27 percent checked that box.*]

Some people collecting or analyzing data may be tempted to use the "finagle factor" to give more emphasis to those facts that *support* their preconceived opinions than to those which *conflict* with their opinions. According to Thomas L. Martin, Jr.:[4]

> *The Finagle Factor allows one to bring* actual *results into immediate agreement with* desired *results easily and without the necessity of having to repeat messy experiments, calculations or designs.* [*When discovered, the Finagle Factor*] *was instantly and immensely popular with engineers and scientists, but found its greatest use in statistics and in the social sciences where actual results so often greatly differ from those desired by the investigator. . . . Thus:*

$$\begin{pmatrix} Desired\ results \\ on\ paper \end{pmatrix} = \begin{pmatrix} Finagle \\ Factor \end{pmatrix} \times \begin{pmatrix} Actual \\ results \end{pmatrix}$$

---

[3] "(Crab) Grass Roots: Questionnaires Sent by Congress Tap Voter Vitriol," *The Wall Street Journal,* Aug. 21, 1975, p. 13.

[4] Thomas L. Martin, Jr., *Malice in Blunderland,* McGraw-Hill Book Company, New York, 1973, p. 6. Martin's Finagle Factor has been referred to by others as the Fudge Factor, the Fake Factor, or the B. S. Factor.

Unlike lies that quickly wear thin, the finagle factor is a very durable weave of logical fallacies and sophistries that often appears and is constantly being recycled into new applications. How can the finagle factor be employed? Thank you for asking. It can be (and often is) used by advertisers. Suppose that you see on television a professional-looking actor saying (with great sincerity) that "8 out of 10 doctors recommend the ingredients found in Gastro-Dismal elixir." Does this message convince you to rush out and buy a bottle? Even assuming that the "8 out of 10" figure is correct, it is likely that the doctors were merely giving their stamp of approval to a number of common ingredients found in many nonprescription products. They probably were not specifically recommending Gastro-Dismal as being any better than other similar brands. (In fact, it is even quite possible that the separate ingredients recommended by most doctors could be put together in some preposterous way by Gastro-Dismal employees so as to be injurious to health.)

One more example of the advertiser's art should be sufficient here. The California Raisin Advisory Board ran the following ad in at least one ladies' magazine a few years ago: "Your husband could dance with you for 11 minutes on the energy he'd get from 49 raisins. Think what would happen if he never stopped eating them." As Stephen Campbell observed, "I can think of a lot of things that might happen to a man if he never stopped eating raisins—and they are all very painful."[5]

Although bias in the form of the finagle factor is generally consciously applied, bias can also appear unintentionally. Sometimes, for example, a researcher is led to faulty conclusions because of unintentionally biased input data. The polls which predicted that Thomas Dewey would defeat Harry Truman in the 1948 election for president, the conclusion of the psychiatrist that most people are mentally unbalanced (based on the input of those with whom he came in contact), the conclusion of the *Literary Digest* that Alf Landon would defeat Franklin Roosevelt for president in 1936—all these examples have in common the fact that bias entered into the picture. In the *Literary Digest* fiasco, for example, the prediction was based on a sample of about 2 million ballots returned (out of a total of 10 million mailed out). Unfortunately for the *Digest* (which ceased to exist in 1937), the ballots had been sent to persons listed in telephone directories and automobile registration records. As any student of history will tell you, those who could afford a telephone and an automobile in 1936 were hardly a cross section of the electorate. Instead, they were among the more prosperous of the voting population, and as a class in 1936 they were predominantly in support of the Republican Landon.

## AGGRAVATING AVERAGES

Let us assume that Ms. Sandy Loam decides that she wants to be a dropout from the city and move to a small farm in the Ozark Mountains where the air and water are clean, where the pace is more relaxed, where she can putter about and grow cucumbers, and where she will have the peace and quiet to write her novel exposing the inhumanity and ribaldry of poets. Sandy finds a farm with a good view (and little else) that suits her needs and is surprised when the realtor tells her that there are 100

---

[5] Stephen K. Campbell, *Flaws and Fallacies in Statistical Thinking*, Prentice-Hall, Inc., Englewood Cliffs, N.J., 1974, p. 174.

farmers in the area and that the average annual income of these farmers is nearly $13,000. Six months later a rally is called to protest the intention of the tax assessor to increase the property taxes in the area. It is pointed out during the meeting that the average income in the area is only $3,000 and that no new taxes can be afforded. Sandy is naturally confused, but since she hasn't made a dime growing cucumbers, she is all too willing to go along with the argument.

How could there be such a drop in average income in just 6 months? The answer, of course, is that nothing has really changed. *Both $13,000 and $3,000 are correct and legitimate averages!* The 100 farmers in the area include 99 whose net income is about $3,000, and one farmer who has invested millions of dollars in a showplace cattle operation that is headquartered atop a mountain and spreads over hundreds of acres. This one farmer nets approximately $1 million annually. Thus, one average—the *arithmetic mean*—is found by figuring the total income (i.e., 99 farmers times $3,000 equals $297,000, and this figure is added to the $1 million of the one-hundredth farmer to give a total of $1,297,000) and dividing by 100 farmers. The arithmetic mean is $12,970, or nearly $13,000. The average of $3,000 is the *median* and represents the earnings of the middle farmer in the group of 100.

In this example, of course, the average of nearly $13,000 is misleading because it distorts the general situation. Yet it is not a lie; it has been correctly computed. The problem of the aggravating average is that the word "average" can apply to several measures of central tendency—e.g., the arithmetic mean, the median, or the *mode*. (The mode is the most commonly occurring value. In our example, it would also be $3,000, but in other examples it could be yet a third figure.) It is not uncommon for one of these averages to be employed in situations in which it is not appropriate, and where it is selected to deliberately mislead the consumer of the information.

## DISREGARDED DISPERSIONS

Suppose that Karl Tell, an economist specializing in nineteenth-century German antitrust matters, is being pressured to coach the track team by the president of the small college where he teaches. Karl isn't too enthused about this prospect, since it will distract him from his study of the robber barons of Dusseldorf, but the president reminds him that he doesn't yet have tenure and that his classes are not in enormous demand. Therefore, Karl does a little checking and finds that the four high jumpers can clear an average of only 4 feet and that the three pole vaulters can manage an average height of only 10 feet. Karl concludes that his first venture into athletic management is likely to result in considerable verbal abuse from both alumni and faculty colleagues. Is Karl correct? Probably, but not because of the data he has gathered. Karl has been the victim of aggravating averages (arithmetic means in this example). Had he checked further, he would have found that one of his four high jumpers consistently clears 7 feet—good enough to win every time in the competition he will face—while the other three can each only manage to stumble over 3-foot heights. Likewise, in the pole vault there is one athlete who vaults 16 feet (with a bamboo pole) and two others who can each manage to explode for only 7 feet. The moral here is really the same as that of the preceding section: *Averages alone often don't adequately describe the true picture.* And we are simply making the further distinction here that *disregarded dispersion* exists when the spread or scatter of the values about the central measure is such that the average tends to mislead. Of course, disregarded dispersions and aggravating averages are usually found acting in concert to confuse and mislead. To summarize, the story is often told of the Chinese warlord who was leading his troops into battle with a rival when he came to a river. Since there were no boats, and since the warlord remembered that he had read somewhere that the average depth of the water in the river was only 2 feet at that time of year, he ordered his men to wade across. After the crossing, the warlord was surprised to learn that a number of his soldiers had drowned. Although the average depth was indeed just 2 feet, in some places it was only a few inches, while in other places it was over the heads of many who became the unfortunate victims of disregarded dispersion.

**THE PERSUASIVE ARTIST**

Statistical tables and charts are often prepared to summarize data, to uncover relationships, and to interpret and communicate quantitative information to those who may be able to use it. When properly prepared, such presentation aids are of real value. Here is a brief description of some of the more *popular types of presentation aids used:*

1   *Statistical tables* efficiently organize classified data in columns and rows so that the user of the tables can quickly find the information needed. Figure 2-1 presents some examples of tables in which the data are classified in a number of ways. In Fig. 2-1*a*, for example, average employment in New York City is classified by specified employment categories and by selected time periods. How are the data classified in Fig. 2-1*b, c,* and *d?*

2   *Line charts* do not present the specific data as well as tables, but they are often able to emphasize relationships more clearly. Quite frequently both a table and a line chart will be used in a presentation, with the chart used during an analysis of the figures in the table. The vertical axis in a line chart is often measured in quantities (e.g., dollars, tons produced) or percentages, while the horizontal axis is often measured in units of time (and thus the line chart may be referred to as a *time-series* chart). The *single*-line chart in Fig. 2-2*a,* for example, shows how state and local governments have been accumulating operating budget surpluses since 1975. Of course, *multiple* series can also be depicted on one line chart, as shown in Fig. 2-2*b* and *c.* The top lines in each represent totals. Note, however, that the three lines in Fig. 2-2*b* are all plotted against the same base line, while in the *component-part* line chart in Fig. 2-2*c* the acreages are added together in such a way that the chart is built up in layers, with the acreage harvested for export being added to the base provided by the acreage harvested for domestic use.

3   *Bar charts* depict by the length of the bars the quantities to be represented. As in the case of line charts, one scale often represents quantities or percentages, while the other scale can show units of time (see Fig. 2-3*a*). And *component-part* bars can also be shown in one chart, as illustrated by Fig. 2-3*b.*

4   *Pie charts* are simply one or more circles divided into sectors, usually for the purpose of showing the component parts of a whole. Single circles can be used, as shown in the pie chart in Fig. 2-4, or several pie charts can be used, for example, for a comparison of changes in the component parts over some period in time.

## New York City nonagricultural employment

Not seasonally adjusted; thousands of persons

| Sector | Average employment January–June 1978 | Average employment January–June 1977 | Change from 1977 to 1978 |
|---|---|---|---|
| Manufacturing | 539.1 | 536.3 | 2.8 |
| Private nonmanufacturing | 2,131.1 | 2,128.4 | 2.7 |
| Construction | 65.5 | 63.1 | 2.4 |
| Finance, insurance and real estate | 414.2 | 413.1 | 1.1 |
| Wholesale and retail trade | 614.5 | 619.2 | −4.7 |
| Transportation and public utilities | 255.0 | 259.8 | −4.8 |
| Services | 780.6 | 771.8 | 8.8 |
| Total private | 2,670.2 | 2,664.7 | 5.5 |
| Government | 501.6 | 492.3 | 9.3 |
| Federal | 83.5 | 84.4 | −0.9 |
| State | 51.0 | 49.3 | 1.7 |
| Local | 367.0 | 358.5 | 8.5 |
| Total nonagricultural | 3,171.8 | 3,157.0 | 14.8 |

Because of rounding, figures may not add to totals.

SOURCE: *Quarterly Review,* Federal Reserve Bank of New York, summer 1978, p. 26.

(a)

## Fortune 500 companies in major cities

1956 and 1977

| City | 1956 | 1977 | Change |
|---|---|---|---|
| New York | 140 | 82 | − 58 |
| Chicago | 47 | 24 | − 23 |
| Pittsburgh | 22 | 14 | − 8 |
| Detroit | 18 | 5 | − 13 |
| Cleveland | 16 | 13 | − 3 |
| Philadelphia | 14 | 7 | − 7 |
| St. Louis | 11 | 13 | + 2 |
| Los Angeles | 10 | 12 | + 2 |
| San Francisco | 8 | 6 | − 2 |
| Boston | 7 | 4 | − 3 |
| Total | 292 | 180 | −113 |

The ten cities are those which had the greatest number of headquarters in 1956, the first year in which this survey was taken.

SOURCE: "The Fortune Directory of the 500 Largest U.S. Industrial Corporations," *Fortune,* July 1957 and May 1978.

(b)

## Estimates and projections of net household formations (millions of households)

| | Total | Age of head: 25–34 years |
|---|---|---|
| 1960–1965 | 4.45 | – |
| 1965–1970 | 5.62 | 1.75 |
| 1970–1975 | 7.21 | 3.32 |
| 1975–1980 | 7.22 | 3.29 |
| 1980–1985 | 6.91 | 2.29 |
| 1985–1990 | 5.85 | 0.76 |

SOURCE: U.S. Bureau of the Census, *Current Population Reports,* ser. P-25, no. 476, "Demographic Projections for the United States," table 7.

(c)

## Credit rating of 25 largest cities, June 30, 1977

| City | Rating | City | Rating |
|---|---|---|---|
| Dallas | Aaa | Seattle | Aa |
| Houston | Aaa | | |
| Indianapolis | Aaa | Baltimore | A 1 |
| Los Angeles | Aaa | Jacksonville | A 1 |
| Milwaukee | Aaa | New Orleans | A 1 |
| San Francisco | Aaa | | |
| | | Cleveland | A |
| Chicago | Aa | Pittsburgh | A |
| Columbus | Aa | St. Louis | A |
| Denver | Aa | | |
| Kansas City | Aa | Boston | Baa |
| Memphis | Aa | Detroit | Baa |
| Phoenix | Aa | Philadelphia | Baa |
| San Antonio | Aa | | |
| San Diego | Aa | New York City | B |

NOTE: Moody's rating scheme assigns the Aaa rating to bonds carrying the smallest degree of investment risk. Bonds rated Aaa and Aa are generally known as "high grade bonds." Ratings then descend in grade through A, Baa, Ba, B, and so on through C. Moody's describes Baa bonds as "medium grade obligations, i.e. they are neither highly protected nor poorly secured"; B bonds "generally lack characteristics of the desirable investment." Those bonds judged strongest in a class have a *1* after the letter rating, as in A 1.

SOURCE: Moody's Investors Service.

(d)

**FIGURE 2-1**

**Statistical tables**

**FIGURE 2-2**
**Line charts**

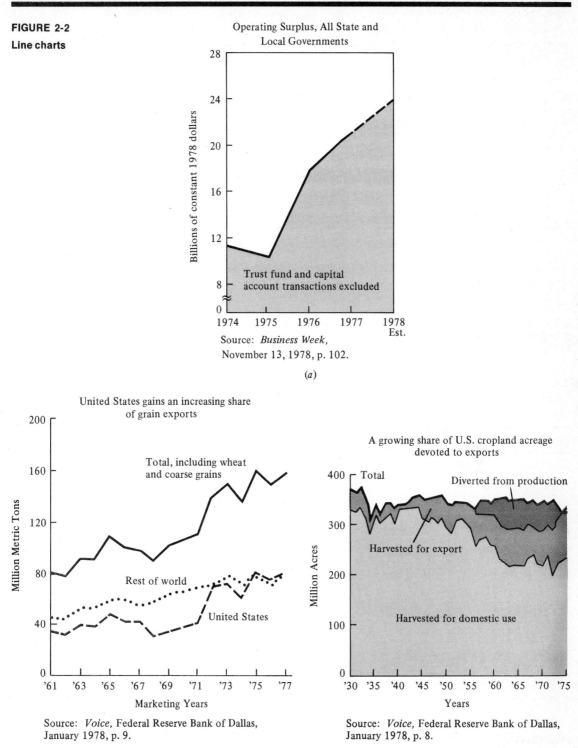

Operating Surplus, All State and
Local Governments

Billions of constant 1978 dollars

Trust fund and capital
account transactions excluded

1974   1975   1976   1977   1978
                                Est.

Source: *Business Week,*
November 13, 1978, p. 102.

(*a*)

United States gains an increasing share
of grain exports

Million Metric Tons

Total, including wheat
and coarse grains

Rest of world

United States

'61  '63  '65  '67  '69  '71  '73  '75  '77

Marketing Years

Source: *Voice,* Federal Reserve Bank of Dallas,
January 1978, p. 9.

(*b*)

A growing share of U.S. cropland acreage
devoted to exports

Million Acres

Total        Diverted from production

Harvested for export

Harvested for domestic use

'30  '35  '40  '45  '50  '55  '60  '65  '70  '75

Years

Source: *Voice,* Federal Reserve Bank of Dallas,
January 1978, p. 8.

(*c*)

**25**    CHAPTER 2/LIARS, #$%*& LIARS, AND A FEW STATISTICIANS

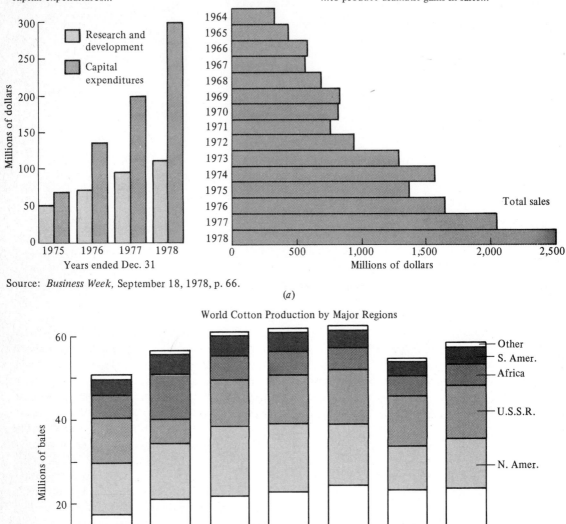

Texts Instruments, Inc. pours on the coal
...in research and development and in
capital expenditures...

...to produce dramatic gains in sales...

Source: *Business Week,* September 18, 1978, p. 66.

(*a*)

World Cotton Production by Major Regions

Source: Adapted from *Economic Review,* Federal Reserve
Bank of Atlanta, May–June 1977, p. 70.

(*b*)

**FIGURE 2-3**

**Bar charts**

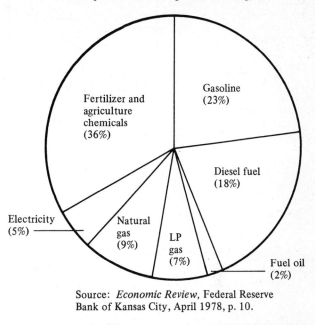

Components of U.S. Agricultural Energy Use

**FIGURE 2-4**

**Pie chart**

Source: *Economic Review,* Federal Reserve Bank of Kansas City, April 1978, p. 10.

So much for ways in which data can be honestly presented. But the purpose of some persuasive artists is to take honest facts and create misleading impressions. How can this be done? There are numerous tricks, but we will limit our discussion here to just a few examples.

Suppose you are running for reelection to a legislative body and during your past 2-year term in office government appropriations for various purposes have increased in your district from $8 million to $9 million. Now, as your fellow politicians know, this is not a particularly good record, but the voters don't need to know that. In fact, you can perhaps turn this possible liability into an asset with the help of a persuasive artist. Figure 2-5*a* would represent one way to present the information honestly. But since your objective is to mislead without actually lying, you prefer instead to distribute Fig. 2-5*b* during your campaign. The difference between Fig. 2-5*a* and *b,* of course, lies in the changing of the vertical scale in the latter figure. (The wavy line correctly indicates a break in the vertical scale, but this is often not considered by unwary consumers of this type of information.) By breaking the vertical scale and by then changing the proportion between the vertical and horizontal scales, you have given us the impression that you have certainly been doing a good job of getting appropriations.[6]

---

[6] We could make Fig. 2-5*b* even more impressive-looking by keeping the vertical scale and *compressing the horizontal* (time) *scale.*

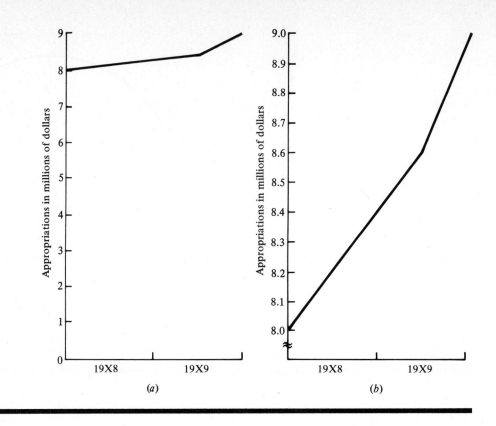

**FIGURE 2-5**

(a)

(b)

Having received favorable comments on your appropriations chart, you decide to employ another trick. New industry has come into your district during the past 2 years, and although there has been some increase in air and water pollution as a result, there has also been an increase in representative average weekly wages of semiskilled workers from $100 per week to $150 per week. Of course, you had little to do with bringing in the new industry, and there is also some disturbing evidence that the fact that wage increases have *followed* the new plants does not necessarily mean that they were *caused* by the new industry, but you see no reason to complicate matters with additional confusing facts.[7] How can you best communicate this wage increase information to your constituents? After discarding several approaches, you decide to use the *pictograph* chart in Fig. 2-6. The *height* of the small money bag represents $100, and the *height* of the large bag is correctly proportioned to represent $150. What is all wrong and quite misleading, however, is the *area* covered by each figure. The space occupied by the larger bag creates a very misleading visual impression. But, of course, that was the intent, wasn't it?

Let us now assume that in spite of your persuasive artwork the voters have

---

[7] We will discuss this matter in the next section.

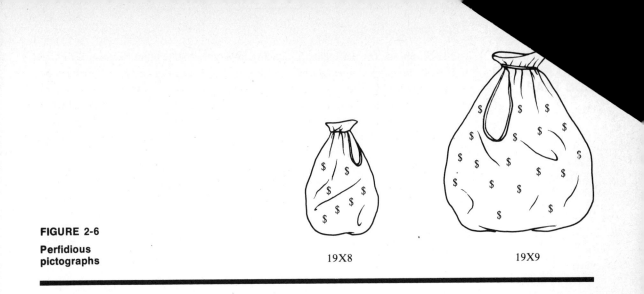

**FIGURE 2-6**

**Perfidious pictographs**

19X8                    19X9

seen fit to throw you out of office in favor of a write-in candidate. However, you are able to find work in your father's manufacturing company. One of the first jobs you are given is to prepare reports for stockholders and the union explaining company progress over the past year. The company has done very well, and profits have amounted to 25 cents of every sales dollar. This could be accurately presented in the form of a picture, as shown in Fig. 2-7a. But a stronger impact can be made on stockholders if the coin used in Fig. 2-7a is shown from the perspective of Fig. 2-7b. Since you don't want the union members to become restless, however, you can show them the profit situation from the perspective of Fig. 2-7c.[8]

**FIGURE 2-7**

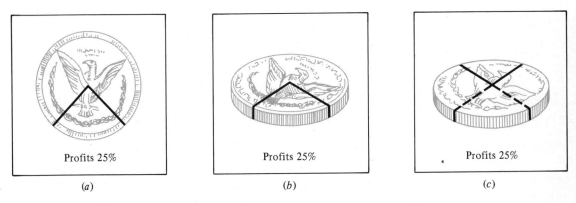

Profits 25%          Profits 25%          Profits 25%

(a)                  (b)                  (c)

[8] For a number of other interesting "persuasive" techniques, see Gene Zelazny, "Grappling with Graphics," *Management Review,* October 1975, pp. 4–16.

The Latin phrase for the logical reasoning fallacy that states that because B *follows* A, B was *caused* by A is *post hoc ergo propter hoc* (i.e., after this therefore because of this). Erroneous cause-and-effect conclusions are often drawn because of the misuse of quantitative methods.[9] In fact, you should always be on your guard when evaluating claims being made by others on the basis of measured relationships among variables being studied. Of course, as the following examples show, some erroneous conclusions are easy to spot:

> *Increased shipments of bananas into the port of Houston have been followed by increases in the national birthrate. Therefore, bananas were the cause of the increase in births.*

> *Ninety-five percent of all those who regularly use marijuana consumed large amounts of milk earlier in their lives. Thus, the early introduction of milk into the diet leads to the use of marijuana.*

> *The average human life-span has doubled in the world since the discovery of the tobacco plant. Therefore, tobacco . . . (this is just too gross to complete).*

But not all examples of the *post hoc* trap are so obvious. Some, in fact, can be subtle. To illustrate, you may recall that a few pages earlier (in our discussion of bias and of the *Literary Digest* poll) we used the following sentence: "Unfortunately for the *Digest* (which ceased to exist in 1937), the ballots had been sent to persons listed in telephone directories and automobile registration records." Did you perhaps conclude from this example that since the *Digest* folded in 1937, the cause was the poor prognostication about the 1936 election? Obviously, the poll didn't help the magazine's reputation, but did it *cause* the publication to go out of business? Isn't it likely that a number of factors combined to bring about this demise?

Let's now conclude our discussion of the *post hoc* trap with a return to the antiseptic world of politics. We have already seen in the preceding section how a politician can claim credit for good events that occur after he or she is elected. Another technique—and one that is often just as misleading—is for a politician to accuse an opponent of being the cause of something bad. For example, *Time* magazine reported that Richard Nixon made the following remarks during the 1968 presidential campaign:[10]

> *Hubert Humphrey defends the policies under which we have seen crime rising ten times as fast as the population. If you want your President to continue a do-nothing policy toward crime, vote for Humphrey. Hubert Humphrey sat on his hands and watched the U.S. become a nation where 50% of the American women are frightened to walk the streets at night.*

These remarks are ironic because of later events and misleading because of the great power attributed to the person occupying the position of vice president.

---

[9] This point will be discussed in greater detail in Chapter 14.

[10] *Time,* Nov. 1, 1968, p. 15.

## ANTICS WITH SEMANTICS

*Failing to define terms* and concepts that are important to a clear understanding of the message, *making improper or illogical comparisons* between unlike or unidentified things, *using an alleged statistical fact or statement to jump to a conclusion* which ignores other possibilities or which is quite illogical, *using jargon and lengthy words* to cloud the message when simple words and phrases are suitable — all these antics with semantics are used to confuse and mislead. The following examples should be adequate to illustrate this unfortunate fact:

1   The example of different unemployment figures produced by different organizations cited in footnote 6, Chapter 1, shows what a failure to define terms can do to understanding. Terms such as "poverty," "population," "living standard," and "cost overhead," to name just a few, are subject to different definitions, and the consumer of the information should be told which is being used.

2   The Federal Trade Commission (FTC) took exception to the advertised claim made by the manufacturers of Hollywood Bread that their product had fewer calories per slice than other breads. According to the FTC, the claim was misleading, since the Hollywood slice was thinner than normal. Actually there was no significant difference in the number of calories when equal amounts of bread were compared.

3   "One of four persons in the world is a Chinese communist. . . . Think about that the next time you listen to Lawrence Welk." What in the world does this mean? "The Egress carburetor is up to 10% less polluting and up to 50% more efficient." Less polluting than what, a steel mill? And more efficient than what, a Boeing 747?

4   "One in ten births is illegitimate. Thus, your estimate of your fellow man is correct 10% of the time." Figures on many activities (including illegitimate births, rapes, marijuana smoking, etc.) are just not reliable, because many cases remain unreported.

5   Representative Ben Grant, in arguing for plain speaking in a proposed new Texas constitution, points out how jargon can be used. If, noted Grant, a man were to give another an orange, he would simply say, "Have an orange." But if a lawyer were the donor, the gift might be accompanied with these words: "I hereby give and convey to you, all and singular, my estate and interests, right, title, claim, and advantages of and in said orange, together with its rind, juice, pulp, and pits, and all rights and advantages therein, with full power to bite, cut, suck, and otherwise eat same. . . ." Alas, the same kind of language may also be used to convey quantitative information.

## THE TREND MUST GO ON!

Another way in which a person may misuse quantitative methods is to assume that because a pattern has developed in the past in a category, that pattern will certainly continue into the future. Such uncritical extrapolation, of course, is foolish. Changes in technology, population, and life-styles all produce economic and social changes that may quickly produce an upturn or a downtrend in an existing social pattern or economic category. The invention of the automobile, for example, brought about significant growth in the petroleum and steel industries and a rapid decline in the production of buggies and buggy whips. Yet, as British editor Norman MacRae has noted, an extrapolation of the trends of the 1880s would show today's cities buried under horse manure.

Of course, the *judicious* projection of past patterns or trends into the future can be a very useful tool for the planner and decision maker, as we shall see in Chapter 13 in our study of time-series analysis. But the failure to apply a generous measure of common sense to extrapolations of past quantitative patterns can lead to erroneous conclusions that people are seriously asked to accept. Perhaps we could be more forgiving of trend projectors who would mislead us if their messages were as entertaining as the following passage from Mark Twain's *Life on the Mississippi:*

> *In the space of one hundred and seventy-six years the Lower Mississippi has shortened itself two hundred and forty-two miles. That is an average of a trifle over one mile and a third per year. Therefore, any calm person, who is not blind or idiotic, can see that in the Old Oölitic Silurian Period, just a million years ago next November, the Lower Mississippi River was upward of one million three hundred thousand miles long, and stuck out over the Gulf of Mexico like a fishing-rod. And by the same token any person can see that seven hundred and forty-two years from now the Lower Mississippi will be only a mile and three-quarters long, and Cairo and New Orleans will have joined their streets together, and be plodding comfortably along under a single mayor and a mutual board of aldermen. There is something fascinating about science. One gets such wholesale returns of conjecture out of such a trifling investment of fact.*

**FOLLOW THE BOUNCING BASE (IF YOU CAN)**

One of the authors of this text (an incredibly old individual) can remember as a small boy going to a movie theater and occasionally seeing a short film that urged the audience to participate in the singing of one or two songs. As the music played, the words of the song were displayed on the screen, and a bouncing ball would indicate to the audience the synchronization of words and music. The audience was asked to sing along by following the bouncing ball. (For the nostalgia buffs, it can be reported that few, if any, did sing along; rather, the time was used to get popcorn and/or attend to other chores.)

And just as old moviegoers often failed to follow the bouncing ball, people today are often confused because of their failure to follow the bouncing base—i.e., *the base period used in computing percentages.* A few examples will illustrate how failure to follow the bouncing base may lead to misunderstanding:

1  A worker is asked to take a 20 percent pay cut from his weekly salary of $300 during a recessionary period. Later, a 20 percent increase is given to the worker. Is he happy? The answer may depend on what has happened to the base. If the *cut* is computed using the earlier period (and the salary of $300) as the base, the reduced pay amounts to $240 ($300 × 0.80). But if the pay *increase* of 20 percent is figured on a base that *has been shifted* from $300 to $240, the worker winds up with a restored pay of $288 (1.20 × $240). Thus, the bouncing base has cost him $12 each week, and this isn't likely to please him.

**2** In Chapter 12 we will study the subject of index numbers, a valuable approach for measuring changes in such economic variables as prices and quantities produced. Unfortunately, index numbers are frequently misused by newspaper reporters. Let us assume, for example, that a price index uses 1967 as the base period and assigns to prices in that period an index number value of 100. Later, in 1973, the index value has risen to 130, and in 1975 the price index is 160. These numbers mean that there was a 30 percent increase in prices between 1967 and 1973 and a 60 percent increase in prices between 1967 and 1975. So far, so good. But now a reporter uses these figures during an article and notes that there has been a *30 percent increase* in prices between 1973 and 1975. It is true that the numbers 130 and 160 represent percentages, and it is also true that there is a difference of 30 *percentage points* between 1973 and 1975. But the *percentage increase* was actually 23.1 percent: $(160 - 130)/130 = 23.1$. In this case, the reporter failed to shift the base to 1973.

**3** Percentage *increases can easily exceed 100 percent.* For example, a company whose sales have increased from $1 million in 1974 to $4 million in 1976 has had a *percentage increase* of 300 percent: ($4 million $-$ $1 million)/$1 million. [Of course, the sales in 1976 *relative to* the sales in 1974 were 400 percent—($4 million $\div$ $1 million) $\times$ 100—and this *percentage relative* figure is sometimes confused with the percentage increase value.] However, *percentage decreases exceeding 100 percent are not possible* if the original data are positive values. For example, *Newsweek* magazine once reported that Mao-Tse Tung had cut the salaries of certain Chinese government officials by 300 percent.[11] Of course, once 100 percent is gone, there isn't anything left. Embarrassed editors later admitted that the cut was 66.67 percent rather than 300 percent.

Although the above examples represent only a few of the types of abuses that may be associated with the use of percentages, they do give you an idea of the importance of following the bouncing base.

## SPURIOUS (AND CURIOUS) ACCURACY

Statistical data based on sample results are often reported in very precise numbers. It is not unusual for several decimal places to be used, and the apparent precision lends an air of infallibility to the information reported. This air of infallibility is frequently enhanced if the data appear on a computer printout. Yet the accuracy image may be false. To illustrate, W. E. Urban, a statistician for the New Hampshire Agricultural Experiment Station in Durham, wrote a letter to the editor of *Infosystems* magazine taking issue with a previously published article. "Your magazine," wrote Urban, "has provided me with an excellent example of impeccable numerical accuracy and ludicrous interpretation which I will save for my statistics classes. With a total sample size of 55, reporting percentages with two decimal places is utter nonsense." The first sample percentage quoted in the article was 31.25, but, as Urban noted, the corresponding percentage in the population from which the sample

---

[11] *Newsweek,* Jan. 16, 1967, p. 6.

was taken was likely to have been anywhere between 12 and 62 percent! As Urban concluded: "I realize that it is painful to throw away all the nice decimals the computer has given us . . . but who are we kidding?" The reply of the editor: "No one. You're right."[12]

Spurious accuracy is not limited to sample results. In 1950, for example, the *Information Please Almanac* listed the number of Hungarian-speaking people at 13,000,000, while in the same year the *World Almanac* placed the number at 8,001,112. Thus, there was a difference of about 5 million people in the estimates of the two well-known almanacs. This isn't particularly surprising. But isn't it curious that the *World Almanac* figure could be so precise? Does it stand to reason that the accuracy could be so great when the figures are well up into the millions?[13] Oskar Morgenstern has summarized this situation very well:[14]

> *It is pointless to treat material in an "accurate" manner at a level exceeding that of the basic errors. The classical case is, of course, that of the story in which a man, asked about the age of a river, states that it is 3,000,021 years old. Asked how he could give such accurate information, the answer was that 21 years ago its age was given as 3 million years.*

## AVOIDING PITFALLS

**Harass them, harass them,**
**Make them relinquish the ball!**
**—Cheer at small but illustrious liberal arts college**

An important function of any course in statistical analysis is to help educated citizens do a better job of distinguishing between valid and invalid uses of quantitative techniques. Thus, information found in most of the chapters that follow should help you to avoid many of the pitfalls discussed in this chapter. Several later chapters, in fact, contain entire sections that point out the pitfalls and limitations associated with various statistical procedures. But even after you finish this book, you will find that it isn't always easy to recognize or cope with statistical fallacies. You must, like the team being encouraged with the cheer printed above, remain on the defense to avoid serious statistical blunders. *To avoid pitfalls, you might ask yourself questions when evaluating quantitative information — questions such as:*

1   *Who is the source of the information you are asked to accept?* Special interests have a way of using statistics to support preconceived positions. Using essentially the same raw data, labor unions might show that corporate profits are very

---

[12] *Infosystems,* August 1972, p. 8.

[13] Albert Sukoff, writing in the March 1973 issue of *Saturday Review of the Society,* observed that "huge numbers are commonplace in our culture, but oddly enough the larger the number the less meaningful it seems to be. . . . Anthropologists have reported on the primitive number systems of some aboriginal tribes. The Yancos in the Brazilian Amazon stop counting at three. Since their word for 'three' is '*poettarrarorincoaroac,*' this is understandable."

[14] Oskar Morgenstern, *On the Accuracy of Economic Observations,* Princeton University Press, Princeton, N.J., 1963, p. 64.

high and thus higher wage demands are reasonable, while the company might make a case to show that profit margins are low and labor productivity is not keeping up with productivity in other industries and in other countries. Also, politicians of opposing parties can use the same government statistics relating to employment, taxation, national debt, welfare spending, budgets, defense appropriations, etc., to draw surprisingly different conclusions to present to the voters.

2  *What evidence is offered by the source in support of the information?* Suspicious methods of data collection and/or presentation should put you on guard. And, of course, you should determine the relevancy of the supporting information to the issue or problem being considered.

3  *What evidence or information is missing?* What *isn't* made available by the source may be more important than what is supplied. If assumptions about trends, methods of computing percentages or making comparisons, definitions of terms, measures of central tendency and dispersion used, sizes of samples employed, and other important facts are missing, there may be ample cause for skepticism.

4  *Is the conclusion reasonable?* Have valid statistical facts or statements been used to support the jump to a conclusion that ignores other plausible possibilities? Does the conclusion seem logical and sensible?

**SUMMARY**

Many people have uncritically accepted statistical conclusions only to discover later that they have been misled. Our purpose in this chapter has *not* been to show you how to misapply statistical procedures so that you may better con fellow human beings. Rather, our purpose has been to alert you to the possibility of misleading statistical information so that you can do a better job of distinguishing between valid and invalid uses of quantitative techniques. Many of the examples selected here have been humorous. But there is nothing amusing about the ways statistics may be used to present biased and prejudiced information that can possibly alter election outcomes or cause a consumer to place unjustified trust in a patent medicine. In the chapters that follow, you will, it is hoped, learn how to avoid the pitfalls that we have discussed here. In the next exciting chapter, for example, you should learn enough about aggravating averages to not be misled by the improper uses of these measures. Don't miss it!

**Important Terms and Concepts**

1  Bias
2  Arithmetic mean
3  Median
4  Mode
5  Dispersion
6  Statistical tables
7  Line charts
8  Bar charts
9  Pie charts
10  Pictograph
11  *Post hoc ergo propter hoc*
12  Base period
13  Percentage relative
14  Spurious accuracy

1 "According to the alumni office, the average Prestige U. graduate, Class of '54, makes $36,123 a year." Comment on this press release.

2 "An independent laboratory test showed that Krinkle Gum toothpaste users report 36 percent fewer cavities." Discuss this advertisement.

3 "Gastro-Dismal elixir has been used by a quarter of a million customers to cure baldness. We have a double-your-money-back guarantee, and only 2 percent of those who used Gastro-Dismal were not helped and asked for a refund." Discuss this advertisement.

4 "There are as many people whose intelligence is above the average as there are people with below-average intelligence." Discuss this statement.

5 "Our firm's income has gone from $5 million to $10 million in just 2 years — an increase of 200 percent." Discuss this statement.

6 "The number of aspirin poisonings makes birth control pills look like one of the safest drugs on the market." Discuss this statement.

7 "Since 66 percent of all rape and murder victims were one-time friends or relatives of their assailants, you are safer at night in a public park with strangers than you are at home." Comment on this remark.

8 "A group of Texas schoolteachers took a history test and failed with an average grade of 60. Thus, Texas schoolteachers are deficient in history." Do you agree?

9 "Last year 760.67 million marijuana cigarettes were smoked in the United States — a flaunting of the law unequaled since Prohibition." Discuss this statement.

10 Why were the polls wrong in 1948 when they predicted that Dewey would defeat Truman for president? (You will have to do outside research to answer this question.)

11 Identify and discuss four popular types of data presentation aids.

12 "Erroneous cause-and-effect conclusions are often drawn because of the misuse of quantitative methods." Discuss this statement.

13 How may antics with semantics be used to confuse and mislead?

14 Is there any distinction between percentage points and percentage increase? Explain.

15 "Percentage decreases exceeding 100 percent are not possible if the original data are positive values." Explain why this is true.

16 What questions might you ask during the evaluation of quantitative information to avoid being misled?

# CHAPTER 3

# STATISTICAL DESCRIPTION: FREQUENCY DISTRIBUTIONS AND MEASURES OF CENTRAL TENDENCY

**LEARNING OBJECTIVES**

After studying this chapter, working the problems, and answering the discussion questions, you should be able to:

☞ Explain how to organize raw data into an array and how to construct and interpret a frequency distribution.

☞ Graphically present frequency distribution data in the form of a histogram or a frequency polygon.

☞ Present an overview of the types of measures that summarize and describe the basic properties of frequency distribution data.

☞ Compute such quantitative measures of central tendency as the mean, median, and mode for ungrouped data and for data organized in a frequency distribution.

☞ Compute a measure used to summarize qualitative data.

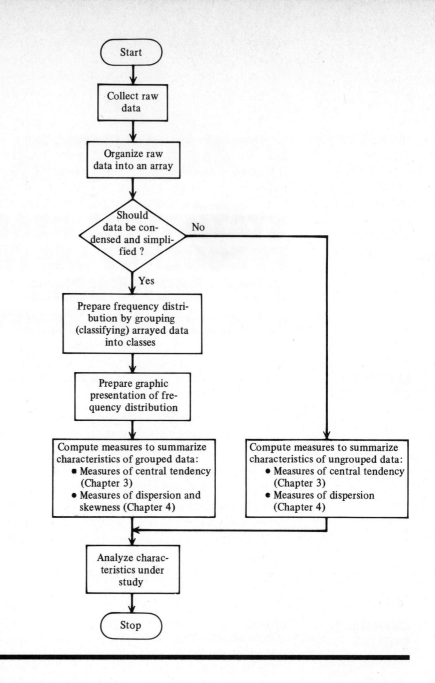

FIGURE 3-1

You may recall from Chapter 1 that "descriptive statistics" is a term applied to the procedures of data collection, classification, summarization, and presentation. In this chapter and in Chapter 4, we are concerned with all these aspects of statistical description. Thus, in this chapter we will (1) briefly consider the *collection and organization of raw (ungrouped) data*, (2) examine procedures for *classifying (and graphically presenting) data in a frequency distribution format*, (3) survey the more popular *types of summary measures used to describe frequency distributions,* and (4) illustrate the procedures used to *compute the most popular measures of central tendency.* A summary of the topics covered in this chapter and in Chapter 4 is presented in Fig. 3-1.

## INTRODUCTION TO DATA COLLECTION AND ORGANIZATION

Since we have already seen in Chapter 1 that *existing data* may be gathered from internal and external sources, and *new original data* may be obtained through the use of personal interviews and mail questionnaires, we need not go into data collection methods in detail here. It should be noted, however, that all the data will have been collected either as a result of *counting* or as the result of the use of some *measuring* instrument. The number of postal service mailboxes in a given geographic area, for example, is determined by counting, while such instruments as the automobile odometer, the service station gas pump, and the bathroom scale provide measurements of distances traveled, gallons of gasoline pumped, and weights of individuals.

The data collected for statistical analysis do not consist of observations that are all identical, since there would be little reason to study such a situation. Rather, data counted or measured for analysis purposes will represent the varying values of a *variable*—i.e., a characteristic that shows variation. A variable for which the values are obtained by counting is called a *discrete variable.* (The number of children per family in a study is an example of a discrete variable.) And a *continuous variable* is one for which the values are measured and recorded to some predetermined degree of accuracy.[1]

Assuming that the data have been collected, let us now turn to the subject of how these facts may be organized in a meaningful way.

## The Raw Data

A listing of the units produced by each of the employees in a plant would probably be of questionable value to a production manager who is trying to obtain an idea of worker productivity, and a sales manager would probably not learn much about sales patterns by combing through the sales invoices issued for a particular time period. To be meaningful, such unorganized raw data should be arranged in some systematic order.

An example of raw sales data is presented in Fig. 3-2. Since we shall be computing various measures to summarize and describe these sales facts later in this chapter and again in Chapter 4, perhaps we should pause here to provide you with some background information. The Boone & Docks Beverage Company was founded by Horatio Boone and Alger Docks during the late 1920s. The company's original product was a potent potable called Sparks and Flashes Forever which owed its

---

[1] Since measured quantities are continuously variable, they give only *approximate* results. The speedometer pointer in an automobile, for example, might give a reading of 45 miles per hour. But if the pointer were lengthened and sharpened, if the speedometer were calibrated more precisely, and if the cable were given closer attention, the reading might be 44 miles per hour. Further refinements and better instruments might give readings of 44.42 miles per hour, 44.4234 miles per hour, etc. Can you explain now why a measurement of 10 inches is only approximate?

## FIGURE 3-2 Raw data

Gallons of Fizzy Cola syrup sold by 50 employees of Boone &
Docks Beverage Company in 1 month

| Employee | Gallons sold | Employee | Gallons sold |
|---|---|---|---|
| P.P. | 95.00 | R.N. | 148.00 |
| S.M. | 100.75 | S.G. | 125.25 |
| P.T. | 126.00 | A.D. | 88.50 |
| P.U. | 114.00 | R.O. | 133.25 |
| M.S. | 134.25 | E.Y. | 95.00 |
| F.K. | 116.75 | Y.O. | 104.50 |
| L.Z. | 97.50 | O.U. | 135.00 |
| F.E. | 102.25 | U.S. | 108.25 |
| A.N. | 110.00 | L.T. | 122.50 |
| R.J. | 125.00 | E.A. | 107.25 |
| O.O. | 144.00 | A.T. | 137.00 |
| U.Y. | 112.00 | R.I. | 114.00 |
| T.T. | 82.50 | N.S. | 124.50 |
| G.H. | 135.50 | I.T. | 118.00 |
| R.I. | 115.25 | N.I. | 119.00 |
| O.S. | 128.75 | G.C. | 117.25 |
| U.S. | 113.25 | A.S. | 93.25 |
| P.O. | 132.00 | N.C. | 115.00 |
| O.R. | 105.00 | Y.A. | 116.50 |
| F.T. | 118.25 | T.N. | 99.50 |
| W.O. | 121.75 | H.B. | 106.00 |
| O.F. | 109.25 | I.E. | 103.75 |
| R.T. | 136.00 | N.F. | 115.25 |
| K.H. | 124.00 | G.U. | 128.50 |
| E.I. | 91.00 | X.N. | 105.00 |

distinctive flavor to a production process that featured four distillations through the radiator of a 1926 Packard. This original product was soon dropped because of what was considered by the founders to be excessive governmental regulation and harassment. Thus, after a period during which they became interested in inspecting the facilities of various correctional institutions. Horatio and Alger changed their product line so that Boone & Docks now markets a rather complete line of soft drink products in bottles and cans. In addition, certain soft drink syrups are sold in restaurants, theaters, and other outlets that mix small amounts of the syrup with carbonated water and sell the result in paper cups. The sales manager is currently interested in seeing how a new Fizzy Cola syrup is selling, and so the raw data in Fig. 3-2 have been gathered. (As you can verify, this unorganized mass of numbers is probably not of much value to the sales manager. And what if there had been 500 values rather than just 50?)

**The Data Array**    Perhaps the simplest device for organizing raw data in a systematic order is the *array* —an arrangement of data items in either an ascending (from lowest to highest value) or descending (from highest to lowest value) order. An array of the Fizzy Cola syrup sales data given in Fig. 3-2 is presented in Fig. 3-3. This array, of course, is in an ascending order.

There are several *advantages in arraying raw data*. *First,* we see in Fig. 3-3 that the sales vary from 82.50 to 148.00 gallons, a *range* of 65.50 gallons. *Second,* it is obvious that the lower one-half of the values are distributed between 82.50 and 115.25 gallons, and the upper 50 percent of the values vary between 115.25 and 148.00 gallons. And *third,* an array *can* show the presence or absence of a large concentration of items around a particular value. (In Fig. 3-3, no single value appears more than twice, but in other arrays there may be a pronounced concentration.)

In spite of these advantages, however, the array is still a rather awkward data organization tool, especially when the number of data items is large. Thus, there often exists the need to compress the data into a more usable form for analysis purposes. The object of the next section is to show how a more compact form of data organization may be developed.

**FIGURE 3-3 Data array**

Gallons of Fizzy Cola syrup
sold by 50 employees of
Boone & Docks Beverage
Company in 1 month

| Gallons sold | Gallons sold |
|---|---|
| 82.50 | 115.25 |
| 88.50 | 116.50 |
| 91.00 | 116.75 |
| 93.25 | 117.25 |
| 95.00 | 118.00 |
| 95.00 | 118.25 |
| 97.50 | 119.00 |
| 99.50 | 121.75 |
| 100.75 | 122.50 |
| 102.25 | 124.00 |
| 103.75 | 124.50 |
| 104.50 | 125.00 |
| 105.00 | 125.25 |
| 105.00 | 126.00 |
| 106.00 | 128.50 |
| 107.25 | 128.75 |
| 108.25 | 132.00 |
| 109.25 | 133.25 |
| 110.00 | 134.25 |
| 112.00 | 135.00 |
| 113.25 | 135.50 |
| 114.00 | 136.00 |
| 114.00 | 137.00 |
| 115.00 | 144.00 |
| 115.25 | 148.00 |

SOURCE: Fig. 3-2.

**FREQUENCY DISTRIBUTIONS**

The purpose of a frequency distribution is to organize the data items into a more compact form without obscuring the essential information contained in the values. This purpose is accomplished by grouping the arrayed data into a relatively small number of classes. Thus, *a frequency distribution (or frequency table) is simply the classification of a group of data items according to some observable characteristic.* In Fig. 3-4, for example, we have organized or grouped the *gallons sold* into *seven classes* and have then indicated the number of employees whose sales have turned up in each of the seven classes. (The term "frequency distribution" comes from this frequency of occurrence of values in the various classes.)

You will notice in Fig. 3-4 that the data are now arranged into a more compact and usable form. A quick glance at the frequency distribution shows, for example, that the sales of about two-thirds of the employees ranged from 100 to 130 gallons (the sales of 33 of the 50 employees are distributed in the middle three classes). In short, Fig. 3-4 gives us a reasonably good overall picture of the sales pattern of Fizzy Cola syrup. Of course, *the reduction or compression of the data has resulted in some loss of detailed information.* We no longer know, for example, exactly how many gallons each employee sold. And we don't know from Fig. 3-4 that the values have a range or spread of exactly 65.50 gallons.[2] On balance, however, the advantages of a well-designed frequency distribution usually outweigh this inevitable loss of detail.

In order to construct a frequency distribution, it is necessary to determine (1) the *number* of classes that will be used to group the data, (2) the *width* of these classes, and (3) the number of observations or *frequencies* in each class. In the fol-

**FIGURE 3-4 Frequency distribution**

Gallons of Fizzy Cola syrup sold by 50 employees of Boone & Docks Beverage Company in 1 month

| Gallons sold | Number of employees (frequencies) |
|---|---|
| 80 and less than 90 | 2 |
| 90 and less than 100 | 6 |
| 100 and less than 110 | 10 |
| 110 and less than 120 | 14 |
| 120 and less than 130 | 9 |
| 130 and less than 140 | 7 |
| 140 and less than 150 | 2 |
| | 50 |

SOURCE: Fig. 3-3.

---

[2] All we know about these matters is that there are two employees in the first class, for example, whose sales were somewhere between 80 and less than 90 gallons and that the range of values is going to be somewhere between 50 and 70 gallons.

lowing section we will briefly examine the first two interrelated considerations. (The last step is a routine transfer of information from an array to a distribution, and so we need not consider it here.)

## Classification Considerations

In constructing a frequency distribution, it is usually desirable to consider the following basic rules or criteria:

1 The *number of classes that will be used* to group the data generally varies between a minimum of 5 and a maximum of 15. The actual number of classes selected depends on such factors as the number of observations being grouped, the purpose for which the distribution is being prepared, and the arbitrary preferences of the analyst. Of course, there would be little or no improvement over the array if we were to group the data in Fig. 3-4 into 22 classes, with each class having a width of 3 gallons. And at the other extreme, valuable detail would be lost if we were to group the data in Fig. 3-4 into only three classes with intervals of 22 gallons each.

2 *Classes must be selected in such a way that (a)* both the smallest and largest data items are included, and *(b)* each item can be assigned to one and only one class— i.e., possible gaps and/or overlaps between successive classes that could cause this rule to be violated must be avoided.

3 Whenever possible, *the width of the classes* —i.e., the size of the *class intervals* — *should be equal.* (It is also often desirable to use class intervals that are multiples of 5, 10, 100, 1,000, etc.) Although unequal class intervals may be needed in frequency distributions in which large gaps exist in the data, such intervals may cause difficulties during the preparation of graphs and the computation of certain descriptive statistical measures. Our Fig. 3-4 has arbitrarily been prepared with seven classes of equal size. How was the interval width of 10 gallons determined? You ask very perceptive questions. The following simple formula was used to estimate the necessary interval:

$$i = \frac{L - S}{c}$$

where $i$ = width of the class intervals
  $L$ = value of the largest item
  $S$ = value of the smallest item
  $c$ = number of classes

Of course, as we have seen in Fig. 3-3, the Fizzy Cola sales data range from a low of 82.50 gallons to a high of 148.00 gallons. Thus,

$$i = \frac{148.00 - 82.50}{7}$$

$$= \frac{65.50}{7}$$

= 9.36, a value close to the convenient class interval size of 10 gallons used in Fig. 3-4

**FIGURE 3-5 Open-ended distribution**

Total income reported by selected families

| Total income | Number of families |
|---|---|
| Under $5,000 | 6 |
| $ 5,000 and under $10,000 | 14 |
| 10,000 and under  15,000 | 18 |
| 15,000 and under  20,000 | 10 |
| 20,000 and under  25,000 | 5 |
| 25,000 and under  30,000 | 4 |
| 30,000 and over | 3 |
| | 60 |

4  Whenever possible, *open-ended classes should be avoided.* Figure 3-5 is an example of an *open-ended distribution.* Although an open-ended class may be needed when a few values are extremely large or small in comparison with the remainder of the more concentrated observations, or when confidential information might possibly be revealed by stating an upper limit,[3] it should be used as sparingly as possible because of graphing problems and because (as we shall see later) it is impossible to compute such important descriptive statistical measures as the arithmetic mean and the standard deviation from an open-ended distribution.

5  When there is a concentration of raw data around certain values, it is desirable to construct the distribution in such a way that these *points of concentration fall at the midpoint of a class interval.* (The reason for this will become apparent later when we consider the computation of the arithmetic mean of data found in a frequency distribution.) In Fig. 3-4, the midpoint of the class "110 and less than 120" is 115 gallons, the lower limit of that class is 110 gallons, and the upper limit is 119.999 . . . gallons. Of course, another analyst could gather additional raw sales data for Fizzy Cola syrup and could then round the sales to the *nearest* gallon. This analyst might then set up a frequency distribution similar to Fig. 3-4 with class intervals of 80 to 89, 90 to 99, 100 to 109, 110 to 119, etc. In this case, the *stated* limits would be only 9 gallons apart, but the size of these class intervals would still be 10 gallons. Why? Because the class "110 to 119" would have a *real lower limit* or *lower boundary* of 109.5 and a *real upper limit* or *upper boundary* of up to 119.5. Thus, the class interval would still have a width of 10 gallons, but the class midpoint in this case would be 114.5 gallons.

After the data are organized into a more compact format, a properly constructed frequency distribution may be used for analysis, interpretation, and communication purposes. It is often possible to prepare a graphical presentation of a frequency dis-

---

[3] For example, placing an upper limit on the data in Fig. 3-5 might tend to reveal the income of an easily identifiable family in a local community.

tribution to achieve one or more of these purposes. How can graphical presentations of frequency distributions be prepared, you eagerly ask? Well it just so happens. . . .

## Graphical Presentation of Frequency Distributions

Although a frequency distribution or a frequency table such as the one shown in Fig. 3-4 is needed if the data are to be used (1) for reference purposes or (2) for the computation of descriptive measures that summarize certain characteristics of the values, a graphical presentation of the data found in a frequency table is more likely to (1) receive the attention of the casual observer and (2) reveal trends or relationships that might be overlooked in a table. The two basic forms of graphical presentation of a numerical distribution are the *histogram* and the *frequency polygon.*

**Histogram** *A histogram is a bar chart of a frequency distribution.* Figure 3-6 is a histogram of the Boone & Docks Fizzy Cola syrup sales data found in the table in Fig. 3-4. As you can see, this histogram simply consists of a set of vertical bars. Values of the variable being measured—in this case gallons of syrup sold—are measured on an arithmetic scale on the horizontal axis. The bars in Fig. 3-6 are of *equal width* and correspond to the *equal class intervals* in Fig. 3-4; the *height* of each bar in Fig. 3-6 corresponds to the *frequency* of the class it represents. Thus, the *area* of a bar above each class interval is proportional to the frequencies represented in that class.[4]

**Frequency polygon** *A frequency polygon is a line chart of a frequency distribution* and is thus an alternative form of graphical presentation. Figure 3-7 is a frequency polygon using the same data and plotted on the same scales as the histogram in Fig. 3-6. (In fact, Fig. 3-6 has been lightly reproduced as background in Fig. 3-7.) As you can see, points are placed at the *midpoints* of each class interval. The *height* of each

**FIGURE 3-6**

**Histogram of frequency distribution of gallons of Fizzy Cola syrup sold by 50 employees of Boone & Docks Beverage Company in 1 month**

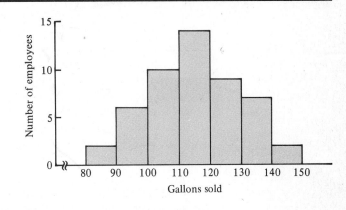

[4] If unequal class intervals were used in a frequency distribution, the *areas of the bars above the various class intervals would still have to be proportional to the frequencies represented in the classes*—e.g., if the third interval is twice as wide as each of the first two, the frequency of the third interval must be divided by two to get the appropriate height for the bar.

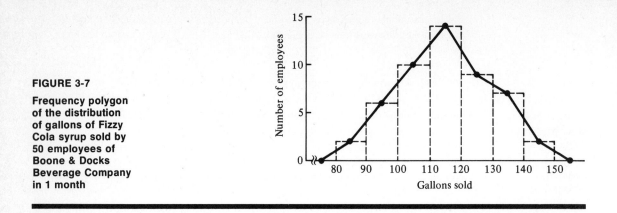

FIGURE 3-7

Frequency polygon
of the distribution
of gallons of Fizzy
Cola syrup sold by
50 employees of
Boone & Docks
Beverage Company
in 1 month

plotted point in Fig. 3-7, of course, represents the *frequency* of the particular class. These points are then connected by a series of straight lines. It is customary to close the polygon at both ends (1) by placing points on the baseline half a class interval to the left of the first class and half a class interval to the right of the last class, and then (2) by drawing lines from the points representing the frequencies in the first and last classes to these baseline points (see Fig. 3-7).

**A comparison** The *advantages of histograms over frequency polygons* are that (1) each individual class is represented by a bar which stands out clearly, and (2) the area of a bar represents the exact number of frequencies in a class interval. However, *the frequency polygon possesses certain advantages*—e.g., it is simpler and has fewer lines than a histogram, and thus it is especially suitable for making comparisons of two or more frequency distributions. Another advantage of the frequency polygon is that if the class intervals in a frequency distribution were continuously reduced in size, and if the number of items in the distribution were continually increased, we would expect the frequency polygon to resemble a smooth curve more and more closely. Thus, for example, if the frequency polygon in Fig. 3-7 represented only a small *sample* of all the available data on Fizzy Cola syrup sales made by hundreds of employees, and if the frequency distribution—i.e., the *population frequency distribution*—that could be prepared to account for all these data were made up of very narrow class intervals, the resulting population frequency distribution curve might be expected to resemble the frequency polygon shown in Fig. 3-8. This bell-shaped *normal curve,* which describes the distribution of many kinds of variables in the physical sciences, the social sciences, medicine, agriculture, business, and engineering, is very important in statistics and will be reintroduced in a number of the following chapters.

**Other considerations** It is sometimes desirable to determine the number of observations that fall above or below a certain value rather than within a given interval. In such cases, the regular frequency distribution may be converted to a *cumulative frequency* distribution, as shown in Fig. 3-9. (As you can see, we have merely arranged the data from Fig. 3-4 in a different format. The eight employees who sold less than 100 gallons, for example, were the two who sold less than 90 plus the six in

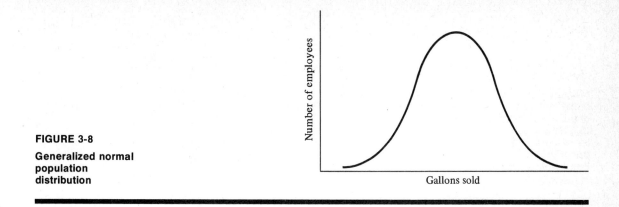

**FIGURE 3-8**

**Generalized normal population distribution**

the class of "90 and less than 100" gallons.) An *ogive* (pronounced "oh jive") is a graphical presentation of a cumulative frequency distribution. The ogive for Fig. 3-9 is shown in Fig. 3-10, and each point represents the number of employees having sales of less than the gallons indicated on the horizontal scale. By adding a percentage scale to the right of the ogive,[5] it is possible to graphically obtain a number of summary measures that we will compute in a later section. For example, if, as shown in Fig. 3-10, we draw a line from the 50 percent point on the percentage scale over to where it intersects with the ogive line, and if we then draw a perpendicular line from this intersection to the horizontal scale, we are able to read the approximate amount of syrup sold by the twenty-fifth employee in the arrayed group of 50. (This value is the *median,* and it will be computed without the use of an ogive later in the chapter.)

**FIGURE 3-9**

Cumulative frequency distribution of gallons of Fizzy Cola syrup sold by 50 employees of Boone & Docks Beverage Company in 1 month

| Gallons sold | Number of employees |
|---|---|
| Less than 80 | 0 |
| Less than 90 | 2 |
| Less than 100 | 8 |
| Less than 110 | 18 |
| Less than 120 | 32 |
| Less than 130 | 41 |
| Less than 140 | 48 |
| Less than 150 | 50 |

---

[5] Since each employee represents 2 percent of the total ($1/50 \times 100$), we simply double the scale on the vertical axis to obtain our percentage scale.

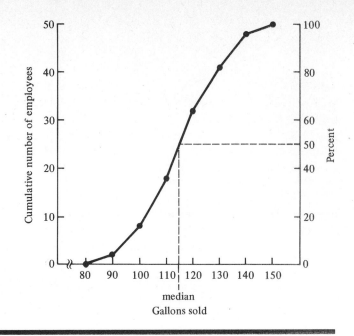

**FIGURE 3-10**

Ogive for the distribution of gallons of Fizzy Cola syrup sold by 50 employees of Boone & Docks Beverage Company in 1 month

**Self-testing Review 3-1**

This is the first in a series of self-testing review sections that will appear from time to time throughout most of the remainder of the book. You are encouraged to pause here to test your understanding of the concepts that have just been presented. *The answers to self-testing review questions are found at the end of the chapter in which they appear.*

Brock and Parse Lee, owners of the Lee Produce Company, are studying the size of the orders placed by customers in an outlying county. In the past week, the following 30 orders have been received:

| | | | | | |
|---|---|---|---|---|---|
| $42.50 | $45.00 | $47.75 | $52.10 | $29.00 | $31.25 |
| 21.50 | 56.30 | 55.60 | 49.80 | 35.55 | 42.30 |
| 43.50 | 34.60 | 65.50 | 45.10 | 40.25 | 58.00 |
| 30.30 | 44.80 | 36.50 | 55.00 | 59.20 | 36.60 |
| 38.50 | 41.10 | 46.00 | 39.95 | 25.35 | 49.50 |

**1  a** Arrange the above data in an ascending array.
   **b** What is the range of values?

**2** Organize the data items according to order size into a frequency distribution having the classes "$20 and under $30," "$30 and under $40," . . . , and "$60 and under $70."

**3  a** Would it have been possible to have used six or seven classes rather than five classes in the above frequency distribution?
   **b** What would have been a reasonable class interval or width if you had prepared a frequency distribution using eight classes rather than five?

**4** Draw a histogram of the frequency distribution prepared in problem **2** above.

**5** Draw a frequency polygon of the frequency distribution prepared in problem **2** above.

## SUMMARY MEASURES OF FREQUENCY DISTRIBUTIONS

In later sections of this chapter (and in Chapter 4) we shall compute various types of measures that summarize and describe the basic properties or characteristics of frequency distribution data.[6] Before we begin these computations, however, it is probably appropriate to pause here long enough to present a brief discussion and graphical overview (using Fig. 3-11) of the summary measures we will encounter.

### Measures of Central Tendency

It has already been noted in Chapters 1 and 2 that the word "average" applies to several measures of central tendency. The purpose of these averages is to summarize in a single value the typical size, middle property, or central location of a set of values. The most familiar average is, of course, the *arithmetic mean,* which is simply the sum of the values of a group of items divided by the number of such items. But, as we saw in the last chapter, the *median* and *mode* are other averages or measures of central tendency that are commonly used. Figure 3-11 shows various possibilities that could exist in different frequency distributions. Suppose in Fig. 3-11*a*, for example, we

**FIGURE 3-11**

**Summary measures of frequency distributions**

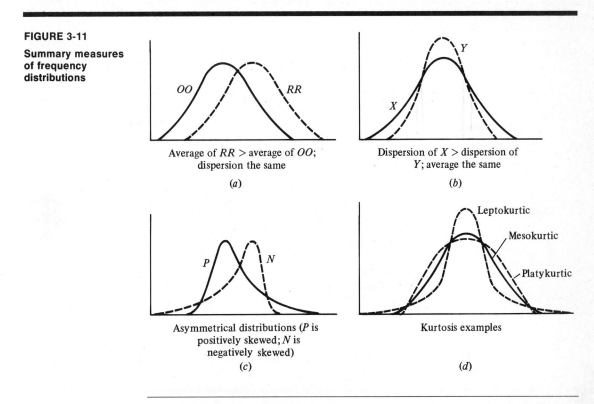

Average of *RR* > average of *OO*; dispersion the same

(*a*)

Dispersion of *X* > dispersion of *Y*; average the same

(*b*)

Asymmetrical distributions (*P* is positively skewed; *N* is negatively skewed)

(*c*)

Kurtosis examples

(*d*)

---

[6] We will also compute many of the same measures using *ungrouped data*—i.e., using data that have not been grouped into frequency distributions.

have the monthly sales distributions of two Boone & Docks products—Opulent Orange (OO) and Ribald Root Beer (RR). Although the spread or scatter of the sales data in each distribution appears to be the same, it is obvious that the average sales of root beer are greater than the average sales of the orange beverage—i.e., the root beer sales are concentrated around a higher value than the orange sales. We will compute measures of central tendency in the latter pages of this chapter.

## Measures of Dispersion

What if two distributions have the same average? Does this mean that there is no difference in the distributions? Perhaps, but then again perhaps not. In Fig. 3-11*b*, distributions X and Y would have the same average, but they are certainly not identical. The difference lies in the degree of *spread, scatter, dispersion,* or *variability* along the horizontal axis of the values in each distribution. Obviously, the dispersion in distribution X is greater than the spread of the values in distribution Y. One simple measure of dispersion is the *range*—i.e., the difference between the highest and lowest values. Other measures that we will be considering are the *average deviation, standard deviation,* and *quartile deviation.* But they all do the same thing: They measure the extent of the dispersion in the distribution. Measures of dispersion will be computed in Chapter 4.

## Measure of Skewness

In Fig. 3-11*a* and *b*, the frequency distribution curves were all *symmetrical.* That is, if you were to draw a perpendicular line from the peaks of these curves to the baseline, you would divide the area of the curves into two *equal* parts. As you can see in Fig. 3-11*c*, however, curves may be *skewed* rather than symmetrical. Skewed curves occur when a few values are much higher or lower than the typical values found in the distribution. For example, distribution P in Fig. 3-11*c* might be the curve resulting from Professor Nastie's first statistics test. Most of the test scores are concentrated around the lower values, although a few curve breakers made extremely high grades. When the extreme values tail off to the *right* (as in distribution P), the curve is said to be *positively skewed.* Distribution N in Fig. 3-11*c*, on the other hand, might be a curve of the test scores obtained by Professor Sweet's statistics students. As you can see, most of the students made high scores (although a few unfortunates had extremely low grades). When extreme values tail off to the *left* (as in distribution N), the curve is said to be *negatively skewed.* A measure of the extent to which a distribution departs from the symmetrical is presented in Chapter 4.

## Measure of Kurtosis

**He learns. He becomes educated.**
**He instigorates knowledge.**

**—The House at Pooh Corner**

It is possible that three distributions (see Fig. 3-11*d*) may be symmetrical, may have the same average value, and may have the same dispersion value.[7] Yet the distributions may still possess different degrees of *kurtosis.* It is beyond the scope of

---

[7] As measured by the standard deviation.

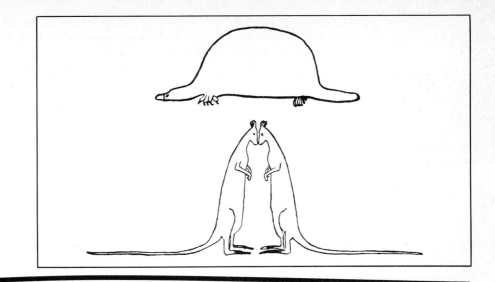

**FIGURE 3-12**

this book to consider the matter of kurtosis any further because little attention is paid to the subject in ordinary statistical analysis. Suffice it to say that a *mesokurtic* curve is generally normally peaked, with moderate-length tails; a *leptokurtic* curve is usually more peaked than normal, with lengthy tails; and a *platykurtic* curve is likely to be squat, with short tails. On the very remote chance that you should wish to remember these outrageous names, you might use the system of William S. Gosset, a famous early statistician who wrote: "I myself bear in mind the meaning of the words by [a] *memoria technica,* where the first figure represents platypus, and the second kangaroos, noted for 'lepping,' though, perhaps, with equal reason they should be hares!" (Figure 3-12 is a reproduction of a sketch supplied by Gosset.)

**MEASURES OF CENTRAL TENDENCY**

Data have a tendency in many cases to congregate about some central value, and this central value is often used as a summary measure to describe the general pattern of the data. If the collected data are limited in number (or if they are to be processed by computer), it may not be worthwhile to condense and/or simplify them by grouping them into a frequency distribution format. On the other hand, as we have seen, if the number of collected data items is large, and if there is a need to compress them into a more humanly usable form, the use of a frequency distribution becomes desirable. In the pages that follow we will first compute measures of central tendency using *ungrouped data,* and then we will look at methods of computing the same measures when the *data are grouped* into a frequency distribution.

**Ungrouped Data**

**The arithmetic mean** When most people use the word "average," they are referring to the arithmetic mean. And when you have added up the examination grades you have made in a subject during the course of a semester and divided by the number of exams taken, you have computed the arithmetic mean. The arithmetic mean is the most commonly used average.

Let us review the computation of the mean by considering the statistics grades made by Peter Parker[8] during one agonizing semester. (The grades have been *arrayed* in a descending order.)

$$
\begin{array}{r}
75 \\
75 \\
61 \\
50 \\
40 \\
25 \\
10 \\
5 \\
\underline{1} \\
342 \text{ Total of all grades}
\end{array}
$$

It is customary to let the capital letter $X$ represent the values of variables (such as Peter's grades). Thus, the tough formula to compute the *mean for ungrouped data* is:

$$\mu = \frac{\Sigma X}{N} \tag{3-1}$$

where $\mu$ (the Greek letter mu) = arithmetic mean
$\Sigma$ (the Greek letter sigma) = "the sum of"
$N$ = number of $X$ items in the listing[9]

Since, in the case of Peter's grades, $\Sigma X$ is 342, Peter's mean semester grade is 38 ($342/N$ or $342/9 = 38$).

**The median** The median is a measure of central tendency that occupies the *middle position* in an *array* of values. (Note that the word "array" has been emphasized; it is necessary to put the data into an ascending or descending order before selecting the median value.) In the example of Peter's grades, the middle value in the array, and thus the median grade, is 40. Although the median in this example deviates by a small amount from the mean, the ultimate result of using either the mean or the median as the semester average grade is the same—Peter doesn't have a clue about the general subject of statistics and has flagged the course. As we saw in the example of farm income in the last chapter, however, one or a few extremely high (or extremely low) values in a series can cause a substantial difference between the mean and the median.

---

[8] A name selected in memory of another loser, Sir Peter Parker, the British naval commander during the Revolutionary Battle of Sullivan's Island outside Charleston, South Carolina. During this battle, while giving orders aboard *HMS Bristol,* Sir Peter had the "unspeakable mortification" to have a cannon ball carry away the seat of his pants. (According to an old ballad, it "propelled him along on his bumpus.")

[9] The symbols used in this formula, and in the other formulas presented in this chapter and in Chapter 4, are appropriate when the data represent *all* the values in a *population.* Thus, the symbol $\mu$ refers to a population mean, and the symbol $N$ refers to the total number of items in a population. As you will see in the first section of Chapter 6 (and in Fig. 6-1), however, if the measures are being computed from *sample* data, different symbols must be used to maintain a distinction between population values and sample values.

What if Peter's instructor had dropped Peter's lowest grade before computing the median? In that event, the middle position in Peter's grade array would have been midway between 40 and 50 — i.e., the median value would then have been 45 (a change that doesn't do a thing for Peter's final grade).

**The mode** The mode, by definition, is the *most commonly occurring value* in a series. Thus, in the example of Peter's grades, the mode would be 75 — an average that appeals to Peter but not to his professor. Although not of much use in our grade example, the mode may be an important measure to a clothing manufacturer who must decide how many dresses of each size should be made. Obviously, the manufacturer will want to produce more dresses in the most commonly purchased size than in the other sizes.

**Self-testing Review 3-2**

1 The parts in this question all pertain to the following ungrouped data:

| (i) | (ii) |
|-----|------|
| 8 | 5 |
| 10 | 4 |
| 10 | 7 |
| 10 | 5 |
| 12 | 2 |
| 16 | 12 |
| 18 | 11 |
| 84 | 6 |
| | 52 |

a Determine the mean, median, and mode for the data in column *i*.
b Determine the mean, median, and mode for the data in column *ii*.
c If the data in column *i* are representative of the sizes of dresses sold during a typical day in a store, which measure of central tendency might be of most interest to the store's dress buyer?

2 Using the data array produced in Self-testing Review 3-1, question **1a,** determine the following measures:
a Arithmetic mean
b Median
c Mode

**Grouped Data**

Approximate values of the arithmetic mean, median, and mode may be computed from data grouped into a frequency distribution format. Let us now use the Boone & Docks Beverage Company data found in Fig. 3-4 to demonstrate these computational procedures.

**The arithmetic mean (direct method)** The direct method of computing the arithmetic mean from a frequency distribution is very similar to the method used to compute the mean from ungrouped data. However, since the compression of the data in frequency distribution form has resulted in the loss of the actual values of the observations in each class in the frequency column, it is necessary to make an assumption about these values. *The assumption (or estimate) made is that every observation in a class has a value equal to the class midpoint.* Thus, in Fig. 3-13, it is assumed that the two employees (*f*) in the first class each sold 85 gallons (*m*) of Fizzy Cola syrup,

**FIGURE 3-13 Computation of arithmetic mean (direct method)**

Gallons of Fizzy Cola syrup sold by 50 employees of Boone & Docks Beverage Company in 1 month

| Gallons sold | Number of employees ($f$) | Class midpoints ($m$) | $fm$ |
|---|---|---|---|
| 80 and less than 90 | 2 | 85 | 170 |
| 90 and less than 100 | 6 | 95 | 570 |
| 100 and less than 110 | 10 | 105 | 1050 |
| 110 and less than 120 | 14 | 115 | 1610 |
| 120 and less than 130 | 9 | 125 | 1125 |
| 130 and less than 140 | 7 | 135 | 945 |
| 140 and less than 150 | 2 | 145 | 290 |
| $N = \Sigma f = 50$ | | | 5760 |

$$\mu = \frac{\Sigma fm}{N} = \frac{5760}{50} = 115.2 \text{ gallons sold}$$

giving a total of 170 gallons ($fm$) sold. Of course, we have the advantage of knowing from Fig. 3-3 that neither employee actually sold 85 gallons, but their total sales of 171 gallons is only 1 gallon over our estimate. And although our assumption in this first class has caused a small *underestimate,* it is quite possible that a similar error occurring in another class may produce a small *overestimate.* Therefore, throughout a properly constructed distribution, the effect may be that most of these errors will be canceled out. For example, we have slightly overestimated the sales of the seven employees in the sixth class because it is assumed that they sold a total of 945 gallons (see the *fm* column in Fig. 3-13). In fact, their sales in Fig. 3-3 amounted to 943 gallons.

The computation of the mean of 115.2 gallons is shown in Fig. 3-13.[10] As you will notice, the *direct method formula for computing the mean for grouped data is:*

$$\mu = \frac{\Sigma fm}{N} \qquad (3\text{-}2)$$

where $f$ = frequency or number of observations in a class

$m$ = midpoint of a class and the assumed value of every observation in the class

$N$ = total number of frequencies or observations in the distribution

In the discussion of classification considerations a few pages earlier, it was pointed out that (1) open-ended classes should be avoided if possible, and (2) points

---

[10] The approximate mean of 115.2 gallons computed in Fig. 3-13 is very close to the true mean of 115.4 gallons, which is found by adding the 50 actual values in Fig. 3-3 (the total is 5,770 gallons) and dividing by 50. Of course, an analyst frequently does not have access to the raw data.

of data concentration should fall at the midpoint of a class interval.[11] Perhaps the reasons for these comments may now be clarified. *First,* the uses of *open-ended distributions are limited by the fact that it is impossible to compute the arithmetic mean from such distributions.* Why is this the case? Because, as you can see in Fig. 3-5, we cannot make any assumption about the income of each of the three families in the "$30,000 and over" class. Since there is no upper limit in this class, *there is no midpoint value* that we can assign as being equal to the total income for each of the three families. And *second,* if the raw data values should be concentrated at the lower (or upper) limits of several classes rather than at the class midpoints, the assumption that we have made in order to compute the approximate value of the mean would be incorrect and could lead to badly distorted results. For example, if the raw data tended to be concentrated around the lower limits of several classes, the computed mean could overstate the true mean by a significant amount.

**The arithmetic mean (shortcut method)** An alternative method of computing the arithmetic mean *when the class intervals are of equal width* is illustrated in Fig. 3-14. Since this method generally involves less-detailed intermediate computations

**FIGURE 3-14 Computation of arithmetic mean (shortcut method)**

Gallons of Fizzy Cola syrup sold by 50 employees of Boone & Docks Beverage Company in 1 month

| Gallons sold | Number of employees ($f$) | $d$ | $fd$ |
|---|---|---|---|
| 80 and less than 90 | 2 | −3 | −6 ⎫ |
| 90 and less than 100 | 6 | −2 | −12 ⎬ −28 |
| 100 and less than 110 | 10 | −1 | −10 ⎭ |
| 110 and less than 120 | 14 | 0 | 0 |
| 120 and less than 130 | 9 | +1 | +9 ⎫ |
| 130 and less than 140 | 7 | +2 | +14 ⎬ +29 |
| 140 and less than 150 | 2 | +3 | +6 ⎭ |
| | $\overline{50}$ | | $\overline{+\ 1}$ |

$$\mu = \mu_a + \left(\frac{\Sigma fd}{N}\right)(i)$$

$$\mu = 115 + (1/50)(10)$$

$$\mu = 115 + 1/5$$

$$\mu = \underline{\underline{115.2}} \text{ gallons sold}$$

---

[11] It was also noted earlier that, whenever possible, class interval widths should be equal. The direct method of computing the mean may be used when *either equal or unequal* widths are found, but the shortcut method presented in the next section can be used *only* when *equal* interval widths are employed.

and thus produces faster results, it is called the *shortcut method.* Basically, *the short-cut approach consists of these steps:*

1   Arbitrarily select the *midpoint* of *any* class to be the *assumed mean* value ($\mu_a$).
2   Determine the number of *class interval* deviations from the selected assumed mean to the midpoints of all other classes in the distribution.
3   Calculate a *correction factor* that will adjust the assumed mean so that the same value found by the use of the direct method will be obtained. (For example, if the assumed mean has been placed *too low,* a *positive correction* factor value will be added to the assumed mean, and if the assumed mean has been *too high,* a *negative correction* factor value will result in a decrease in the assumed mean.)

The formula that accomplishes these steps is:

$$\mu = \mu_a + \left(\frac{\Sigma fd}{N}\right)(i) \tag{3-3}$$

where $\mu_a$ = assumed mean placed at the *midpoint* of *any* class[12]
$\quad$ $f$ = frequencies or number of observations in a class
$\quad$ $d$ = deviations from the *assumed mean* to the middle of each class in *class interval units*
$\quad$ $N$ = total number of frequencies in the distribution
$\quad$ $i$ = size of the class interval

What was that you just said? "If this is a so-called 'shortcut,' I will take the long (and direct) way around. . . ." This isn't an unusual reaction, but please be patient while we trace through the use of formula (3-3) in Fig. 3-14. As you will notice, we have arbitrarily given the assumed mean a value of 115, which is the midpoint of the class "110 and less than 120." The values of $N$(50) and $i$ (10) are found simply by inspecting the distribution. Thus, only the numerator of $\Sigma fd$ in the fractional part of formula (3-3) remains to be determined. Since, by definition, the values of $d$ are deviations from the assumed mean of 115 to the midpoints of all the classes in class interval units, the midpoint of the class "110 and less than 120" would obviously be zero deviations from the assumed mean of 115. And since the midpoints of the classes "100 and less than 110" and "120 and less than 130" are *one class interval* from the assumed mean, the $d$ values for those classes are $-1$ and $+1$, respectively. The other values in the $d$ column in Fig. 3-14 are found in the same way.

---

[12] The word "midpoint" has been emphasized here for good reason. As we will see, in computing most other descriptive measures for frequency distributions the first symbol in many formulas refers to the *lower limit* of a class rather than to the class midpoint. Thus, when students are asked to compute several of these measures, they sometimes forget and also place the assumed mean at the lower limit of a class. Just make sure that *you* are not one of those who will commit this error if you are asked to take an exam on these measures. The word "any" has also been emphasized. The use of any class midpoint for the assumed mean will result in the correct answer, but *the computations are simplified* if you select the class (usually near the center of the distribution) which you think has the best chance of actually including the mean value.

Therefore, to find $\Sigma fd$, it is only necessary to (a) multiply the f values for each class by the corresponding d values to get the values in the fd column, and (b) total the negative and positive values in the fd column to arrive at the algebraic sum. (In Fig. 3-14 the value of $\Sigma fd$ is + 1, but this total will be a negative value if the assumed mean has been placed too high.) Once $\Sigma fd$ has been found, all that remains is to perform the calculations shown in Fig. 3-14 in order to compute the mean. (You will notice that the computed mean of 115.2 gallons in Fig. 3-14 is exactly the same as the computed value in Fig. 3-13.)

**The median** As we have seen, the median is the value that occupies the *middle position* in an array of values. However, since the actual values of the observations that go to make up a frequency distribution are lost when the distribution is constructed, it is only possible to approximate the median value from grouped data. We can illustrate this approximation process by referring to Fig. 3-15. Figure 3-15*a* shows what it is we are looking for when we compute the median for our Boone & Docks example problem. We might assume from the data given in the distribution in Fig. 3-15*b* that the employee whose sales quantity was lowest (call him employee 1) sold approximately 80 gallons, and we might guess that the highest-selling employee (call her employee 50) sold approximately 149.9 gallons. What we are looking for, however, is the approximate quantity sold by the middle (or twenty-fifth) employee.

The *first* step in computing this median value is to locate the *median class*—i.e., the class in our example which contains the twenty-fifth or median worker in the group of 50 employees. Although this is a simple matter, the optional *cumulative frequencies* column in Fig. 3-15*b* may be useful. As you can see, the 18 employees in the first three classes sold less than 110 gallons, and the 32 employees in the first four classes sold less than 120 gallons. Therefore, the twenty-fifth employee must be one of the 14 in the fourth or median class.

The *next* step is to determine which one of the 14 is the median employee. (You have probably already figured it out, but for the sake of others let's explain this step.) If 18 have been accounted for in the first three classes, and if we are looking for number 25, that employee must be the *seventh* one in the group of 14 in the fourth class (that is, $25 - 18 = 7$). In other words, in our example it just so happens that the median employee is found seven-fourteenths or one-half of the way through the median class.[13]

The *final* step is to compute the median value by *interpolating* within the median class. In order to perform this final step, *it is assumed that the sales of the 14 employees in the median class are evenly distributed throughout the class.*[14] Given this assumption, it is easy to see that if our median employee happens to be the middle one in the class, his or her sales should be halfway through the class interval. In short, the median value should be 115 gallons sold by the twenty-fifth employee.

---

[13] This is just a coincidence. The median observation could have been anywhere in the median class.

[14] This assumption of an even distribution of values throughout the median class is seldom likely to be exactly correct, and so the computed median will only be an approximation.

**FIGURE 3-15a What are we looking for?**

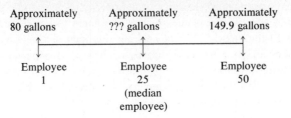

**FIGURE 3-15b Computation of the median**

Gallons of Fizzy Cola syrup sold by 50 employees of
Boone & Docks Beverage Company in 1 month

| Gallons sold | Number of employees ($f$) | Cumulative frequencies ($CF$) |
|---|---|---|
| 80 and less than 90 | 2 | 2 |
| 90 and less than 100 | 6 | 8 |
| 100 and less than 110 | 10 | 18 |
| 110 and less than 120* | 14 | 32 |
| 120 and less than 130 | 9 | 41 |
| 130 and less than 140 | 7 | 48 |
| 140 and less than 150 | 2 | 50 |
| | 50 | |

$$Md = L_{md} + \left(\frac{N/2 - CF}{f_{md}}\right)(i)$$

$$= 110 + \left(\frac{50/2 - 18}{14}\right)(10)$$

$$= 110 + (7/14)(10)$$

$$= 110 + 5$$

$$= \underline{\underline{115.0}} \text{ gallons}$$

* Median class.

The formula for computing the median given below is simply a formal presentation of the preceding paragraphs.

$$Md = L_{md} + \left(\frac{N/2 - CF}{f_{md}}\right)(i) \tag{3-4}$$

where $Md$ = median

$L_{md}$ = lower limit of the median class

$N$ = total number of frequencies in the distribution

$CF$ = cumulative frequencies in all classes up to, but not including, the median class

$f_{md}$ = frequency of the median class

$i$ = size of the interval of the median class

The computation using formula (3-4) is shown in Fig. 3-15b. You will also notice that the computed median of 115.0 gallons and the median located on the ogive in Fig. 3-10 are the same.[15]

**The mode** The mode is, by definition, the most commonly occurring value. But since actual values of observations are unknown in frequency distributions, the mode must also be approximated. It is assumed that the most commonly occurring value in a frequency distribution will be found in the largest class and directly under the peak of a frequency polygon. Thus, the class in the distribution with the largest number of frequencies is the *modal class*. Figure 3-16 demonstrates how a mode may be *graphically* approximated from a histogram, and the following formula shows how a mode may be approximated in mathematical terms:

$$Mo = L_{Mo} + \left(\frac{d_1}{d_1 + d_2}\right)(i) \qquad (3\text{-}5)$$

where $Mo$ = mode
$\qquad L_{Mo}$ = lower limit of the modal class
$\qquad d_1$ = difference between the frequency of the modal class and the frequency of the class immediately preceding it in the distribution
$\qquad d_2$ = difference between the frequency of the modal class and the frequency of the class immediately following it in the distribution
$\qquad i$ = size of the interval of the modal class

Using the data from the Boone & Docks example problem in Fig. 3-15b, we would calculate the mode as follows:

$$Mo = 110 + \left(\frac{4}{4 + 5}\right)(10)$$
$$= 110 + (40/9)$$
$$= 110 + 4.44$$
$$= 114.44 \text{ gallons sold}$$

**FIGURE 3-16**

**Graphical representation of the mode**

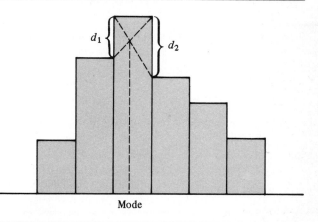

Mode

[15] You can verify from Fig. 3-3 that the true median is 115.25; therefore, our approximation is quite close.

CHAPTER 3/FREQUENCY DISTRIBUTIONS AND MEASURES OF CENTRAL TENDENCY

**Other measures** In addition to the mean, median, and mode, there are other specialized measures of central tendency that are occasionally used. The *weighted arithmetic mean,* for example, is a modification of the measure we have been computing that assigns *weights* or indications of *relative importance* to the values to be averaged. Thus, if you get grades of 83 and 87 on hourly statistics exams and a grade of 95 on the final, if the hourly exams each carry a weight of 25 percent of your semester grade, and if the final counts 50 percent of your semester grade, your weighted mean or semester average will be

$$\frac{83(25) + 87(25) + 95(50)}{100} = 90 \qquad \text{(congratulations!)}$$

In addition to the weighted arithmetic mean, there are two other lesser means that might be mentioned. Although it is beyond the scope of this book to go into the computation of these measures, there are certain circumstances under which they should be used. The *geometric mean,* for example, is the measure that should be employed to average ratios or rates of change expressed in positive numbers, and the *harmonic mean* is a measure that should be used to average time rates under certain conditions.

**Summary of Comparative Characteristics**

There is no general rule that will always identify the proper measure of central tendency to use. In a perfectly *symmetrical distribution,* the issue of which average to use may be simplified by the fact that the *arithmetic mean, median, and mode have the same value* (see Fig. 3-17a). But if the data are such that a skewed distribution results, the values of the three measures are different. In a *positively skewed* distribution, for example, the *mode* will remain under the peak of the curve and will have the smallest value; the *mean* will be influenced by the relatively limited number of extremely large values, will be pulled out from under the peak of the distribution in the direction of those extreme values, and will have the largest value; and the value of the *median* will lie between the values of the mode and the mean (see Fig. 3-17b). In a *negatively skewed* distribution, the *mode* will have the largest value and will still be found under the peak of the curve; the *mean* will have the smallest value because some extremely small data items will have been used in its computation; and, as always, the *median* will lie between the mode and the mean (see Fig. 3-17c).[16]

In selecting the proper average to use, the characteristics of each measure must be considered, and the type of data available must be evaluated. A summary of comparative characteristics for each measure is given below.

---

[16] In fact, it has been observed that in moderately skewed continuous distributions, the median will be located approximately two-thirds of the distance toward the mean from the mode. This approximation can be useful in checking the reasonableness of computed values of the three measures. In our Boone & Docks example problem, the distance between the mean of 115.2 gallons and the mode of 114.44 gallons is 0.76 gallon. If we add two-thirds of this 0.76 gallon to the mode of 114.44, we arrive at an approximation of the median of 114.95, which verifies the reasonableness of the median computation of 115.0 gallons.

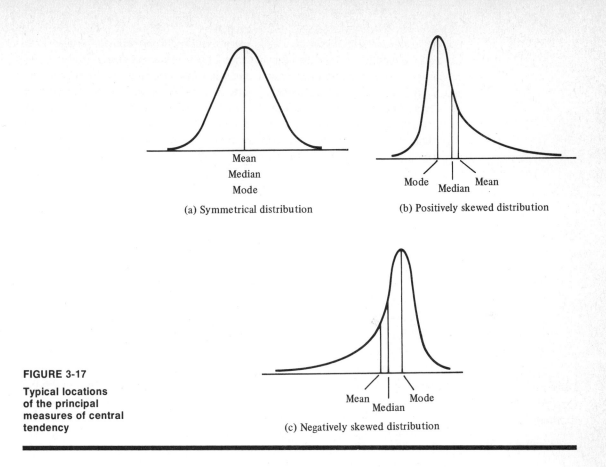

FIGURE 3-17

Typical locations
of the principal
measures of central
tendency

(a) Symmetrical distribution

(b) Positively skewed distribution

(c) Negatively skewed distribution

**The arithmetic mean** Some of the more important characteristics of the mean are:

1   *It is the most familiar and most widely used measure.* Long explanations of its meaning are thus not usually required.

2   *It is a computed measure whose value is affected by the value of every observation.* A change in the value of any observation will change the mean value; however, the mean value may not be the same as any of the observation values.

3   *Its value may be distorted too much by a relatively few extreme values.* Because it is affected by all the values of the variable, the mean (as we saw in Chapter 2) can lose its representative quality in badly skewed distributions.

4   *It cannot be computed from an open-ended distribution in the absence of additional information.* This point has been discussed earlier.

5   *It is the most reliable average to use when sample data are being used to make inferences about populations.* As we will see in Chapters 6 and 7, the mean of a sample of observations taken from a population may be used to estimate the value of the population mean.

**6** *It possesses two mathematical properties that will prove to be important in subsequent chapters.* The *first* of these properties is that the sum of the differences between data items and the mean of those items will be zero—i.e., $\Sigma (X - \mu) = 0$. And the *second* property is that if these differences between data items and the mean are *squared,* the sum of the squared deviations will be a *minimum value*—i.e., $\Sigma (X - \mu)^2$ = minimum value. To illustrate these properties suppose we have the following observations: 2, 3, 4, 7, and 9. The mean of these items is 25/5 or 5. These two properties of the mean may now be demonstrated as follows:

| $X$ | $(X - \mu)$ | $(X - \mu)^2$ |
|-----|-------------|----------------|
| 2 | −3 | 9 |
| 3 | −2 | 4 |
| 4 | −1 | 1 |
| 7 | +2 | 4 |
| 9 | +4 | 16 |
| 25 | 0 | 34 |

As you can see, $\Sigma (X - \mu)$ *must* equal zero (this is really a definition of the mean); and $\Sigma (X - \mu)^2$ in this case is 34. If we were to use any other value in place of the true mean of 5 in our example and follow through with the same procedure of determining and then squaring the deviations of the actual values from this other value, the result would be a total *greater than* our minimum value[17] of 34. (Of course, there isn't anything profound to note about the number 34 itself; a different set of data would almost certainly produce a minimum value of some other number.)

**The median** Some of the important characteristics of the median are:

**1** *It is easy to define and easy to understand.* The computation and interpretation of the median, as we have seen, is not difficult.

**2** *It is affected by the number of observations, but not by the value of these observations.* Thus, extremely high or low values will not distort the median.

**3** *It is frequently used in badly skewed distributions.* The median will not be affected by the size of the values of extreme items, and so it is a better choice than the mean when a distribution is badly skewed.

**4** *It may be computed in an open-ended distribution.* Since the median value is located in the median class interval, and since that interval is virtually certain of not being open-ended, the median may be determined.

**5** *It is generally less reliable than the mean for statistical inference purposes.* In the statistical inference chapters in Part 2 we will use the mean exclusively as the measure of central tendency.

---

[17] If, for example, we were to substitute the number 4 for our true mean of 5 and follow through with the procedure of determining and then squaring differences, $\Sigma (X - 4)^2$ would be 39. What would be the result if 6 were substituted for the true mean of 5?

**The mode** Some of the characteristics of the mode are:

1 *It is generally a less popular measure than the mean or median.*

2 *It may not exist in some sets of data, or there may be more than one mode in other sets of grouped data.* A distribution with two peaks (i.e., a *bimodal* distribution) should probably be reclassified into more than one distribution.

3 *It can be located in an open-ended distribution.*

4 *It is not affected by extreme values in a distribution.*

**Self-testing
Review 3-3**

1 Using the frequency distribution constructed in Self-testing Review 3-1, question **2**, determine the following measures:
  a The arithmetic mean   43
  b The median   44.4
  c The mode   44

  *med   mean   mode*

2 Compare the mean and median for the Lee Produce Company data computed in question **2a** and **b** of Self-testing Review 3-2 with the same measures computed in problem **1a** and **b** above. How do you account for the differences?

3 a From the information obtained in question **1** above, is the distribution for the Lee Produce Company negatively or positively skewed?
  b Why?   +ve

**SUMMARIZING
QUALITATIVE
DATA**

The primary purpose of this chapter has been to discuss *quantitative* measures of central tendency that may be used to summarize data. Before concluding this chapter, however, we should briefly mention the key measure used to summarize the relative frequency with which a particular characteristic occurs. This measure, used to summarize *qualitative* data, is the *percentage* or *proportion.*

If, for example, a machine produces 250 parts and a quality control check reveals that 21 of the parts are defective, the percentage of defective parts is (21/250) (100) or 8.4 percent.[18] An analytical approach (to be discussed in Part 2) using this percentage of defective parts might be used by a production manager as the basis for taking corrective action. Similarly, a campaign manager whose candidate receives 540 votes in a poll of 1,200 voters prior to an election—i.e., (540/1,200) (100) or 45 percent—may use this qualitative summary measure for analysis and planning purposes, and a television executive whose program is watched in 252 of

---

[18] We will summarize qualitative data in terms of percentages rather than proportions in this book because percentages are more frequently used in everyday discussion. A proportion, of course, is obtained simply by moving the percentage decimal point two places to the left. Thus 8.4 percent is .084 in proportion terms.

the 900 homes surveyed (i.e., in 28 percent of the homes) at the time the program is aired may also use this qualitative measure for decision-making purposes.

**SUMMARY**

Statistical description begins with the collection of raw data. To be meaningful, however, unorganized raw data should be arranged in some systematic order. One simple device for organizing raw data is the data array. From an array, it is possible to compute measures to summarize characteristics of ungrouped data. But when the number of data items is large, there often exists the need to compress the data into the more compact form of a frequency distribution. A number of basic rules or criteria for constructing a frequency distribution were presented in this chapter. Two basic forms of graphical presentation of the information found in a frequency distribution — the histogram and the frequency polygon — were also introduced in this chapter.

The basic properties or characteristics of frequency distribution data may be summarized and described by measures of central tendency, dispersion, skewness, and kurtosis. In this chapter we have concentrated on the computation of measures of central tendency such as the mean, median, and mode for both ungrouped and grouped data. A summary of the comparative characteristics of each of these measures has also been presented. In the next chapter we will look at the methods used to compute measures of dispersion and skewness. A summary of the characteristics of measures of dispersion will also be included.

**Important Terms and Concepts**

1  Variable
2  Discrete variable
3  Continuous variable
4  Array
5  Frequency distribution
6  Class interval
7  Frequencies
8  Class midpoint
9  Open-ended distribution
10  Histogram
11  Frequency polygon
12  Normal curve
13  Cumulative frequency distribution
14  Ogive
15  Arithmetic mean
16  Median
17  Mode

18  Ungrouped data
19  Grouped data
20  Dispersion
21  Skewness
22  Positively skewed
23  Negatively skewed
24  Symmetrical distribution
25  Kurtosis
26  $\mu = \dfrac{\Sigma X}{N}$
27  $\mu = \dfrac{\Sigma fm}{N}$
28  Direct method
29  Shortcut method
30  $\mu = \mu_a + \left(\dfrac{\Sigma fd}{N}\right)(i)$
31  Assumed mean
32  Median class

**33** $Md = L_{md} + \left(\dfrac{N/2 - CF}{f_{md}}\right)(i)$

**34** $Mo = L_{mo} + \left(\dfrac{d_1}{d_1 + d_2}\right)(i)$

**35** Weighted arithmetic mean

**36** $\Sigma (X - \mu) = 0$

**37** $\Sigma (X - \mu)^2 = $ minimum value

**Problems**

**1** The following scores were made by Professor Shirley A. Meany's accounting students on a test:

| | | | | |
|----|----|----|----|----|
| 68 | 52 | 49 | 56 | 69 |
| 74 | 41 | 59 | 79 | 81 |
| 42 | 57 | 60 | 88 | 87 |
| 47 | 65 | 55 | 68 | 65 |
| 50 | 78 | 61 | 90 | 85 |
| 65 | 66 | 72 | 63 | 95 |

  **a** Arrange the above grades in an ascending array.
  **b** What is the range of the values?
  **c** Compute the arithmetic mean, median, and mode for these grade values.

**2 a** Organize the data items from problem 1 above into a frequency distribution having the classes "40 and under 50," "50 and under 60," . . . , and "90 and under 100."
  **b** Compute the mean (by both the direct and shortcut methods), the median, and the mode for this frequency distribution.
  **c** Compare the values obtained in problem **2b** with the values found in problem **1c** and explain any discrepancies.
  **d** Is the distribution of test scores positively or negatively skewed? Explain your answer.

**3** The following distribution gives the miles traveled by 100 Lee Produce Company route trucks for the year 197x.

| Miles traveled | Number of trucks |
|---|:---:|
| 5,000 and under 7,000 | 5 |
| 7,000 and under 9,000 | 10 |
| 9,000 and under 11,000 | 12 |
| 11,000 and under 13,000 | 20 |
| 13,000 and under 15,000 | 24 |
| 15,000 and under 17,000 | 14 |
| 17,000 and under 19,000 | 11 |
| 19,000 and under 21,000 | 4 |

  **a** Construct a histogram and a frequency polygon of this mileage distribution.
  **b** Construct an ogive, and graphically locate the value of the median.
  **c** Compute the mean by the direct method and by the shortcut method.
  **d** Compute the median and interpret its meaning.
  **e** Compute the mode and interpret its meaning.
  **f** Is the distribution skewed? Explain your answer.

**4** The following data represent the annual earnings of families in two small New England communities, Languor and Friskyville:

| Annual earnings | Number of families | |
|---|---|---|
| | Languor | Friskyville |
| $ 5,000 and under $ 8,000 | 5 | 2 |
| 8,000 and under 11,000 | 40 | 20 |
| 11,000 and under 14,000 | 73 | 32 |
| 14,000 and under 17,000 | 52 | 58 |
| 17,000 and under 20,000 | 22 | 35 |
| 20,000 and under 23,000 | 8 | 30 |
| 23,000 and under 26,000 | – | 15 |
| 26,000 and over | – | 8 |
| | 200 | 200 |

**a** Compute the arithmetic mean annual earnings for each community.
**b** Compute the median annual earnings for each community.
**c** Compute the modal annual earnings for each community.
**d** Now that you have summarized the distributions, compare and analyze the economic situations in these two communities.

*Note:* If it is not possible to perform all the operations in any or all of the above parts, please indicate your reasons for omitting the operation(s).

**5** In order to prepare a government report, a university must determine the percentage of men and women faculty members in its several schools and colleges. The faculty data are as follows:

| School/College | A | B | C | D |
|---|---|---|---|---|
| Men | 148 | 64 | 12 | 102 |
| Women | 32 | 42 | 26 | 48 |

What is the percentage of women faculty members in each school or college?

**6** In his last game, the kicker for Gridiron University had punts of 42, 38, 51, 48, and 45 yards. Compute the arithmetic mean and median.

**7** The following distributions show the salaries paid to male and female employees by Chauvinists, Inc.

| Annual salaries | Number of employees | |
|---|---|---|
| | Male | Female |
| $ 6,000 and under $ 9,000 | 2 | 10 |
| 9,000 and under 12,000 | 5 | 25 |
| 12,000 and under 15,000 | 20 | 9 |
| 15,000 and under 18,000 | 18 | 6 |
| 18,000 and under 21,000 | 15 | 3 |
| 21,000 and under 24,000 | 8 | 2 |
| 24,000 and under 27,000 | 5 | 0 |
| 27,000 and under 30,000 | 2 | 0 |
| | 75 | 55 |

**a** Compute the arithmetic mean for each distribution.

**b** Compute the median for each distribution.

**c** Compute the mode for each distribution.

**Topics for Discussion**

**1** Distinguish between a discrete variable and a continuous variable.

**2** "Since measured quantities are continuously variable, they give only approximate results." Discuss this comment.

**3** What are the possible advantages to be obtained from arraying raw data?

**4** What are the implications of the fact that the reduction of data into a frequency distribution form results in some loss of detailed information?

**5** What basic rules or criteria should be observed in constructing a frequency distribution?

**6** "Whenever possible, the width of the classes should be equal." Discuss this statement.

**7** "Whenever possible, open-ended classes should be avoided." Discuss this statement.

**8** Why should points of data concentration be placed at the midpoint of a class interval?

**9** **a** What is a histogram?
**b** What is a frequency polygon?
**c** What is an ogive?

**10** What are the advantages and disadvantages of histograms and frequency polygons?

**11** Discuss the types of summary measures that may be used to describe the characteristics of frequency distributions.

**12** **a** What basic assumption is necessary in order to compute the arithmetic mean from grouped data?
**b** What basic assumption is necessary in order to compute the median from grouped data?
**c** What basic assumption is made in order to compute the mode from grouped data?

**13** "If the raw data tended to be concentrated around the lower limits of several classes, the computed mean could overstate the true mean by a significant amount." Explain this comment.

**14** **a** Discuss the typical locations of the mean, median, and mode in a negatively skewed distribution.
**b** Discuss the typical locations of these three measures in a positively skewed distribution.

**15** Summarize the comparative characteristics of the mean, median, and mode.

**16** Identify and discuss the two mathematical properties of the arithmetic mean.

**17** "A quantitative measure of central tendency tells us how much; a qualitative summary measure tells us how many." Discuss this statement.

**3-1**

**1  a**  The data array is:

| | |
|---|---|
| $21.50 | $43.50 |
| 25.35 | 44.80 |
| 29.00 | 45.00 |
| 30.30 | 45.10 |
| 31.25 | 46.00 |
| 34.60 | 47.75 |
| 35.55 | 49.50 |
| 36.50 | 49.80 |
| 36.60 | 52.10 |
| 38.50 | 55.00 |
| 39.95 | 55.60 |
| 40.25 | 56.30 |
| 41.10 | 58.00 |
| 42.30 | 59.20 |
| 42.50 | 65.50 |

**b**  The range is $44.00 ($65.50 − $21.50).

**2**

| Size of order | Number of orders |
|---|---|
| $20 and under $30 | 3 |
| 30 and under  40 | 8 |
| 40 and under  50 | 12 |
| 50 and under  60 | 6 |
| 60 and under  70 | 1 |
| | 30 |

**3  a**  Yes, the distribution could have had more classes, but because of the limited number of frequencies, five classes adequately present the data.

**b**  $i = \dfrac{L - S}{c} = \dfrac{\$65.50 - \$21.50}{8} = \dfrac{\$44.00}{8} = \$5.50$, or an interval of $6

**4**  See the figure presented below.

**5**  See the frequency polygon superimposed on the figure below.

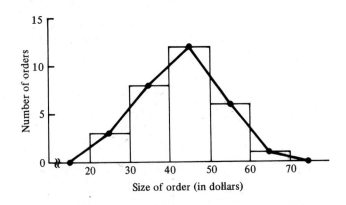

**3-2**

**1** **a** $\mu = 84/7 = 12$; $Md = 10$; $Mo = 10$

**b** $\mu = 52/8 = 6.5$; $Md = 5.5$ (You didn't forget to array the data, did you?); $Mo = 5$

**c** The mode

**2** **a** $\mu = \$1,298.40/30 = \$43.28$

**b** $Md = \$43.00$ (the middle position in the array)

**c** There is no mode, since no value occurs more than once.

**3-3**

**1** **a** $\mu = \$45 + \left(\dfrac{-6}{30}\right)(10)$

$= \$45 - \$2$

$= \$43.00$ by the shortcut method

or

$\mu = \$1,290/30 = \$43.00$ by the direct method

**b** $Md = \$40 + \left(\dfrac{15 - 11}{12}\right)(10)$

$= \$40 + \left(\dfrac{4}{12}\right)(10)$

$= \$43.33$

**c** $Mo = \$40 + \left(\dfrac{4}{4 + 6}\right)(10)$

$= \$40 + \left(\dfrac{4}{10}\right)(10)$

$= \$44.00$

**2** The differences are due to the fact that the above measures computed from a frequency distribution are approximations of the true values. The true mean, for example, is the $43.28 found in question **2a** of Self-testing Review 3-2. But since we did not have the actual values when we computed the mean from the frequency distribution, our computed value of $43.00 is only approximately correct.

**3** **a** Negatively skewed.

**b** Because the mean is the smallest value and the mode is the largest value.

# CHAPTER 4

# STATISTICAL DESCRIPTION: MEASURES OF DISPERSION AND SKEWNESS

**LEARNING OBJECTIVES**

After studying this chapter, working the problems, and answering the discussion questions, you should be able to:

☞ Explain the reasons for measuring absolute dispersion, relative dispersion, and skewness.

☞ Compute such measures of absolute dispersion as the range, the average deviation, and the standard deviation for ungrouped data.

☞ Compute the standard deviation and the quartile deviation for data organized in a frequency distribution.

☞ Explain the meaning of, and some of the characteristics of, the measures of absolute dispersion discussed in this chapter.

☞ Compute (and explain the purpose of) the coefficient of variation and a coefficient of skewness.

**CHAPTER OUTLINE**

MEASURES OF ABSOLUTE DISPERSION
  Ungrouped Data
  Self-testing Review 4-1
  Grouped Data
  Summary of Comparative
    Characteristics
  Self-testing Review 4-2

MEASURE OF RELATIVE DISPERSION
  Self-testing Review 4-3

MEASURE OF SKEWNESS
  Self-testing Review 4-4

SUMMARY

IMPORTANT TERMS AND CONCEPTS

PROBLEMS

TOPICS FOR DISCUSSION

ANSWERS TO SELF-TESTING REVIEW
  QUESTIONS

The measures of central tendency discussed in the last chapter are generally not, by themselves, sufficient to describe and summarize adequately the data being studied. In addition, *measures of dispersion* (or *variability* or *spread*) are needed. As the words "dispersion," "variability," and "spread" suggest, the measures that summarize this characteristic indicate to what extent the individual items in a series are scattered about an average size.

There are at least *two reasons for measuring dispersion*. The *first* reason is to form a judgment about the reliability of the average value. For example, if there is a large amount of scatter among the items in a series, the average size used to summarize the values may not be at all representative of the data being studied. (Of course, this thought is not new to us, for we saw in Chapter 2 the dangers of disregarded dispersions.) And a *second* reason for measuring dispersion is to learn the extent of the scatter so that steps may be taken to *control* the existing variation. For example, a tire manufacturer tries to produce a product that will have a long average mileage life. But the manufacturer is also interested in producing a tire of a *uniform* high quality so that there will not be a wide spread in tire mileage results that will alienate customers. (You are very pleased with the 40,000 miles you got from a set of these tires, but I am really chapped by the 12,000 miles I got with the same tires.) By measuring the existing variation, the manufacturer may see a need to improve the uniformity of the product through better inspection and quality control procedures.

In addition to measuring dispersion, it may be desirable to measure the extent to which a distribution departs from the symmetrical—i.e., the *degree of skewness* that is present—in order to also form an opinion about the representativeness of various descriptive measures. Thus, in this chapter we will consider the following topics: (1) *measures of absolute dispersion*, (2) a *measure of relative dispersion*, and (3) a *measure of skewness*.

## MEASURES OF ABSOLUTE DISPERSION

Measures of absolute dispersion, like measures of central tendency, are expressed in the *units of the original observations* (e.g., gallons sold, dollars earned, miles driven, test grades) and may be computed for *ungrouped data* as well as for data *grouped* into frequency distributions.

## Ungrouped Data

Three common measures of dispersion are often computed from ungrouped data. These measures are the *range*, the *average deviation*, and the *standard deviation*.

**The range** The range is the simplest and crudest measure of dispersion and is merely the difference between the highest and lowest values in an array. The range is used to report the movement of stock prices over a period of time, and weather reports typically state the high and low temperature readings for a 24-hour period. Since we discussed the range in the last chapter, we need not consider it further here.

**The average deviation** Let us assume that after his recent academic ordeal (discussed in the last chapter) our friend Peter Parker decides to get away and go to a seaside resort where he can romp, play, and drown his sorrows (but, it is hoped, not himself). A fellow student who works part-time at a travel agency tells Peter that her agency is making group travel arrangements for two suitable resorts. At the urging of Peter, his friend reports that the mean age of the unmarried females signed to go to resort A is 19, while the mean age of those going to resort B is 31. Peter quickly signs up to go to resort A and goes home to pack. The actual ages of the unmarried females going to each resort are as follows:

| Ages of unmarried females going to resort A | Ages of unmarried females going to resort B |
|---|---|
| 2 ⎫ | 18 |
| 2 ⎬ triplets | 19 |
| 2 ⎭ | 19 |
| 4 | 19 |
| 5 | 19 |
| 7 | 19 |
| 10 | 20 |
| 11 | 20 |
| 11 | 45 |
| 34 | 45 |
| 35 | 46 |
| 35 | 47 |
| 50 | 48 |
| 58 | 50 |
| 266   Total age | 434   Total age |

$$\mu = 266/14 = 19 \qquad \mu = 434/14 = 31$$

Had Peter looked beyond the mean age, it is likely that he would have made a different decision. What Peter really wanted was a mean age of 19 *and very little spread or scatter* of the individual ages about the mean. In short, Peter would have preferred a very small measure of dispersion to go along with the mean of 19. (Of course, Peter made a grade of 5 on the test covering measures of dispersion.) One measure of dispersion which would have alerted Peter to the spread of ages is the *average* (or *mean*) *deviation*.

To compute the average deviation it is necessary to (1) compute the mean of the observations being studied, (2) determine the absolute deviation—i.e., the deviation without regard to the algebraic sign—of each observation from this mean, and (3) compute the average (mean) of these absolute deviations. The appropriate formula is:

$$AD = \frac{\Sigma |X - \mu|}{N} \qquad (4\text{-}1)$$

where $AD$ = average deviation

$\qquad X$ = values of the observations

$\qquad \mu$ = mean of the observations

$\qquad |\;\;|$ = algebraic signs of the deviations are to be ignored[1]

$\qquad N$ = total number of observations

---

[1] If the signs of these deviations were not ignored, the sum of these deviations would always equal zero. Why? Because you will recall that the first mathematical property of the mean is $\Sigma (X - \mu) = 0$. Thus, it would be impossible to compute the average deviation unless absolute values were used.

**FIGURE 4-1 Computation of the average (mean) deviation**

Peter Parker passed a pack of pulchritude

| Ages of females going to resort B (1) | Mean age (2) | $\|X - \mu\|$ (1) − (2) |
|---|---|---|
| 18 | 31 | 13 |
| 19 | 31 | 12 |
| 19 | 31 | 12 |
| 19 | 31 | 12 |
| 19 | 31 | 12 |
| 19 | 31 | 12 |
| 20 | 31 | 11 |
| 20 | 31 | 11 |
| 45 | 31 | 14 |
| 45 | 31 | 14 |
| 46 | 31 | 15 |
| 47 | 31 | 16 |
| 48 | 31 | 17 |
| 50 | 31 | 19 |
| 434 | | 190 |

$$\mu = \frac{434}{14} = 31$$

$$AD = \frac{\Sigma |X - \mu|}{N} = \frac{190}{14} = \underline{13.57} \text{ years}$$

Figure 4-1 demonstrates the use of formula (4-1) in computing the average deviation for the ages of unmarried females going to resort B. The average deviation or spread in this example is 13.57 years. (Would the average deviation have been larger or smaller if we had computed it for those going to resort A?)

Unlike the range, the average deviation takes every observation into account and shows the average scatter of the data items about the mean; however, it is still relatively simple to understand and compute. Unfortunately, the procedure of ignoring the algebraic signs limits its use in further calculations.

**The standard deviation** The standard deviation is also used in conjunction with the mean and is the most widely used measure of dispersion. Like the computation of the average deviation, the computation of the standard deviation is based on, and is representative of, the deviations of the individual observations about the mean of those values. And another similarity with the average deviation is that as the actual observations become more widely scattered about their mean, the standard deviation will become larger and larger.[2] Unlike the computation of the average deviation,

---

[2] Alternatively, if all the observations in a series were identical in value, i.e., if there were no spread or scatter of values about the mean, the standard deviation would be zero.

**FIGURE 4-2 Computation of the standard deviation (ungrouped data)**

Peter Parker passed a pack of pulchritude

| Ages of females going to resort B $(X)$ (1) | Mean age $(\mu)$ (2) | $(X - \mu)$ (1) − (2) (3) | $(X - \mu)^2$ [(1) − (2)]² (4) |
|---|---|---|---|
| 18 | 31 | −13 | 169 |
| 19 | 31 | −12 | 144 |
| 19 | 31 | −12 | 144 |
| 19 | 31 | −12 | 144 |
| 19 | 31 | −12 | 144 |
| 19 | 31 | −12 | 144 |
| 20 | 31 | −11 | 121 |
| 20 | 31 | −11 | 121 |
| 45 | 31 | 14 | 196 |
| 45 | 31 | 14 | 196 |
| 46 | 31 | 15 | 225 |
| 47 | 31 | 16 | 256 |
| 48 | 31 | 17 | 289 |
| 50 | 31 | 19 | 361 |
| 434 | | 0 | 2,654 |

$$\mu = \frac{\Sigma X}{N} \qquad \sigma = \sqrt{\frac{\Sigma(X - \mu)^2}{N}}$$

$$= \frac{434}{14} = 31 \qquad = \sqrt{\frac{2,654}{14}}$$

$$= \sqrt{189.57}$$

$$= \underline{\underline{13.8}} \text{ years}$$

however, the computation of the standard deviation does not ignore the algebraic signs of the deviations about the mean. Let us now work through an example problem to illustrate the procedure for computing the standard deviation.

Figure 4-2 shows the calculation of the standard deviation for the unmarried females going to resort B. The steps involved in the computation are:

1 The arithmetic mean of the data is computed. (We have seen that it is 31.)

2 The mean is subtracted from each of the individual ages in column 1. (See column 3.)

3 The deviations of the individual ages about the mean (column 3) are squared (see column 4) and totaled. This total, $\Sigma(X - \mu)^2$, is the second mathematical property of the mean discussed in Chapter 3; it is a minimum value. Appendix 5, at the back of the book, has been provided to help you quickly look up the squares of numbers from 1 through 1,000.

4 The mean of the deviations in column 4 is computed. This value is called the *variance*. (Take another look at the steps required to compute the variance, for it

is an important statistical measure in its own right. In Chapter 10, we will be computing the variances of a number of samples as a part of an *analysis of variance* procedure that is used to test whether the arithmetic means of a number of populations are likely to be equal.) Unfortunately, however, although the variance measures the average amount of variability that exists about the mean of the original values, it is not expressed in the units of the original data. That is, the variance in our example is 189.57, but this value represents the average variability of *squared* ages. Thus, to obtain a measure of dispersion expressed in terms of the original values, the following final step is needed.

5   *The standard deviation is computed by taking the square root of the variance.* (Appendix 5, at the back of the book, will also help you quickly look up square root values.) As you can see in Fig. 4-2, the standard deviation for our example is 13.8 years of age. (The standard deviation will always be larger than the average deviation for the same data because the squaring of deviations places more emphasis on the extreme values.) Would the standard deviation figure have been larger or smaller if we had computed it for the unmarried females going to resort A?

The appropriate formula for the above steps is:[3]

$$\sigma = \sqrt{\frac{\Sigma(X - \mu)^2}{N}} \tag{4-2}$$

where $\sigma$ (small Greek letter sigma) = standard deviation
$\qquad\qquad\quad X$ = values of the observations
$\qquad\qquad\quad \mu$ = mean of the observations
$\qquad\qquad\quad N$ = total number of observations[4]

---

[3] An alternative formula that you may find easier to use when the number of observations is large is:

$$\sigma = \sqrt{\frac{\Sigma X^2 - N(\mu)^2}{N}}$$

Squaring the values in column 1 of Fig. 4-2 and then adding these squared values will produce a total of 16,108. The standard deviation will then be computed as follows:

$$\sigma = \sqrt{\frac{16,108 - 14(31)^2}{14}}$$

$$= \sqrt{\frac{16,108 - 14(961)}{14}}$$

$$= \sqrt{\frac{2,654}{14}}$$

$$= \underline{13.8} \text{ years}$$

[4] The symbols used in this formula (and in the formulas for the standard deviation for grouped data that will be presented later) are appropriate when the data represent *all* the values in a *population*. As you will see in the first section of Chapter 6 (and in Fig. 6-1), however, if the standard deviation is being computed from sample data, different symbols must be used to maintain a distinction between population values and sample values.

1  Using the ages of the unmarried females going to resort A in the Peter Parker fiasco, compute the following measures:
   a   The range
   b   The average deviation
   c   The variance
   d   The standard deviation

2  The following statistics grades were made by Peter Parker last semester and were presented in Chapter 3:

| Grades | | $X-\mu$ | $X-\mu^2$ |
|---|---|---|---|
| 75 | 38 | 37 | 1369 |
| 75 | | 37 | 1369 |
| 61 | | 23 | 529 |
| 50 | | 12 | 144 |
| 40 | | 2 | 4 |
| 25 | | -13 | 169 |
| 10 | | -28 | 784 |
| 5 | | -33 | 1089 |
| 1 | | -37 | 1369 |
| 342 | | | 6826 |

Compute the following:
   a   The average deviation
   b   The variance   758
   c   The standard deviation   27.53

## Grouped Data

When data are grouped into a frequency distribution format, the average deviation is seldom computed. Rather, the primary measures of dispersion are the standard deviation, which is used along with the mean for descriptive purposes, and the quartile deviation, which, together with the median, is used to describe those distributions whose characteristics either cannot or should not be represented by the mean and the standard deviation. In the following demonstrations of the procedures for computing approximations of the standard and quartile deviations, we shall once again use the Boone & Docks Beverage Company data found in Chapter 3, Fig. 3-4.

**The standard deviation (direct method)** The direct method of computing the standard deviation from a frequency distribution is very similar to the method used to compute the measure from ungrouped data. *The direct method formula used to compute the standard deviation* in Fig. 4-3 is:

$$\sigma = \sqrt{\frac{\Sigma f(m - \mu)^2}{N}} \tag{4-3}$$

where $f$ = frequency or number of observations in a class
      $m$ = midpoint of a class and the assumed value of every observation in the class
      $N$ = total number of frequencies or observations in the distribution

As you can see in Fig. 4-3, the standard deviation for the Boone & Docks Company data is 14.63 gallons. And did you also notice in Fig. 4-3 that the procedure to compute the standard deviation by the direct method involves several columns and a

## FIGURE 4-3 Computation of standard deviation (direct method)

Gallons of Fizzy Cola syrup sold by 50 employees of
Boone & Docks Beverage Company in 1 month

| Gallons sold | Number of employees | Class midpoints | | Deviation $(m - \mu)$ | $(m - \mu)^2$ | $f(m - \mu)^2$ |
|---|---|---|---|---|---|---|
| | $(f)$ | $(m)$ | $(fm)$ | | | |
| 80 and less than 90 | 2 | 85 | 170 | −30.2 | 912.04 | 1824.08 |
| 90 and less than 100 | 6 | 95 | 570 | −20.2 | 408.04 | 2448.24 |
| 100 and less than 110 | 10 | 105 | 1050 | −10.2 | 104.04 | 1040.40 |
| 110 and less than 120 | 14 | 115 | 1610 | −0.2 | 0.04 | 0.56 |
| 120 and less than 130 | 9 | 125 | 1125 | 9.8 | 96.04 | 864.36 |
| 130 and less than 140 | 7 | 135 | 945 | 19.8 | 392.04 | 2744.28 |
| 140 and less than 150 | 2 | 145 | 290 | 29.8 | 888.04 | 1776.08 |
| | 50 | | 5760 | | | 10,698.00 |

$$\mu = \frac{\Sigma fm}{N} = \frac{5760}{50} = 115.2 \text{ gallons} \qquad \sigma = \sqrt{\frac{\Sigma f(m - \mu)^2}{N}}$$

$$= \sqrt{\frac{10,698.00}{50}}$$

$$= \sqrt{213.96}$$

$$= \underline{\underline{14.63}} \text{ gallons}$$

number of tedious calculations?[5] If the class intervals in the distribution are of equal width, you can eliminate some of the monotonous calculations by using the short-cut method presented below; if unequal widths are used, you have no choice but to use a direct method.

---

[5] An alternative direct method formula that you might find easier to use is:

$$\sigma = \sqrt{\frac{\Sigma f(m)^2 - N(\mu)^2}{N}}$$

Squaring each of the class midpoint values in column 2 of Fig. 4-3, multiplying each of these squared values by the corresponding class frequencies found in column 1 of Fig. 4-3, and then adding the products of this multiplication step will produce a total of 674,250. The standard deviation will then be computed as follows:

$$\sigma = \sqrt{\frac{674,250 - 50(115.2)^2}{50}}$$

$$= \sqrt{\frac{674,250 - 663,552}{50}}$$

$$= \sqrt{\frac{10,698}{50}}$$

$$= \underline{14.63} \text{ gallons}$$

CHAPTER 4/STATISTICAL DESCRIPTION: MEASURES OF DISPERSION AND SKEWNESS

**FIGURE 4-4 Computation of standard deviation (shortcut method)**

Gallons of Fizzy Cola Syrup sold by 50 employees of
Boone & Docks Beverage Company in 1 month

| Gallons sold | Number of employees (f) | d | fd | fd² |
|---|---|---|---|---|
| 80 and less than 90 | 2 | −3 | −6⎤ | 18 |
| 90 and less than 100 | 6 | −2 | −12⎬ −28 | 24 |
| 100 and less than 110 | 10 | −1 | −10⎦ | 10 |
| 110 and less than 120 | 14 | 0 | 0 | 0 |
| 120 and less than 130 | 9 | +1 | 9⎤ | 9 |
| 130 and less than 140 | 7 | +2 | 14⎬ +29 | 28 |
| 140 and less than 150 | 2 | +3 | 6⎦ | 18 |
| | 50 | | +1 | 107 |

$$\sigma = i\sqrt{\frac{\Sigma fd^2}{N} - \left(\frac{\Sigma fd}{N}\right)^2}$$

$$= 10\sqrt{107/50 - (1/50)^2}$$

$$= 10\sqrt{2.14 - (0.0004)}$$

$$= 10\sqrt{2.1396}$$

$$= 10\,(1.463)$$

$$= \underline{14.63}\text{ gallons}$$

**The standard deviation (shortcut method)** Another method of computing the standard deviation *when the class intervals are of equal width* is illustrated in Fig. 4-4. Just as in the calculation of the mean by the shortcut method (see Fig. 3-14 in Chapter 3), the procedure here consists of (1) arbitrarily selecting the midpoint of any class to be the assumed mean value, (2) determining the deviations from the assumed mean to the middle of each class in class interval units, (3) multiplying the frequency ($f$) in each class by the class deviation ($d$) from the assumed mean, and (4) multiplying these $fd$ values by $d$ again to get $fd^2$ values for each class. If all this seems confusing to you, just remember that Fig. 4-4 is very similar to Fig. 3-14 (used to compute the mean by the shortcut method), with the sole difference being that the values in the $d$ and $fd$ columns have been multiplied to get the *one additional $fd^2$* column. Thus, if you have already computed the mean for a distribution by the short-cut method, all that is required is to add the $fd^2$ column to your work sheet and apply the following *shortcut formula:*

$$\sigma = i\sqrt{\frac{\Sigma fd^2}{N} - \left(\frac{\Sigma fd}{N}\right)^2} \qquad (4\text{-}4)$$

where $i$ = size of the class interval
$f$ = frequency in a class
$d$ = deviations from the assumed mean to the middle of each class in class interval units
$\Sigma fd^2$ = sum of $fd$ times $d$ for each class, and *not* $\Sigma f$ times $\Sigma d^2$

The standard deviation of 14.63 gallons found in Fig. 4-4 is, of course, exactly the same result computed in Fig. 4-3.

**Interpreting the standard deviation** As we saw in the last chapter in Fig. 3-8, when a large number of values are analyzed, they are often found to be distributed or scattered about their arithmetic mean in a symmetrical way. The standard deviation is a particularly important measure of dispersion because of its relationship with the mean in such a bell-shaped or normal distribution. Although this relationship will be considered in much greater detail in the next chapter, it is appropriate here to show briefly how the standard deviation may be used with the mean to indicate the proportions of the observations in the distribution that fall within specified distances from the mean.

Suppose, for example, that a large number of people were given an IQ test, that the resulting raw scores were organized into a frequency distribution, that the frequency polygon prepared from the distribution was symmetrical or normal in shape, and that an arithmetic mean of 100 points and a standard deviation of 10 points were computed. In such a situation, the mean IQ score would be directly under the peak of the curve, and the following relationships would exist: (1) *68.3 percent* of the test scores would fall within *one* standard deviation of the mean—i.e., 68.3 percent of the people would have test scores between 90 and 110 points; (2) *95.4 percent* of the test scores would fall within *two* standard deviations of the mean—i.e., slightly over 95 percent of those taking the test would have scores between 80 and 120 points; and (3) *virtually all* (99.7 percent) the test scores would fall within *three* standard deviations of the mean (scores between 70 and 130).[6] Figure 4-5 illustrates these relationships.

These precise relationships that exist between the mean and the standard deviation in a normal distribution may also be used for analysis purposes with distributions that are only approximately normal. Thus, we are now in a position to

---

[6] These very precise percentage figures are obtained from Appendix 2 at the back of the book. The use of Appendix 2 will be explained in the next chapter.

**FIGURE 4-5**

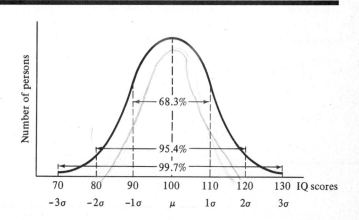

CHAPTER 4/STATISTICAL DESCRIPTION: MEASURES OF DISPERSION AND SKEWNESS

interpret the meaning of the standard deviation of 14.63 gallons in our Boone & Docks example distribution, since that distribution is approximately normal. We can conclude that approximately the middle two-thirds of the 50 employees sold syrup quantities between $\mu \pm 1 \sigma$, that is, between 115.20 gallons $\pm$ 14.63 gallons (or from 100.57 to 129.83 gallons). Furthermore, about 95 percent of the employees sold syrup quantities between $\mu \pm 2\sigma$, or between 85.94 and 144.46 gallons.[7] Of course, all the employees sold syrup quantities between $\mu \pm 3\sigma$.

**The quartile deviation** Like the range, the quartile deviation is a measure that describes the existing dispersion in terms of the *distance* between selected observation points. With the range, of course, the observation points were simply the highest and lowest values. In determining the quartile deviation, however, we will compute an *interquartile range* that will include approximately[8] the *middle 50 percent* of the values in the distribution. Thus, our observation points, as illustrated in Fig. 4-6, will be at the *first* ($Q_1$) and *third* ($Q_3$) quartile positions. The interquartile range will then simply be the distance or difference between $Q_3$ and $Q_1$. The first quartile position is simply the point which separates the *lower* 25 percent of the values from the upper 75 percent, and the third quartile position is the point which separates the *upper* 25 percent of the values from the lower 75 percent. Thus, the lower and upper 25 percent of the values are not considered in the computation of the quartile deviation (and open-ended distributions present no problems). *The second quartile value is just another name for the median.*[9] And as a matter of fact, in computing the values of $Q_3$ and $Q_1$ in order to find the interquartile range, we will be following an interpolation procedure that is completely analogous to the procedure used in the last chapter to compute the median. So let's get started on the quartile deviation by once again using the Boone & Docks Company data.[10]

---

[7] You can verify from the data array in Fig. 3-3 that 66 percent of the employees sold between 100.57 and 129.83 gallons and that 96 percent of them sold between 85.94 and 144.46 gallons.

[8] Once again, since the actual values of the observations are lost when a distribution is constructed, it is only possible to approximate the interquartile range.

[9] Also, the median is the 5th decile, the 50th percentile . . . it's enough to make you cry.

[10] By now, seeing this distribution again may also be enough to make you cry!

**FIGURE 4-6**

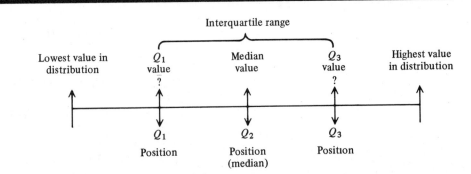

**FIGURE 4-7 Computation of the quartile deviation**

Gallons of Fizzy Cola syrup sold by 50 employees of
Boone & Docks Beverage Company in 1 month

| Gallons sold | Number of employees ($f$) | Cumulative frequencies ($CF$) |
|---|---|---|
| 80 and less than 90 | 2 | 2 |
| 90 and less than 100 | 6 | 8 |
| 100 and less than 110 | 10 | 18 |
| 110 and less than 120 | 14 | 32 |
| 120 and less than 130 | 9 | 41 |
| 130 and less than 140 | 7 | 48 |
| 140 and less than 150 | 2 | 50 |
| | 50 | |

$$Q_1 = L_{Q_1} + \left(\frac{N/4 - CF}{f_{Q_1}}\right)(i) \qquad Q_3 = L_{Q_3} + \left(\frac{3N/4 - CF}{f_{Q_3}}\right)(i)$$

$$= 100 + \left(\frac{50/4 - 8}{10}\right)(10) \qquad = 120 + \left(\frac{150/4 - 32}{9}\right)(10)$$

$$= 100 + \left(\frac{12.5 - 8}{10}\right)(10) \qquad = 120 + \left(\frac{37.5 - 32}{9}\right)(10)$$

$$= 100 + 4.5 \qquad\qquad = 120 + 6.11$$

$$= 104.5 \text{ gallons} \qquad\qquad = 126.11 \text{ gallons}$$

Interquartile range $= Q_3 - Q_1$

$$= 126.11 - 104.5$$

$$= 21.61 \text{ gallons}$$

Quartile deviation $= \dfrac{Q_3 - Q_1}{2}$

$$= \frac{21.61}{2} \text{ gallons}$$

$$= \underline{\underline{10.81}} \text{ gallons}$$

The computation of the quartile deviation is illustrated in Fig. 4-7. The *formula
for computing the $Q_1$ value is:*

$$Q_1 = L_{Q_1} + \left(\frac{N/4 - CF}{f_{Q_1}}\right)(i) \qquad (4\text{-}5)$$

where $Q_1 =$ first quartile value

$L_{Q_1} =$ lower limit of the first quartile class—i.e., the first class whose cumulative frequency exceeds $N/4$ observations

$CF =$ cumulative frequencies in all classes up to, but not including, the *first
quartile class*

$f_{Q_1} =$ frequency of the first quartile class

$i =$ size of the interval of the first quartile class

And *the formula for computing the $Q_3$ value is:*

$$Q_3 = L_{Q_3} + \left(\frac{3N/4 - CF}{f_{Q_3}}\right)(i) \qquad (4\text{-}6)$$

where $Q_3$ = third quartile value

$\quad L_{Q_3}$ = lower limit of the third quartile class — i.e., the first class whose cumulative frequency exceeds $3N/4$ observations

$\quad CF$ = cumulative frequencies in all classes up to, but not including, the *third quartile class*

$\quad f_{Q_3}$ = frequency of the third quartile class

$\quad i$ = size of the interval of the third quartile class

As you will notice in Fig. 4-7, the computed values of $Q_1$ and $Q_3$ are 104.5 gallons and 126.11 gallons. Thus, approximately the middle 50 percent of the Boone & Docks employees sold syrup quantities between 104.5 and 126.11 gallons. Stated another way, the interquartile range $(Q_3 - Q_1)$ of 21.61 gallons indicates that the sales of the central 50 percent of the employees varied within the approximate value of 21.61 gallons.

The *quartile deviation (QD) is simply one-half the interquartile range* and is sometimes referred to as the *semiinterquartile range*. That is,

$$QD = \frac{Q_3 - Q_1}{2} \qquad (4\text{-}7)$$

Obviously, the smaller the $QD$, the greater the degree of concentration of the middle half of the observations in the distribution. If a distribution is normal in shape, exactly 50 percent of the values will be found in the range of the median $\pm 1QD$ because the values of $Q_1$ and $Q_3$ will be equal distances from the median. This relationship can also be used for analysis purposes with distributions that are only approximately normal. Thus, in our Boone & Docks example, we can conclude that approximately the middle 50 percent of the employees sold syrup quantities between $Md \pm 1QD$, that is, between 115.0 gallons $\pm$ 10.81 gallons (or from 104.19 to 125.81 gallons).[11]

**Summary of Comparative Characteristics**

There is no general rule that will always identify the proper measure of absolute dispersion to use. In selecting the appropriate measure, the characteristics of each must be considered, and the type of data available must be evaluated. A summary of comparative characteristics for each measure discussed in the preceding pages is given below.

**The range** Some of the characteristics of the range are:

1 *It is the easiest measure to compute.* Since its calculation involves only one subtraction, it is also the easiest measure to understand.

---

[11] You can verify from the data array in Fig. 3-3 that the middle 50 percent of the employees actually sold between 104.75 and 125.81 gallons.

**2** *It emphasizes only the extreme values.* Because the more typical items are completely ignored, the range may give a very distorted picture of the true dispersion pattern.

**The average deviation** Some of the characteristics of the average deviation are:

**1** *It gives equal weight to the deviation of every observation.* Thus, it is more sensitive than measures such as the range or quartile deviation that are based on only two values.

**2** *It may be computed from the median as well as from the mean.* Although formula (4-1) called for the computation of the deviations about the mean of the values, the median could also have been used to compute an average deviation.

**3** *It is an easy measure to compute.* It is also not difficult to understand.

**4** *It is not influenced as much by extreme values as the standard deviation.* The squaring of deviations in the calculation of the standard deviation places more emphasis on the extreme values.

**5** *Its use is limited in further calculations.* Because the algebraic signs are ignored, the average deviation is not as well suited as the standard deviation for further computations.

**The standard deviation** Included among the characteristics of the standard deviation are:

**1** *It is the most frequently encountered measure of dispersion.* Because of the mathematical properties it possesses, it is more suitable than any other measure of dispersion for further analysis involving statistical inference procedures. We shall use the standard deviation extensively in the chapters in Part 2.

**2** *It is a computed measure whose value is affected by the value of every observation in a series.* A change in the value of any observation will change the standard deviation value.

**3** *Its value may be distorted too much by a relatively few extreme values.* Like the mean, the standard deviation can lose its representative quality in badly skewed distributions.

**4** *It cannot be computed from an open-ended distribution in the absence of additional information.* As formula (4-2) shows, if the mean cannot be computed, neither can the standard deviation.

**The quartile deviation** Some of the characteristics of the quartile deviation are:

**1** *It is similar to the range in that it is based on only two values.* As we have seen, these two values identify the range of the middle 50 percent of the values.

**2** *It is easy to define and easy to understand.* (All right, so the computations were a bit tedious. But the concept of the quartile deviation wasn't hard to understand, was it?)

**3** *It is frequently used in badly skewed distributions.* The quartile deviation will not be affected by the size of the values of extreme items, and so it may be preferable to the average or standard deviation when a distribution is badly skewed.

**4** *It may be computed in an open-ended distribution.* Since the upper and lower 25 percent of the values are not considered in the computation of the quartile deviation, an open-ended distribution presents no problem.

**Self-testing Review 4-2**

**1** The data for the following frequency distribution were first presented in Self-testing Review 3-1, and the mean, median, and mode for this distribution of orders placed by customers of the Lee Produce Company were computed in Self-testing Review 3-3.

| Size of order | Number of orders |
|---|---|
| $20 and under $30 | 3 |
| 30 and under 40 | 8 |
| 40 and under 50 | 12 |
| 50 and under 60 | 6 |
| 60 and under 70 | 1 |
| | 30 |

Determine the following measures:
**a** The standard deviation
**b** The quartile deviation

**2** Analyze the above distribution by the use of
**a** The mean and the standard deviation
**b** The median and the quartile deviation

**MEASURE OF RELATIVE DISPERSION**

The standard deviation and the other gauges of dispersion we have studied in the preceding pages are measures of *absolute dispersion.* That is, they are expressed in the units of the original observations. Thus, one distribution of the annual earnings of a group of fire fighters may have a standard deviation of $400, and another distribution of the annual earnings of a group of plumbers may have a standard deviation of $800. Or, to use another example, one distribution may have a standard deviation of 14.63 gallons, while another has a standard deviation of $9.80. If we desire to *compare* the dispersions of the first two distributions, can we conclude that the distribution with the $800 standard deviation has twice the variability of the one with the $400 standard deviation? Or can we conclude that the distribution with the 14.63-gallon standard deviation is more widely dispersed than the one with the $9.80 measure? (Can we even logically compare gallons and dollars?) The answer to these questions is that we can't conclude anything by simply comparing the measures of absolute dispersion. Rather, what is needed for comparison purposes is a measure of the degree of *relative dispersion* that exists in the distributions being studied.

The most popular measure of relative dispersion is the *coefficient of variation (CV)*, which is simply the standard deviation of a distribution expressed as a percentage of the mean of the distribution. That is,

$$CV = \frac{\sigma}{\mu} \ (100) \qquad\qquad\qquad \text{(4-8)}$$

Thus, if the mean for the distribution of annual earnings of the fire fighters is \$10,000, the coefficient of variation is computed as follows:

$$CV = \frac{\$400}{\$10,000} \ (100) = 4 \text{ percent}$$

And let us assume that the mean annual earnings for the group of plumbers is \$22,000. The coefficient of variation for this distribution is then 3.64 percent: $CV =$ (\$800/\$22,000)(100). Therefore, the distribution with the standard deviation of \$800 not only *does not* have twice the variability of the one with the \$400 standard deviation; it actually has *less* relative dispersion. In other words, the annual earnings received by the group of plumbers are slightly more uniform than the earnings received by the fire fighters.

**Self-testing Review 4-3**

1 Compute the coefficient of variation for the Boone & Docks example problem.

2 Compute the coefficient of variation for the Lee Produce Company data discussed in Self-testing Review 4-2.

3 Which of these distributions has the greater relative dispersion?

**MEASURE OF SKEWNESS**

You will remember from Fig. 3-17 that on the frequency polygon of a symmetrical distribution, the mean, median, and mode are all located directly under the peak of the curve. You will also recall that as a distribution departs from the symmetrical — i.e., as it becomes skewed — these measures of central tendency are separated (see Fig. 3-17b and c), with the mode remaining under the peak of the curve and the mean moving the greatest distance out in the direction of the tail of the distribution. These relationships between measures of central tendency are utilized in the following *coefficient of skewness* formula, which gives the direction (negative or positive) as well as an indication of the degree of skewness ($Sk$):

$$Sk = \frac{3(\mu - Md)}{\sigma} \qquad\qquad\qquad \text{(4-9)}$$

When a distribution is symmetrical, formula (4-9) will give a value of zero, because the mean and median are equal, and the numerator of the formula will thus be zero. As the mean and median become separated, however, the coefficient of skewness will move from zero toward either a negative or positive value of one. (It will seldom exceed $\pm 1$.) Obviously, the closer the value is to zero, the less skewed the distribution will be. The value is not, of course, expressed in terms of any units of measure. The coefficient of skewness for our Boone & Docks example distribution is computed as follows:

$$Sk = \frac{3(115.2 - 115.0)}{14.63} = +.041$$

This value of +.041 represents a very small degree of positive skewness. (If the distribution had been skewed to the left, the value of the mean would have been smaller than the median and the sign of the numerator of formula (4-9) would have been negative.)

**Self-testing Review 4-4**

1 Compute the coefficient of skewness for the Lee Produce Company distribution. (The values of the measures needed are found in earlier self-testing review sections.)

2 Interpret the value computed in question **1.**

**SUMMARY**

Measures of dispersion are needed to form a judgment about the reliability of the average value and to learn the extent of the scatter of the observations so that steps may be taken to control the existing variation. Such measures may be computed for ungrouped data as well as for data grouped into frequency distributions. Among the measures commonly computed for ungrouped data are the range, the average deviation, and the standard deviation; the most popular measures used with grouped data are the standard deviation and the quartile deviation. Methods for computing each of these measures have been presented in this chapter. In addition, a summary of the comparative characteristics for each of these measures has been prepared. The standard deviation is the most important measure, but there are times when it either cannot or should not be used. The quartile deviation, for example, might be the measure of choice in an open-ended or badly skewed distribution.

If there is a need to compare the dispersions of two or more distributions, a measure of the degree of relative dispersion that exists should be used. The coefficient of variation is well suited for comparison purposes. In addition to measuring dispersion, it may be desirable to measure the extent to which a distribution departs from the symmetrical in order to form an opinion about the representativeness of various descriptive measures. The coefficient of skewness may be used for this purpose.

**Important Terms and Concepts**

1 Absolute dispersion

2 Range

3 Average deviation

4 Standard deviation

5 Variance

6 $\mu \pm 1\sigma$

7 $\mu \pm 2\sigma$

8 $\mu \pm 3\sigma$

9 Interquartile range

10 Quartile deviation

11 First quartile

12 Third quartile

13 Second quartile (median)

14 Relative dispersion

15 Coefficient of variation

16 Coefficient of skewness

17 $AD = \dfrac{\Sigma|X - \mu|}{N}$

18 $\sigma = \sqrt{\dfrac{\Sigma(X - \mu)^2}{N}}$

19 $\sigma = \sqrt{\dfrac{\Sigma f(m - \mu)^2}{N}}$

**20** $\sigma = i\sqrt{\dfrac{\Sigma fd^2}{N} - \left(\dfrac{\Sigma fd}{N}\right)^2}$

**22** $CV = \dfrac{\sigma}{\mu}(100)$

**21** $QD = \dfrac{Q_3 - Q_1}{2}$

**23** $Sk = \dfrac{3(\mu - Md)}{\sigma}$

**Problems**

**1** The following scores were made by a few of Professor Shirley A. Meany's accounting students who dared to take an optional makeup test:

72
65
43
50
68
62

Compute the following measures for these test scores:
**a** The range
**b** The average deviation
**c** The variance
**d** The standard deviation

**2** The following distribution represents the test scores made by Professor Meany's students in problem **1**, Chapter 3:

| Test scores | Number of students |
| --- | --- |
| 40 and under 50 | 4 |
| 50 and under 60 | 6 |
| 60 and under 70 | 10 |
| 70 and under 80 | 4 |
| 80 and under 90 | 4 |
| 90 and under 100 | 2 |

Compute the following measures for this distribution:
**a** The standard deviation (by the direct method)
**b** The standard deviation (by the shortcut method)
**c** The quartile deviation
**d** The coefficient of variation
**e** The coefficient of skewness

*Note:* The mean and median were to be computed in problem **2b,** Chapter 3.

**3** Analyze the distribution in problem **2** above by the use of
**a** The mean and the standard deviation
**b** The median and the quartile deviation

**4** Using the mileage data in problem **3,** Chapter 3, compute the following measures:
**a** The standard deviation
**b** The quartile deviation
**c** The coefficient of variation
**d** The coefficient of skewness

**5** Compare the distributions in the preceding problems **2** and **4,** and indicate which has the greater relative dispersion.

**6** Using the annual earnings data for families in Languor and Friskyville found in problem **4,** Chapter 3, perform the following operations:
   **a** Compute the standard deviation for each community.
   **b** Compute the quartile deviation for each community.
   **c** Compute the coefficient of variation for each community.
   **d** Compute the coefficient of skewness for each community.
   **e** Compare and analyze the economic situations in these two communities.
   *Note:* If it is not possible to perform all the operations in any or all of the above parts, please indicate your reasons for omitting the operation(s).

**7** Compute the average deviation and the standard deviation for the yardage figures given in problem **6,** Chapter 3.

**8** Using the annual salary data of Chauvinists, Inc., found in problem **7,** Chapter 3, perform the following operations:
   **a** Compute the standard deviation for both male and female employees.
   **b** Compute the quartile deviation for both male and female employees.
   **c** Compute the coefficient of variation for each group.
   **d** Compute the coefficient of skewness for each group.
   **e** Compare and analyze the economic situations of each employee group.

**Topics for Discussion**

**1** Why is it necessary to measure dispersion?

**2** Why must the algebraic signs of the deviations about the mean be ignored when computing the average deviation?

**3** How does the computation of the standard deviation for ungrouped data differ from the computation of the average deviation?

**4** Define "variance," and discuss its limitations as a measure of absolute dispersion.

**5** When must the direct method of computing the standard deviation for grouped data be used?

**6** Discuss the relationship that exists between the mean and the standard deviation in a normal distribution.

**7** Discuss the relationship that exists between the median and the quartile deviation in a symmetrical distribution.

**8** "An interpolation procedure just like the one used to compute the median is used to calculate the values of $Q_1$ and $Q_3$." Discuss this statement.

**9** Discuss the important characteristics of:
   **a** The range
   **b** The average deviation
   **c** The standard deviation
   **d** The quartile deviation

**10** What is the purpose of the coefficient of variation?

**11** What is the purpose of the coefficient of skewness?

**4-1**

1   **a**   Range = 56 years

    **b**   $AD = 234/14 = 16.71$ years

    **c**   $\sigma^2 = 4{,}860/14 = 347.14$

    **d**   $\sigma = \sqrt{347.14} = 18.63$ years

2   **a**   $AD = 222/9 = 24.67$ points

    **b**   $\sigma^2 = 6{,}826/9 = 758.4$

    **c**   $\sigma = \sqrt{758.4} = 27.54$ points

**4-2**

1   **a**   $\sigma \text{ (direct method)} = \sqrt{\dfrac{2880}{30}} = \sqrt{96} = \$9.80$

    or

$$\sigma \text{ (shortcut method)} = 10\sqrt{\frac{30}{30} - \left(\frac{-6}{30}\right)^2}$$

$$= 10\sqrt{1.00 - .04}$$

$$= 10\sqrt{.96}$$

$$= \$9.80$$

    **b**   $Q_1 = 30 + \left(\dfrac{7.5 - 3}{8}\right)(10)$          $Q_3 = 40 + \left(\dfrac{22.5 - 11}{12}\right)(10)$

           $= \$35.62$                     $= \$49.58$

      $QD = \dfrac{\$49.58 - \$35.62}{2} = \$6.98$

2   **a**   Since this distribution may be considered approximately normal, Brock and Parse Lee may conclude that about the middle two-thirds of their orders from the outlying county are within one standard deviation of the mean, that is, between $\$43.00 \pm \$9.80$ (or from $\$33.20$ to $\$52.80$).

    **b**   And about the middle 50 percent of the orders are between $Md \pm 1\,QD$, that is, between $\$43.33 \pm \$6.98$ (or from $\$36.35$ to $\$50.31$).

**4-3**

1   $CV = (14.63/115.2)(100) = 12.70$ percent

2   $CV = (\$9.80/\$43.00)(100) = 22.79$ percent

3   There is a much greater degree of dispersion in the Lee Produce Company distribution.

**4-4**

1   $Sk = \dfrac{3(\$43.00 - \$43.33)}{\$9.80} = -.10$

2   There is a slight degree of negative skewness in the Lee Produce Company data, but not enough to prevent the use of the mean and the standard deviation as representative descriptive measures.

# SAMPLING IN THEORY AND PRACTICE

Sometimes he thought sadly to himself, "Why?" and sometimes he thought, "Wherefore?" and sometimes he thought, "Inasmuch as which?"
— Winnie-the-Pooh

If complete knowledge were readily available, many of this world's problems would be solved, since complete knowledge is synonymous with certainty. Instead of relying on complete knowledge, however, decision makers must often resign themselves to using inferences based on sample results. Alas, they are destined to make statements using a vocabulary of "perhaps," "likely," "maybe," "almost surely," and "probably."

Fortunately for the decision maker, the science of statistics provides techniques for gathering and analyzing sample data in an objective manner. Statistical principles will not provide absolute certainty in decisions, but these principles will reduce the level of uncertainty which would have existed without any sample data. Statistical techniques *do not guarantee* results; they merely elevate a person from the status of complete ignorance to the status of definable doubt. Statistical laws merely bestow upon the individual the privilege of saying "maybe."

In this part of the text we will cover an area formally known as *inferential statistics*. Specifically, in Chapter 7 we will discuss methods of *objectively making estimates*. For example, these methods may be applicable in estimating the average salary of cab drivers in New York City, or in estimating the percentage of voters who favor a political candidate. And in Chapters 8 through 11, we will discuss *hypothesis-testing* techniques to be used in specified situations. Thus, you may want to test the assumption that the average salary of New York City cab drivers is $19,000 per year, or you may want to determine the validity of a candidate's statement that 60 percent of the voters support her bid for office.

But before we walk, we must learn to crawl. Before covering the material in Chapters 7 through 11, we must understand the foundations or validity of these methods. Chapters 5 and 6, in a sense, explain why there is justification for the methods you will be using in the later chapters of this part. The intent of these early chapters is to let you see why we can sometimes be confident in our estimates and conclusions.

Chapter 5 covers the area of *probability*. Since complete certainty is non-existent in decision making, everything must be placed in the context of likely outcomes. For example, in deciding on whom to bet to win more football games, we would probably decide to bet on Notre Dame rather than on Aunt Alice's School of the Dance because of likely outcomes. Chapter 6 ties together the principles of probability and sampling techniques. If we are to make inferences based on sample results, we should have some idea of the likelihood of being correct with our inferences. Chapters 5 and 6 discuss likelihood, and likelihood applied in the context of sampling.

The chapters included in Part 2 are:

# CHAPTER 5

# PROBABILITY AND PROBABILITY DISTRIBUTIONS

**LEARNING OBJECTIVES**

After studying this chapter, working the problems, and answering the discussion questions, you should be able to:

☞ Define probability and explain how probabilities may be classified.

☞ Correctly use addition and multiplication rules to perform probability computations.

☞ Explain what a probability distribution is and compute the probabilities in a binomial probability distribution.

☞ Use the table of areas for the standard normal probability curve (Appendix 2) to determine probabilities for a normal distribution.

**CHAPTER OUTLINE**

MEANING AND TYPES OF PROBABILITY

PROBABILITY COMPUTATIONS
  Addition Rule for Mutually Exclusive
    Events
  Addition Rule When Events Are Not
    Mutually Exclusive
  Multiplication Rule for Independent
    Events
  Multiplication Rule for Dependent
    Events
  Self-testing Review 5-1

PROBABILITY DISTRIBUTIONS

THE BINOMIAL DISTRIBUTION
  Combinations—A Brief Digression

Back to the Binomial
  Self-testing Review 5-2

THE NORMAL DISTRIBUTION
  Areas under the Normal Curve
  Using the Table of Areas
  Computing Normal Curve Probabilities
  Self-testing Review 5-3

SUMMARY

IMPORTANT TERMS AND CONCEPTS

PROBLEMS

TOPICS FOR DISCUSSION

ANSWERS TO SELF-TESTING REVIEW
  QUESTIONS

There are many instances when we must make decisions under conditions of uncertainty. A retailer must decide how much inventory to stock; a manager must decide whether or not to market a new product; a bridge player must decide which opponent to finesse for the queen.

In each of these cases, the individual makes the decision on the basis of what he or she thinks will happen. The retailer will look at past sales records to determine what demand is likely to be and then set inventory policy accordingly. With regard to the marketing of the new product, the manager's decision will be positive if she thinks the chances are good for success and negative if she feels the chances are poor. The bridge player will assess the bidding and other factors to determine which of his opponents is more likely to have the queen.

Each of these decisions, then, is made on the basis of what the decision maker thinks is likely to occur. In other words, a decision is based on the *probability* that a particular event will occur. Probability is also important in many statistical analyses. Thus, in this chapter we shall be concerned with the (1) *methods of computing probabilities*, (2) *concept of the probability distribution*, and (3) *characteristics of the binomial and normal probability distributions*.

## MEANING AND TYPES OF PROBABILITY

*Probability may be defined as the relative frequency of occurrence of an event after a very large number* (actually an unlimited number) *of trials*. In other words, it is the *proportion* of times an event can be expected to occur. Since we are speaking of proportions, probability measurements are stated as fractions between 0 and 1. A probability of 0 means an event can *never* occur, and a probability of 1 means an event will *always* occur.

If you were given a penny to flip (which is about all a penny is good for these days), you would likely believe that the probability of getting a head is .5. This does not mean, of course, that if you flipped the penny twice you would have to get one head. It does mean, though, that if you flipped the penny a very large number of times, the proportion of heads would approach .5.

You might wish to try an experiment along these lines: Flip a penny 10 times and keep a record of the number of heads; then flip the penny 10 more times and keep a cumulative record of the number of heads obtained in the 20 trials. If this experiment is continued for a total of 100 or 200 flips, you will find the proportions of heads getting ever closer to .5. Not only is this an instructive exercise; being somewhat time-consuming, it achieves the socially desirable result of keeping you out of various dens of iniquity.

*Probabilities may be classified in several ways.* The classification of particular interest to our discussion is according to the time of determination. In this sense, there are two basic types of probabilities, *a priori* and *empirical*.

An *a priori* probability is one that can be determined without experimentation. For example, in rolling a single die (one-half of a pair of dice) there are six possible outcomes. Unless someone has slipped you a loaded die, each outcome is equally likely to occur. Therefore, the probability of throwing any given number is 1/6.

An *empirical* probability is determined from observation and experimentation. This type of probability is very important to many types of decisions. The quantity of inventory a retailer should stock, the number of nurses a hospital should employ, and the rates that should be charged by an insurance company are all determined on the basis of empirical probabilities. Also, if you are still flipping that penny, have now flipped it 10,000 times, and the probability of heads, let us assume, continues to remain at about .65, you may be justified in revising your *a priori* probability expectation and concluding, perhaps, that you have been given an unfair or loaded coin.

## PROBABILITY COMPUTATIONS

**Anybody can win unless there happens to be a second entry.**

**—George Ade**

Computations involving probabilities follow two basic rules: the *addition rule* and the *multiplication rule*. The first has to do with the probability that *one or the other* of two events will occur, and the second with the probability that they will *both* occur.

## Addition Rule for Mutually Exclusive Events

Two events are said to be *mutually exclusive* if the occurrence of one event precludes the occurrence of the other event. For example, when you flip a penny a single time, if you get a head it is obviously impossible to get a tail, and vice versa. Therefore, these two possible outcomes are mutually exclusive.

When two events are mutually exclusive, the probability that *one or the other* of the two events will occur is the *sum of their separate probabilities*. For two events, A and B, this addition rule may be stated as follows:[1]

$$P(A \text{ or } B) = P(A) + P(B)$$

For example, if you roll a single die, the probability of getting either a 1 or a 2 would be computed as follows:

$$P(1 \text{ or } 2) = P(1) + P(2)$$
$$= 1/6 + 1/6$$
$$= 2/6$$

## Addition Rule When Events Are Not Mutually Exclusive

If two events are not mutually exclusive, it is possible for both events to occur. For example, if you draw a card from a deck and if you want either an ace or a spade, it will be possible for you to draw the ace of spades. Therefore, these two events, ace and spade, are not mutually exclusive. In a case like this, our addition rule must be stated as:

$$P(A \text{ or } B) = P(A) + P(B) - P(A \text{ and } B)$$

The probability of getting either an ace or a spade from a deck of cards will be:

$$P(\text{ace or spade}) = P(\text{ace}) + P(\text{spade}) - P(\text{ace and spade})$$
$$= 4/52 + 13/52 - 1/52$$
$$= 16/52$$

---

[1] The symbols may be interpreted as follows: The probability that either A or B will occur is equal to the probability that A will happen plus the probability that B will happen.

## Multiplication Rule for Independent Events

In determining a *joint probability,* that is, the probability that both events will occur, it is necessary to know if the two events are independent. Two events are said to be *independent* if the occurrence or nonoccurrence of one event in no way affects the probability that the other event will occur. This may be illustrated by assuming you have one red die and one green die and you wish to know the probability of throwing a 2 with this pair of dice. This would mean, of course, throwing a 1 on the red die and a 1 on the green die. The probability of throwing a 1 on the red die is 1/6 and will be 1/6 regardless of the result on the green die. Since the probabilities of getting a 1 on the green die or a 1 on the red die are not affected by the result on the other die, these events are said to be independent.

*If two events are independent, the probability that they will both occur is the product of their separate probabilities.* This may be stated as:

$$P(A \text{ and } B) = P(A) \times P(B)$$

For the problem of throwing a 2 with the dice, the probability would be:

$$P(1 \text{ on red and } 1 \text{ on green}) = P(1 \text{ on red}) \times P(1 \text{ on green})$$
$$= 1/6 \times 1/6$$
$$= 1/36$$

Another illustration of independent events is sampling with replacement. Let us consider a bowl containing 10 poker chips, 6 red and 4 white. A chip is drawn, its color is noted, and it is replaced in the bowl; then a second chip is drawn. The probability that the second chip will be red or the probability that it will be white is not affected by the result of the first draw. Therefore, the probability that a sample of two with replacement will result in two red chips is:

$$P(\text{two red}) = P(\text{red on first draw}) \times P(\text{red on second draw})$$
$$= .6 \times .6$$
$$= .36$$

## Multiplication Rule for Dependent Events

If two events are dependent rather than independent — i.e., if the probability of occurrence of one event is *dependent or conditional* on the occurrence or nonoccurrence of the other event — then the probability that they will both occur is:[2]

$$P(A \text{ and } B) = P(A) \times P(B/A)$$

Let us again consider our previous example of the bowl containing six red and four white poker chips. A chip is drawn, and then a second chip is drawn (and the first chip is not replaced). The probability that the second chip will be red or the probability that it will be white depends on the result of the first draw. The probability that a sample of two drawn in this fashion will result in two red chips is:

$$P(\text{two red}) = P(\text{red on first draw}) \times P(\text{red on second draw/red on first draw})$$
$$= 6/10 \times 5/9$$
$$= 30/90$$

---

[2] The term "$P(B/A)$" is a dependent or conditional probability and is read "the probability of B given A."

**1** An urn contains eight red marbles, seven green marbles, and five white marbles. What is the probability that a marble selected at random will be:
  **a** Either red or green?
  **b** Either green or white?

**2** If one card is drawn from a deck of playing cards, what is the probability that it will be either a face card or a heart?

**3** An urn contains 14 red marbles and 6 green marbles. If sampling is done with replacement, what is the probability that a sample of two will contain:
  **a** Two red marbles?
  **b** Two green marbles?

**4** An urn contains 12 red marbles and 8 green marbles. If sampling is done without replacement, what is the probability that a sample of two will contain:
  **a** Two red marbles?
  **b** Two green marbles?

## PROBABILITY DISTRIBUTIONS

A *probability distribution* is simply a complete listing of all possible outcomes of an experiment, together with their probabilities. This concept can be illustrated by the probabilities of throwing various numbers with a pair of dice. This example will also allow us to further illustrate the application of the addition and multiplication rules discussed earlier.

In a previous section we computed the probability of throwing a 2 with a pair of dice to be 1/36. Let us now consider the probability of throwing a 3. We can obtain this result either by getting a 1 on the red die and a 2 on the green die or by getting a 2 on the red die and a 1 on the green die. The probability of a 1 on the red die and a 2 on the green die is computed as follows:

$$P(1 \text{ on red and } 2 \text{ on green}) = P(1 \text{ on red}) \times P(2 \text{ on green})$$
$$= 1/6 \times 1/6$$
$$= 1/36$$

The probability of 2 on red and 1 on green will be:

$$P(2 \text{ on red and } 1 \text{ on green}) = P(2 \text{ on red}) \times P(1 \text{ on green})$$
$$= 1/6 \times 1/6$$
$$= 1/36$$

We wish to know the probability that one or the other of these events will occur. Therefore, we now use the addition rule to determine the probability of throwing a 3 with the pair of dice:

$$P(3) = P(1 \text{ on red and } 2 \text{ on green}) + P(2 \text{ on red and } 1 \text{ on green})$$
$$= 1/36 + 1/36$$
$$= 2/36$$

The probabilities of throwing a 4, 5, 6, 7, 8, 9, 10, 11, or 12 may be computed in the same way. The complete probability distribution is given in Fig. 5-1.

It should be noted that the total of the probabilities in Fig. 5-1 is 1. This will be true for all probability distributions. Since the distribution is a complete listing of

**FIGURE 5-1**

Probability distribution of results of throwing a pair of dice

| Result | Probability |
|--------|-------------|
| 2 | 1/36 |
| 3 | 2/36 |
| 4 | 3/36 |
| 5 | 4/36 |
| 6 | 5/36 |
| 7 | 6/36 |
| 8 | 5/36 |
| 9 | 4/36 |
| 10 | 3/36 |
| 11 | 2/36 |
| 12 | 1/36 |
| Total | 36/36 = 1 |

*all* outcomes of the experiment, the probability that one or the other of these outcomes will occur is a certainty.

There are several probability distributions that are important in understanding and applying statistical methods. We shall now examine two of these, the binomial distribution and the normal distribution, in some detail.

**THE BINOMIAL DISTRIBUTION**

The *binomial distribution describes the distribution of probabilities when there are only two possible outcomes for each trial of an experiment.* For example, when flipping a penny, there are only two possible outcomes, heads or tails. Therefore, the probability distribution showing the probabilities of zero, one, two, three, and four heads in four flips of the penny would be a binomial distribution.

This distribution also describes the probabilities associated with many *more practical* problems. When a manufacturer is trying to determine the quality of a product, each item inspected is either good or defective. In an election poll, each person interviewed either plans to vote for a particular candidate or does not plan to vote for that candidate. In a market survey, each person interviewed either plans to buy a new car this year or does not plan to buy a new car this year.

**Combinations — A Brief Digression**

There was once a brainy baboon,
Who always breathed down a bassoon,
For he said, "It appears
That in billions of years
I shall certainly hit on a tune."

— **Sir Arthur Eddington**

Before discussing the computation of binomial probabilities, it is desirable here to consider briefly the subject of *combinations*.[3] Let us assume that you have one card identified by the letter A, another by the letter B, and a third by the letter C. How many different combinations of two cards can be made *without regard to the order in which the cards appear?* That is, how many combinations of three cards taken two at a time are possible? We can answer this terribly difficult question by simply listing the possible combinations: AB, AC, and BC.

The general formula for the number of combinations of $n$ things taken $r$ at a time — or $_nC_r$ — is:

$$_nC_r = \frac{n!}{r!(n-r)!}$$

The symbol $n!$ (or *n factorial*)[4] means the product of

$$n(n-1)(n-2)(n-3) \ldots [n-(n-1)]$$

Thus, 6! equals $6 \times 5 \times 4 \times 3 \times 2 \times 1$, or 720. We noted above that three combinations of two could be made of the three cards A, B, and C. Verifying this result by our formula gives:

$$_3C_2 = \frac{3!}{(2!)(3-2)!}$$

$$= \frac{3 \times 2 \times 1}{2 \times 1 \times 1}$$

$$= 3$$

And if we wish to know the number of combinations of seven things taken three at a time, we find:

$$_7C_3 = \frac{7!}{(3!)(7-3)!}$$

$$= \frac{7 \times 6 \times 5 \times 4 \times 3 \times 2 \times 1}{(3 \times 2 \times 1)(4 \times 3 \times 2 \times 1)}$$

$$= 35$$

**Back to the Binomial**

In dealing with binomial probabilities, we may speak of the two possible outcomes as successes and failures. A *success* is simply the outcome for which we wish to find the probability distribution, such as heads on a penny or a defective item from a production line. The probability of success in *any one trial* may be identified here as $p$, and the probability of failure in the *same single trial* may be labeled $q$. Thus, the sum of $p$ and $q$ is one in all cases. The probability of $r$ successes in $n$ trials is

$$P(r) = (_nC_r)(p)^r(q)^{n-r}$$

We can illustrate the computation of binomial probabilities with some examples.

---

[3] The subject of *permutations* is often discussed in statistics texts at about this time. We have elected to delete the topic here; however, problem 9 at the end of the chapter briefly explains how permutations differ from combinations.

[4] It should be remembered that 0! is, by definition, equal to 1.

**Example 1** Assuming that a penny is flipped four times, what is the probability of getting *zero* heads? For this problem $p$ is the probability of getting heads on any one flip of the penny and is equal to 1/2. The probability of not getting heads, $q$, is also 1/2. The number of trials, $n$, is 4, and the number of successes for which we want the probability, $r$, is 0. The probability of zero heads in four flips is:

$$P(0) = {}_4C_0 \; (1/2)^0 \; (1/2)^4$$

$$= \frac{4!}{0! \; 4!} \; (1) \; (1/16)$$

$$= 1/16$$

Of course, in addition to the possibility of no heads when a penny is flipped four times, there are also the possibilities of getting one head, two heads, three heads, or four heads. The probabilities for these other possible outcomes are shown below:

$$P(1) = {}_4C_1 \; (1/2)^1 \; (1/2)^3$$

$$= \frac{4!}{1! \; 3!} \left(\frac{1}{2}\right) \left(\frac{1}{8}\right)$$

$$= (4) \; (1/16)$$

$$= 4/16$$

$$P(2) = {}_4C_2 \; (1/2)^2 \; (1/2)^2$$

$$= \frac{4!}{2! \; 2!} \left(\frac{1}{4}\right) \left(\frac{1}{4}\right)$$

$$= \frac{4 \times 3 \times 2 \times 1}{2 \times 1 \times 2 \times 1} \left(\frac{1}{16}\right)$$

$$= 6/16$$

$$P(3) = {}_4C_3 \; (1/2)^3 \; (1/2)^1$$

$$= \frac{4!}{3! \; 1!} \left(\frac{1}{8}\right) \left(\frac{1}{2}\right)$$

$$= 4/16$$

$$P(4) = {}_4C_4 \; (1/2)^4 \; (1/2)^0$$

$$= \frac{4!}{4! \; 0!} \left(\frac{1}{16}\right) \; (1)$$

$$= 1/16$$

These probabilities are summarized in Fig. 5-2, and a histogram of the probability distribution is given in Fig. 5-3.

**Example 2** The output of a production process is 10 percent defective. What is the probability of selecting *exactly two* defectives in a sample of *five*? For this problem, $p = .1$ and $q = .9$. Thus, the solution is:

## FIGURE 5-2

Probability distribution for the number of heads in four flips of a penny

$$4C_3 \left(\tfrac{1}{2}\right)^3 \left(\tfrac{1}{2}\right)^{\!1}$$

| Number of heads | Probability |
|---|---|
| 0 | 1/16 |
| 1 | 4/16 |
| 2 | 6/16 |
| 3 | 4/16 |
| 4 | 1/16 |
| Total | 16/16 = 1 |

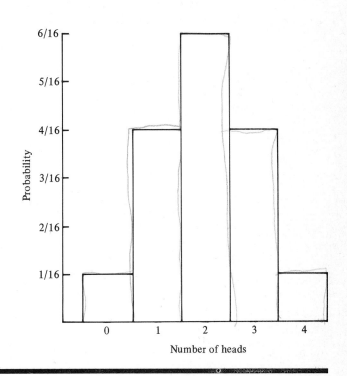

## FIGURE 5-3

**Probability distribution of the number of heads in four flips of a penny**

$$P(2) = {}_5C_2 \, (.1)^2 \, (.9)^3$$

$$= \frac{5!}{2! \, 3!} \, (.01) \, (.729)$$

$$= \frac{5 \times 4 \times 3 \times 2 \times 1}{2 \times 1 \times 3 \times 2 \times 1} \, (.00729)$$

$$= (10) \, (.00729)$$

$$= .0729$$

$$P(2)$$
$$= {}_5C_2 (.1)^2(.9)^3$$

FIGURE 5-4

Probability distribution for the
number of defectives in a sample
of five

| Number of defectives | Probability |
| --- | --- |
| 0 | .59049 |
| 1 | .32805 |
| 2 | .07290 |
| 3 | .00810 |
| 4 | .00045 |
| 5 | .00001 |
| Total | 1.00000 |

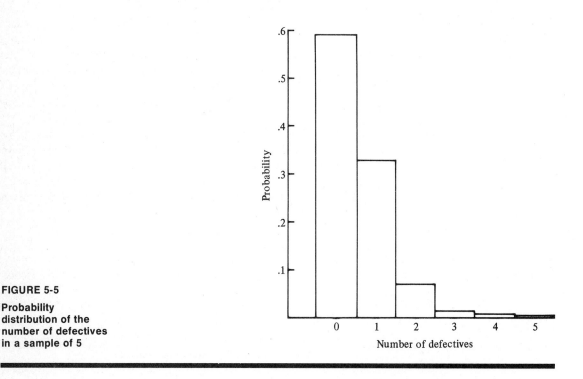

**FIGURE 5-5**

**Probability
distribution of the
number of defectives
in a sample of 5**

The probabilities for *all* the possible outcomes in this sample of five are given in
Fig. 5-4, and this probability distribution is illustrated in Fig. 5-5.

A comparison of Figs. 5-3 and 5-5 shows a marked difference in shape. The
probability distribution for the number of heads obtained in four flips of a penny is
symmetrical because *p* and *q* are equal. However, in Fig. 5-5, where *p* and *q* are not
equal, we find a marked skewness.

Now that you have seen how binomial probabilities are calculated, we will let
you in on a little secret: You can often look up the appropriate probability values by

using a table designed for this purpose. Such a table is presented in Appendix 1 in the back of the book.

**Self-testing Review 5-2**

1  In six flips of a penny, what is the probability of getting:
   a  Exactly three heads?
   b  Exactly four heads?
   c  At least four heads?

2  If 30 percent of a population own their own homes, what is the probability that a sample of seven from this population will contain:
   a  Exactly two homeowners?
   b  Exactly four homeowners?
   c  At least five homeowners?

## THE NORMAL DISTRIBUTION

The binomial distribution is an example of a *discrete* distribution, that is, a distribution in which there is a finite number of values a variable can take. When we are concerned with a variable, such as time, that can have an infinite number of values, the probability distribution is *continuous*. The most important of the continuous distributions is the *normal distribution*. (An expression of its importance—and its shape—was presented by W. J. Youden of the National Bureau of Standards in the manner shown in Fig. 5-6a.)

As indicated in Fig. 5-6a, the normal curve is symmetrical and, because of its appearance, is sometimes called a bell-shaped curve. It is actually not a single curve but a family of curves. A particular normal curve is defined by its mean, $\mu$, and its standard deviation, $\sigma$. Figure 5-6b shows three normal curves with the same mean, but with different standard deviation values; Fig. 5-6c shows three normal curves with the same standard deviation, but with different mean values.

## Areas under the Normal Curve

Probabilities for continuous probability distributions are represented by areas under the curve; that is, *the probability that the variable will take on a value between a and b is the area under the curve between two vertical lines erected at points a and b.* For example, if the breaking strength of a material is normally distributed with a mean of 110 pounds and a standard deviation of 25 pounds, the probability that a piece of this material will have a breaking strength between 110 and 120 pounds will be the area under the curve covered by this interval, as shown in Fig. 5-7.

To determine the area covered by an interval under the normal curve, we use a *table of areas* developed for this purpose. It is, of course, impossible to construct tables for every normal curve. However, it is possible to construct a table for what is known as a *standardized normal curve*. This table can then be used to determine probabilities for any normal distribution.

In order to understand the use of the table of areas, *it is necessary to understand the relationship of the standard deviation to the normal curve.* An interval of a given number of standard deviations from the mean will cover the same area in any normal curve. Thus, the interval from 50 to 70 for a normal curve with a mean of 50 and a standard deviation of 20 *will cover the same area* as the interval from 170

FIGURE 5-6

Normal distributions

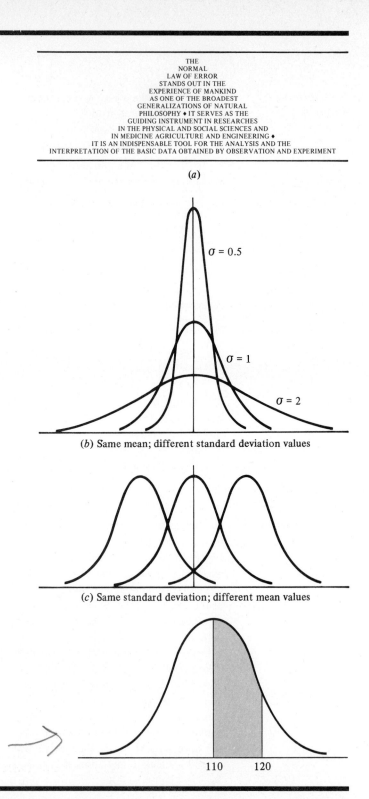

THE
NORMAL
LAW OF ERROR
STANDS OUT IN THE
EXPERIENCE OF MANKIND
AS ONE OF THE BROADEST
GENERALIZATIONS OF NATURAL
PHILOSOPHY ♦ IT SERVES AS THE
GUIDING INSTRUMENT IN RESEARCHES
IN THE PHYSICAL AND SOCIAL SCIENCES AND
IN MEDICINE AGRICULTURE AND ENGINEERING ♦
IT IS AN INDISPENSABLE TOOL FOR THE ANALYSIS AND THE
INTERPRETATION OF THE BASIC DATA OBTAINED BY OBSERVATION AND EXPERIMENT

(*a*)

$\sigma = 0.5$

$\sigma = 1$

$\sigma = 2$

(*b*) Same mean; different standard deviation values

(*c*) Same standard deviation; different mean values

FIGURE 5-7

Probability of
breaking strength
between 110 and
120 pounds

110    120

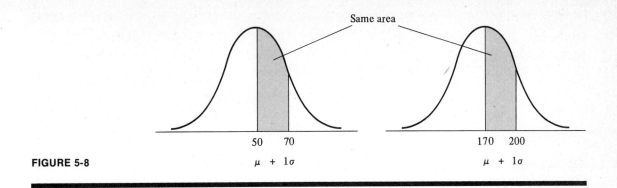

**FIGURE 5-8**

to 200 in a normal curve with a mean of 170 and a standard deviation of 30, since both of these intervals cover a distance of one standard deviation from the mean (see Fig. 5-8).

**Using the Table of Areas**

The table of areas under the normal curve (see Appendix 2 at the back of the book) shows the area between the mean and a given number of standard deviations from the mean. The symbol $Z$ is used to represent this number of standard deviations. Thus, $Z$ may be defined as the *difference* between any variable value ($x$) and the mean of all $x$ values ($\mu$) divided by the standard deviation ($\sigma$). Or,

$$Z = \frac{x - \mu}{\sigma}$$

How is the table of areas used? Thank you for asking. Let us assume that we have a normal distribution with a mean of 50 and a standard deviation of 20. Let us also assume that one of the values of the variable from which the mean of 50 was computed is 75. How many $Z$ values (or standard deviations) from the mean will 75 fall? The answer is:

$$Z = \frac{x - \mu}{\sigma}$$

$$= \frac{75 - 50}{20}$$

$$= 1.25$$

Thus, a value of 75 will lie 1.25 standard deviations to the *right* of the mean of 50. Similarly, the $Z$ value for 25 will be:

$$Z = \frac{x - \mu}{\sigma}$$

$$= \frac{25 - 50}{20}$$

$$= -1.25, \text{ or } 1.25 \text{ standard deviations to the } left \text{ of the mean of 50}$$

**105**

The first column of the table of areas in Appendix 2 gives values of $Z$ to one decimal place. The remaining columns give the second decimal place. To find the area for $Z = 1.25$, go down the column to 1.2 and across to the column headed .05. The area given by the table is .3944.

Since the normal curve is symmetrical, what is true for one-half of the curve is true for the other half. Therefore, the table is made up for only one-half of the curve, and the area for $Z = -1.25$ is also .3944. All right, you may say, but what is the meaning of this .3944? This question is answered in the next section.

**Computing Normal Curve Probabilities**

It is probably appropriate here to pause just long enough to emphasize two precautions. *First,* remember that $Z$ is the number of standard deviations *from the mean.* The interval from $Z = 1$ to $Z = 2$ is one standard deviation wide, but it *will not* cover the same area as the area between the mean and one $Z$ value. *Second,* remember that the normal curve is a probability distribution, and thus the total area under the curve is 1. Therefore, the area covered by one-half the curve is .5.

To determine the probability of getting a value within a particular interval, *the following procedure should be used:*

1  Determine the $Z$ value for each limit of the interval.

2  From the table of areas, determine the *area* for each $Z$ value.

3  If both limits of the interval are on *opposite sides* of the mean, *add* the areas determined in the previous step. If the limits are on the *same side* of the mean, *subtract* the smaller area from the larger one.

Let us illustrate this procedure with several examples — all of which deal with the life expectancy of light bulbs whose lifetimes are normally distributed with a mean life of 750 hours and with a standard deviation of 80 hours.

**Example 1** What is the probability that a light bulb will last between 750 hours and 830 hours (see Fig. 5-9)?

$$\text{For 830 hours, } Z = \frac{830 - 750}{80}$$

$$= 1.00$$

$$\text{For } Z = 1.00, \text{ the area} = .3413$$

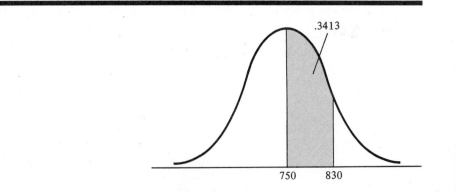

**FIGURE 5-9**

**Probability of a light bulb lasting between 750 and 830 hours**

.3413

750    830

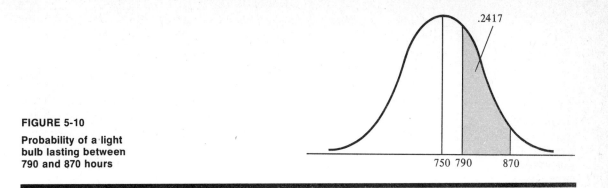

**FIGURE 5-10**

**Probability of a light bulb lasting between 790 and 870 hours**

The probability that a light bulb will last between 750 and 830 hours is thus .3413 (the shaded area in Fig. 5-9).

**Example 2** What is the probability that a light bulb will last between 790 hours and 870 hours (see Fig. 5-10)?

$$\text{For 790 hours, } Z = \frac{790 - 750}{80}$$

$$= .50$$

$$\text{For 870 hours, } Z = \frac{870 - 750}{80}$$

$$= 1.50$$

Then, for $Z = 1.50$, the area $= .4332$
and for $Z = .50$, the area $\quad = \underline{.1915}$
and the probability is $\qquad \quad .2417$

The *difference* between the two areas is .2417, which is both the shaded area in Fig. 5-10 and the probability that a bulb will last between 790 and 870 hours.

**Example 3** What is the probability that a light bulb will last between 730 hours and 850 hours (see Fig. 5-11)?

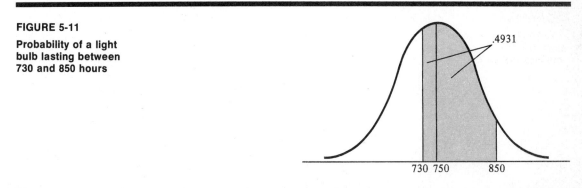

**FIGURE 5-11**

**Probability of a light bulb lasting between 730 and 850 hours**

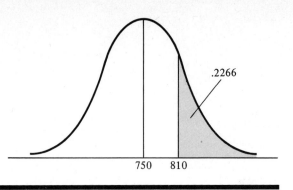

FIGURE 5-12

Probability of a light
bulb lasting more
than 810 hours

$$\text{For 730 hours, } Z = \frac{730 - 750}{80}$$

$$= -.25$$

$$\text{For 850 hours, } Z = \frac{850 - 750}{80}$$

$$= 1.25$$

Then, for $Z = -.25$, the area = .0987
and for $Z = 1.25$, the area  = .3944
and the probability is          .4931

The *sum* of the two areas is .4931, which is both the shaded area of Fig. 5-11 and the probability that a bulb will last between 730 and 850 hours.

**Example 4** What is the probability that a light bulb will last more than 810 hours (see Fig. 5-12)?

$$\text{For 810 hours, } Z = \frac{810 - 750}{80}$$

$$= .75$$

For $Z = .75$, the area = .2734

Thus, if .2734 is the area between 750 and 810 hours, the area *beyond* 810 hours must be the *difference* between .5000 (the total area greater than the mean of 750 hours) and .2734. In short, the probability that a bulb will last more than 810 hours (and the shaded area in Fig. 5-12) is .2266.

**Example 5** What is the probability of a light bulb lasting less than 770 hours (see Fig. 5-13)?

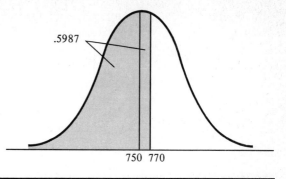

**FIGURE 5-13**

**Probability of a light bulb lasting less than 770 hours**

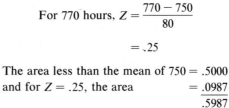

$$\text{For } 770 \text{ hours, } Z = \frac{770 - 750}{80}$$

$$= .25$$

The area less than the mean of $750 = .5000$
and for $Z = .25$, the area $\quad = \underline{.0987}$
$\qquad\qquad\qquad\qquad\qquad\quad .5987$

The answer is the *sum* of the areas (and probabilities) or .5987.

**Example 6** What is the probability that a light bulb will last more than 670 hours (see Fig. 5-14)?

$$\text{For } 670 \text{ hours, } Z = \frac{670 - 750}{80}$$

$$= -1.00$$

The area greater than the mean of $750 = .5000$
and for $Z = -1.00$, the area $\quad = \underline{.3413}$
$\qquad\qquad\qquad\qquad\qquad\qquad .8413$

The *sum* of the areas (and probabilities) is .8413, and this is the answer.

**FIGURE 5-14**

**Probability of a light bulb lasting more than 670 hours**

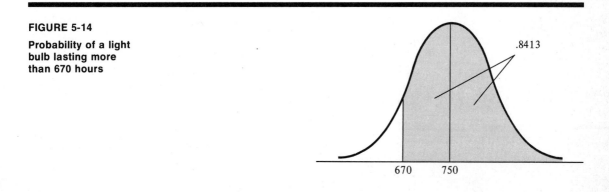

1 The breaking strength of a material is normally distributed with a mean of 90 pounds and a standard deviation of 20 pounds. What is the probability that a piece of this material will have a breaking strength:

   a  Between 90 and 114 pounds?

   b  Between 95 and 110 pounds?

   c  Between 80 and 110 pounds?

   d  Greater than 70 pounds?

   e  Greater than 100 pounds?

## SUMMARY

In this chapter we have covered the basic concepts of probability and have looked in some detail at the binomial and normal distributions. These concepts will play an important part in the following chapters dealing with sampling. Since probabilities are often determined from either the binomial or the normal distribution, an understanding of the material presented in this chapter is essential to an understanding of the sampling concepts introduced in the next chapter. (You will be delighted to know, however, that there are no light bulb problems in the next chapter.)

## Important Terms and Concepts

1  Probability

2  *A priori* probability

3  Empirical probability

4  Mutually exclusive events

5  Addition rule

6  Independent events

7  Multiplication rule

8  Dependent events

9  Conditional probability

10  Probability distribution

11  Binomial distribution

12  Combinations

13  Normal distribution

14  Areas under the normal curve

15  $Z$ value

16  Continuous distribution

## Problems

1  What is the probability of throwing a 1 or 2 with a single die?

2  If a single card is drawn from a deck of playing cards, what is the probability that it will be either a black card or an ace?

3  A box of parts contains eight good items and two defective items.

   a  What is the probability that in a sample of two, taken with replacement, both items will be good?

   b  What is the probability that they will both be defective?

4  A box of parts contains seven good items and three defective items.

   a  What is the probability that in a sample of two, taken without replacement, both items will be good?

   b  What is the probability that they will both be defective?

5  If a penny is flipped seven times, what is the probability of getting:

   a  Exactly four heads?

   b  Exactly five heads?

   c  Four or five heads?

**6** If 60 percent of a population are Democrats, what is the probability that a sample of six from this population will contain:

**a** Exactly four Democrats?

**b** Exactly five Democrats?

**c** Four or five Democrats?

**d** At least four Democrats?

**7** The lifetimes of batteries produced by a firm are normally distributed with a mean of 110 hours and a standard deviation of 10 hours. What is the probability that a battery will last:

**a** Between 110 and 115 hours?

**b** Between 107.5 and 120 hours?

**c** Between 112 and 123 hours?

**d** Between 90 and 102 hours?

**e** More than 113 hours?

**8** The scores of students on a standardized test are normally distributed with a mean of 300 and a standard deviation of 40. What is the probability that a student will score:

**a** Between 310 and 330?

**b** Between 280 and 340?

**c** Less than 320?

**d** More than 260?

**e** More than 380?

**9** *Permutations.* We saw in the chapter that *order was not important* in considering the number of possible combinations—i.e., the card combination **AB** was considered the same as **BA**. *Permutations differ from combinations in that order is important*—i.e., **AB** is one permutation, and **BA** is a second permutation. The formula for the number of permutations of *n* things taken *r* at a time is:

$$nPr = \frac{n!}{(n-r)!}$$

**a** In how many different ways can a football fan enter a stadium by one gate and leave by a different gate if there are 15 gates in the stadium?

**b** In how many different ways can a committee of 3 be chosen from a group of 10 deans? (Are you sure this is a permutation?)

**10** There are 50 students in the M.B.A. program at Systems Tech. In this class, 20 students are taking statistics, 15 are taking finance, and 10 are taking both statistics and finance. If a student is chosen at random,

**a** What is the probability that he or she is taking either statistics or finance?

**b** What is the probability that he or she is not taking statistics?

**11** The diameters of a particular group of parts are normally distributed with a mean of 2 inches and a standard deviation of 0.2 inches. If a part is chosen at random, what is the probability that it will have a diameter:

**a** Between 1.8 and 2.1 inches?

**b** Between 1.5 and 1.7 inches?

**c** Greater than 2.2 inches?

**d** Less than 2.3 inches?

**Topics for Discussion**

1 Explain the difference between *a priori* probability and empirical probability.

2 "In determining a joint probability, it is necessary to know if the two events are independent." Discuss this statement.

3 Define a probability distribution, and give an example of how one could be created.

4 Why is the formula for combinations incorporated into the formula for computing binomial probabilities?

5 "The binomial distribution is an example of a discrete distribution." Explain this statement.

6 "Probabilities for continuous probability distributions are represented by areas under the curve." Discuss this statement.

7 a What is a $Z$ value?
  b How is a $Z$ value used?

8 Discuss the procedure for computing normal curve probabilities.

**Answers to Self-testing Review Questions**

5-1

1 a .75
  b .6

2 22/52 or .4231

3 a .49
  b .09

4 a 132/380 or .34737
  b 56/380 or .14737

5-2

1 a .3125
  b .2344
  c The answer here is the *sum of* the probabilities of getting four heads, five heads, and six heads. That is, the probability of getting at least four heads is .2344 + .0938 + .0156, or .3438.

2 a .3177
  b .0972
  c The sum of the probabilities of five, six, and seven homeowners is .0250 + .0036 + .0002, or .0288.

5-3

1 a .3849
  b .2426
  c .5328
  d .8413
  e .3085

# CHAPTER 6

# SAMPLING CONCEPTS

**LEARNING OBJECTIVES**

After studying this chapter, working the problems, and answering the discussion questions, you should be able to:

☞ Understand and appreciate (a) the purpose and importance of sampling and (b) the advantages made possible by sampling.

☞ Describe the types of samples that may be selected.

☞ Trace through the steps that are required to (a) produce a sampling distribution of sample means, (b) compute the mean of this sampling distribution, and (c) compute the standard deviation of this sampling distribution.

☞ Define the Central Limit Theorem and explain the relationship that exists between the standard error of the mean and the size of the sample.

☞ Trace through the steps necessary to (a) produce a sampling distribution of sample percentages, (b) compute the mean of this sampling distribution, and (c) compute the standard deviation of this sampling distribution.

**CHAPTER OUTLINE**

POPULATION, SAMPLE, SYMBOLS: A REVIEW
Self-testing Review 6-1

IMPORTANCE OF SAMPLING
Self-testing Review 6-2

ADVANTAGES OF SAMPLING
Cost
Time
Accuracy of Sample Results
Other Advantages
Self-testing Review 6-3

**Theirs not to reason why,**
**Theirs but to do and die.**
> **—Excerpt from "The Charge of the Light Brigade"**
> **by Alfred, Lord Tennyson**

Many college students whose degrees require a course in the area of inferential statistics appear to have adopted the above quote as their own approach to learning statistics. These hapless students usually plunge gallantly but imprudently into inferential statistics without a firm understanding of sampling concepts, an understanding which would eliminate much of their confusion. In preparation for exams, they gamely struggle to memorize formulas, forsaking any attempt to understand the basic nature of the formulas, and as a result their test grades become interesting statistics from which the instructor makes inferences about their intelligence. (It's no wonder that many students take pleasure in mispronouncing statistics as "sadistics.")

In the following chapters you will be introduced to statistical inference. But first *you should understand why* it is valid to use sample results to make estimates and decisions about population characteristics. This chapter will attempt to provide you with an understanding of the basic nature of inferential statistics by presenting the theoretical and intuitive bases for estimation and hypothesis testing.

More specifically, in this chapter (after briefly *reviewing* some population and sample *definitions*) we will examine (1) the *importance of* and *advantages of sampling,* (2) some methods of *sample selection,* and (3) the extremely important concepts associated with *sampling distributions of means and percentages*. These latter topics represent the cornerstone of statistical inference, and in these sections we will discuss *why* it is possible to infer a population characteristic with a sample characteristic.

Sample characteristics are interrelated with population characteristics, and an understanding of this interrelationship should help you avoid enlisting in the ranks of the gallant but doomed soldiers of the Light Brigade.

## POPULATION, SAMPLE, SYMBOLS: A REVIEW

On Tuesday when it hails and snows
The feeling in me grows and grows
That hardly anybody knows
If those are these and these are those.

**—Winnie-the-Pooh**

Usually, the term "population" brings to mind a large mass of people who reside in a geographic area. In statistics the term has a broader meaning. *A population is the total of any kind of units under consideration by the statistician.* These individual units may be items such as business firms, credit accounts, chickens, transistors, and even people. A population is *finite* if the total is a limited, specific number. An *infinite* population is unlimited in size.

*A sample is any portion of the population selected for study.* Consider, for example, the American League as a population of baseball teams. If the Boston Red Sox and the Texas Rangers were selected for a statistical study of the American League, these two teams would be considered a sample.

Whether a group of items is a population or whether a group of items is a sample, a statistician usually describes the group with measures such as a total number, an average, a standard deviation, and the like. For example, a group of students may be described by the total number of students, the average grade, and a standard deviation of grades. If a particular measure describes a population, it is a parameter. *A parameter is a characteristic of a population.* If a particular measure describes a group of items which is a sample, the measure is known as a statistic. *A statistic is a characteristic of a sample.*

### FIGURE 6-1

Distinctions between a population and a sample

| Area of distinction | Population | Sample |
|---|---|---|
| Definition | Defined as a total of the items under consideration by the researcher | Defined as a portion of the population selected for study |
| Characteristics | Characteristics of a population are parameters | Characteristics of a sample are statistics |
| Symbols | Greek letters or capitals<br>$\mu$ = population mean<br>$\sigma$ = population standard deviation<br>$N$ = population size<br>$\pi$ = population percentage | Roman letters<br>$\bar{x}$ = sample mean<br>$s$ = sample standard deviation<br>$n$ = sample size<br>$p$ = sample percentage |

You may think that we are "slicing the baloney awfully thin" with all these distinctions in terminology, but the terminology is so important that there is justification for any apparent overemphasis in presentation. As you will see, sample results are generalized to describe the population; i.e., *statistics are used to estimate parameters*. The distinctions made now between parameters and statistics minimize the danger of confusion in later chapters.

Statisticians maintain the distinction between parameters and statistics through the use of different symbols. *Greek letters are usually used to denote parameters, while lowercase Roman letters denote sample statistics*. Figure 6-1 illustrates the common symbols. The population mean is designated by $\mu$ (mu), while the sample mean is denoted by $\bar{x}$. The standard deviation of the population is indicated by $\sigma$ (sigma), while the sample standard deviation is represented by $s$. And the population percentage is represented by $\pi$ (pi), while the sample percentage is designated by $p$.

**Self-testing Review 6-1**

1   Suppose we are interested in performing a study of a university, and there are four possible groups from which we could collect data. Would each of the following groups be the population of the university or only a sample?
   a   All students enrolled at the university
   b   All students enrolled in a psychology course
   c   All students enrolled in the business school
   d   All students in every division of the university

2   In an earlier example, the American League was considered a population. Is it possible to consider the American League a sample? If so, under what circumstances?

3   A workers' union has a membership of 300 persons. Data were collected from 25 of them, and their average age was 39. The average age of the entire union membership was therefore estimated to be approximately 39. A subsequent polling of all members indicated the true average age was 42.
   a   What figure is a parameter?
   b   What figure is a statistic?
   c   The sample statistic 39 was used to estimate the _____.

**IMPORTANCE OF SAMPLING**

Sampling occurs frequently in the course of daily events and should not be viewed as just a concept employed solely by statisticians. Although the samplings in daily life do not have the sophistication of formal statistical studies, they do serve a fundamental purpose of providing information for judgments. Here are a few examples.

1   A homemaker tastes a spoonful of the soup which she is preparing for supper. She wants to know if the soup has an acceptable flavor.

2   A prospective car buyer test-drives an automobile in order to judge whether or not it is a potential lemon.

3   Pieces of ore are analyzed to determine the potential of a new mine.

The examples could go on *ad infinitum* (and *ad nauseam*), but let's look at the rationale for sampling.

*The purpose of sampling is to provide sufficient information so that inferences may be made concerning the characteristics of the population.* Many times it is not possible to study an entire population to determine the true characteristics. The homemaker cannot taste the whole pot of soup to determine if it is really acceptable. The test driver cannot drive the car for 3 years to find out if it will eventually be a lemon. The entire mine cannot be analyzed to be sure about its potential. Consequently, sampling provides convenient information as a basis for judgments about the population.

*The goal in sampling is to select a portion of the population which is maximally representative of the characteristics of the population.* If we are to make a judgment about a population from sample results, we want the sample results to be as representative of the population as possible. Suppose we want to perform a sociological study of a town. Not only should the sample contain enough people to represent the upper-income group, the middle-income group, and other income groups; it should also be representative of such characteristics as age, education, and race. Unless a sample is similar to the population, there can be no reliability in estimates based on sample results.

Unfortunately, it is extremely difficult, if not nearly *impossible,* to have a sample *completely* representative of the population. It would be unreasonable to expect a sample result to have *exactly* the same value as some population characteristic because sampling error is always present.

Although sampling error will always exist, this does not mean that sample results are useless. Statisticians learn to cope with error. If sampling error can be objectively assessed, the precision of estimates can also be objectively assessed. Statisticians are often willing to give up the benefits of a complete enumeration for advantages to be gained from sampling. The advantages of sampling allow a certain degree of sampling error to be tolerated.

**Self-testing Review 6-2**

1  "Sampling is a concept which can be best illustrated by statisticians." Comment on this statement.

2  What is the purpose of sampling?

3  What is considered the goal of sampling?

4  If there are any errors in sampling, the results of the sample have little use. True or false? Why?

5  How can the precision of an estimate be determined?

**ADVANTAGES OF SAMPLING**

Complete information acquired through a census is generally desirable. If every individual unit in a population is examined, we would be extremely confident in describing the population. But, as in many situations, what you want is not necessarily what you can get. Census data are a luxury item in many situations and are thus often not available for studying a population. Sampling rather than census taking is us ally the rule rather than the exception. Imagine the case of the homemaker tasting the whole pot of soup to make sure about the need for spices. After constantly sipping, there would be no soup left for the family! In the following paragraphs we will elaborate on some of the *major advantages of sampling*.

**Cost**

Any data-gathering effort will require money expenditures for such things as mailings, interviewing, and the tabulation of data. The more data to be handled, the higher the costs will likely be. Consider consumer surveys of the United States: If every citizen were to be polled, the cost would run easily into millions of dollars. Any benefits derived from census data might be negated by the costs. For example, a national food company might want to improve a product to increase sales. The company could survey every potential customer, but it is very likely that the costs of a census would wipe out any additional revenues generated from an improved product. Any time a sample can be taken with less expenditure than that required for a census, cost becomes an acceptable (although not sufficient) reason for sampling.

**Time**

The oft-quoted maxim "Time is money" characterizes many important business decisions. Speed in decisions is often crucial, since many profitable business opportunities pass quickly. Assume you are the owner of a company which has an innovative idea for a better mousetrap, and your top advisers have indicated to you that rival firms are also racing to build a better mousetrap. Being the first company to have a better mousetrap on the market may lead to tremendous sales revenues, but do you actually have a better mousetrap? Will the public beat a path to your company's door for your innovative idea, or will the new mousetrap fail to appeal to the public? Obviously a census requires too much time, which is a valuable commodity. The answer lies in sampling, since it can produce adequate information with less consumption of time.

**Accuracy of Sample Results**

Sometimes the results of a small sample provide information that is almost as accurate as that resulting from a complete census. How is this possible? Remember that the object in sampling is to achieve representation of the population characteristics. There are sampling methods which produce samples that are highly representative of the population. In these situations, larger samples will not produce results which are *significantly* more accurate. Let's again consider the situation of the homemaker. If she has stirred the soup well before sampling, two sips should be sufficient to tell her about the entire pot. Any additional sips will only serve to decrease the volume of soup available for supper. (It's fortunate that our homemaker isn't brewing whiskey.)

**Other Advantages**

Destructive tests are often employed in testing product quality. For example, a company may be interested in the tensile strength of a truckload of iron bars which it has received. In order for the tensile strength to be tested, the iron bars must be subjected to pressure until they break. Obviously, not all the bars can be tested, or else the company would have a truckload of broken iron bars.

Many times the resources may be available for a census, but the nature of the population requires a sample. Suppose we are interested in the number of humpback whales left in the world. Environmental organizations may be willing to sponsor our project, but migration movements, births, and deaths prevent complete enumeration. One approach to the problem is to sample a small area of the ocean and use the results to make a projection.

**Self-testing Review 6-3**

1 "Complete information is always desirable, but sample results may sometimes be almost as accurate." Comment on this statement.

2 When will costs justify sampling over a census? Is cost alone a sufficient reason for a sample?

## SAMPLE SELECTION

We have emphasized that a sample should be as representative of the population as possible. The more representative a sample is, the more confidence we can have in our estimations. In this section of the chapter we will present the more common methods employed in sample selection. *A variety of methods exist because there is no one best method. An appropriate method for sample selection is determined by the nature of the population and the skill of the researcher.*

## Judgment Samples

Sample selection is sometimes based on the opinion of one or more persons who feel sufficiently qualified to identify items for a sample as being characteristic of the population. *Any sample based on someone's expertise about the population is known as a judgment or purposive sample.* As an example, consider the seasoned political campaign manager who intuitively designates certain voting districts as reliable indicators or estimators of public opinion of his candidate. The sample of voting districts is based on the campaign manager's expertise and involves no complex statistical computations.

A judgment sample is convenient, but this convenience is also a disadvantage. Since the judgment sample is not determined by any statistical techniques, it is difficult to assess objectively the degree of representativeness. This difficulty of objective assessment leaves an uncomfortable uncertainty in any estimation based on the sample results. This does not mean, however, that a judgment sample should never be used. The quality of a judgment sample depends on the researcher's expertise, but experience may serve as a valuable tool in surveys.

## Probability Samples

A *probability sample* is one in which the *probability of selection of each element in the population is known prior to sample selection.* Unlike the use of a judgment sample, the use of a probability sample will result in evaluations of population characteristics that may be objectively assessed. In this section we will briefly discuss *three of the major types of probability samples.*

**Simple random** Simple random sampling is a method of sample selection in which each possible sample combination has an equal probability of occurrence, and *each* *item in the population has an equal chance of being included* in the sample. Suppose we have a basketball team with 10 players and we want to estimate the average score of each player per game. Assume further that we would like to select a simple random sample of size 3. With 10 players and a sample of 3, the number of possible combinations of 3 players is

$$_{10}C_3 = \frac{10!}{3! \; 7!} = 120$$

Therefore, each combination must have a 1/120 chance of being selected, and each ball player must have a 3/10 chance of being in the sample.

In order to assure that sample selection is left *completely* to chance, a table of random numbers should be employed in the selection of a simple random sample. Since a complete theoretical explanation of a table of random numbers is beyond the scope of this book, it is sufficient for you to know that each digit of the table is determined *solely by probability* and that each digit and each sequence of digits in a table of random numbers has an equal chance of occurrence. Appendix 3 in the back of the book contains random numbers.

To illustrate the use of random numbers in simple random sampling, suppose we have a list of 200 homemakers eligible for a consumer survey and we want a sample of 20. We could obtain a simple random sample in the following manner:

1  Assign each and every homemaker a number from 000 to 199. Each homemaker should have her own unique number. The first homemaker would have 000, the second would have 001, and so forth.

2  Next consult a table of random numbers. Figure 6-2 is a sample table of random numbers.

3  It is *essential* that we establish a *systematic way* of selecting a sequence of digits from the table so that no bias will enter into the selection of digits. In our case, we need a sequence of three digits. Let's say that our pattern of selection will be the last three digits of each block of numbers and that we will work down a column.

4  Select a random number in the systematic manner and *match the random number assigned to a homemaker.* For example, homemaker number 124 will be selected first, and then homemaker number 109 will be added to the sample. If we have a random number we cannot use, as in the case of 379, we will proceed to the next random number, 194, and continue this process of selection until there is a sample of 20 homemakers.

**FIGURE 6-2**

An example of a table of random numbers

| | | | |
|---|---|---|---|
| 5124 | 0746 | 6296 | 9279 |
| 5109 | 1971 | 5971 | 1264 |
| 4379 | 6296 | 8746 | 5899 |
| 8194 | 3721 | 4621 | 3634 |

**Stratified** If a population is divided into *relatively homogeneous groups or strata* and a sample is drawn from *each group* to produce an overall sample, this is known as *stratified sampling*. Stratified sampling is usually performed when there is *a large variation within the population and the researcher has some prior knowledge of the structure of the population* on which he can stratify the population. The sample results from each stratum are weighted and calculated with the sample results of other strata to provide an overall estimate.

As an illustration, suppose our population is a university student body and we want to estimate the average annual expenditures of a college student for nonschool items. Assume we know that because of different life-styles, the older students spend more than the younger students, but there are fewer older students than younger students because of some dropout factor. To account for this variation in life-style and group size, the population of students can easily be stratified into freshmen, sophomores, juniors, and seniors. A sample can be taken from each stratum and each result weighted to provide an overall estimate of average nonschool expenditures.

**Cluster** A *cluster sample* is a sample in which the individual units are *groups or clusters of single items.* It is *always assumed* that the individual items within each cluster are representative of the population. Consumer surveys of large cities often employ cluster sampling. The usual procedure is to divide a map of the city into small blocks, each block containing a cluster of households to be surveyed. *A number of clusters are selected for the sample,* and all the households in a cluster are surveyed. A distinct benefit of cluster sampling is savings in cost and time. Less energy and money are expended if an interviewer stays within a specific area rather than traveling across stretches of the city.

**Self-Testing Review 6-4**

1 Which, if any, sampling method is applicable to all situations?

2 Is a probability sample more representative of a population than a judgment sample? Why is it or isn't it?

3 Why is a probability sample more desirable than a judgment sample?

4 Assume we have a population of 10 and we wish to take a simple random sample of 2. What are the chances of selection for each sample combination? What are the chances of each individual item to be selected for the sample?

5 What is the basic assumption in cluster sampling?

6 What must we know about a population in order to obtain a stratified sample?

**SAMPLING DISTRIBUTION OF MEANS—A PATTERN OF BEHAVIOR**

The sample mean only approximates the population mean and is very rarely the equivalent of the population mean. Suppose the average income of a probability *sample*[1] of city residents is $6,251. We could venture to say that the *approximate* value of the population mean is $6,251. Intuitively, we know that the chances are very slim that the sample mean equals the population mean. A different sample of residents would most likely yield a different mean, such as $6,282, while another

---

[1] Unless otherwise indicated, the word "sample" will always refer to a probability sample in the pages that follow.

sample might yield a mean of $6,249. *This variation in sample means is known as sampling variation.*

In stating that a sample mean is an approximation of the population mean, we have made an intuitive assumption that the sample mean is related in some manner to the population mean. We intuitively assume that the value of the sample mean *tends toward* the value of the population mean. As we shall see later in this chapter, our intuition is correct; the population characteristics determine the range of values a sample mean may take.

Let us assume that we have a *population* of 15 cards numbered from 0 through 14. And let us further assume that we want to select random samples of size 6 from this population. The number of *possible* samples that *could* be selected is computed as follows:

$$_{15}C_6 = \frac{15!}{6! \ 9!} = 5,005 \text{ possible samples}$$

*One* of these 5,005 possible samples would consist of the cards numbered 2, 4, 6, 8, 10, and 12; a *second* possible sample selection would be made up of the cards numbered 1, 4, 3, 7, 8, and 13; and a *third* possible sample would comprise the cards numbered 14, 0, 7, 10, 9, and 8. (You can figure out the other 5,002 possible samples next summer at your leisure.) The arithmetic *means* of these 3 possible samples, in the order presented, are 7, 6, and 8. If there are 5,005 possible samples, there are, of course, 5,005 possible sample means. And if we were to select all the 5,005 possible samples, compute the mean of each of these samples, and arrange the 5,005 sample means in a frequency distribution, this distribution would be called a *sampling distribution of means*. Thus, a sampling distribution of means is the distribution of the arithmetic means of all the possible random samples of size *n* that could be selected from a given population.

If we are not careful at this point, we can run into some semantic difficulties, so let's pause here to consider the three fundamental types of distributions in Fig. 6-3. There is nothing new about Fig. 6-3a; it is merely the frequency distribution of whatever population happens to be under study and could, of course, take many shapes. The mean and standard deviation of the *population distribution* ($\mu$ and $\sigma$) are quite familiar to us. And there is really nothing very new about the distributions in Fig. 6-3b; they are simply the frequency distributions of some of the samples that could be selected from a population. As is true in any frequency distribution, the individual values (in a sample) are distributed about the mean of those values ($\bar{x}$). The sample standard deviation (s) is the measure of dispersion in these *sample distributions*, which, like the population distribution, can take many shapes. Theoretically, of course, there could be as many sample distributions as there are possible samples that could be taken in a given situation—e.g., if there could be 5,005 possible samples, there could be 5,005 possible sample distributions. Finally, in Fig. 6-3c, we come to the *sampling distribution of the means*—a distribution that *is* new to us, and one that *should not be confused with a sample distribution* (even though the terms are confusingly similar). In the sampling distribution in Fig. 6-3c, the possible sample mean values are distributed about the mean of the sampling distribution (sometimes called the grand mean). The mean of the sampling distribution, you will remember from Fig. 6-3c, is identified by the symbol $\mu_{\bar{x}}$, and the standard deviation of the sampling distribution is identified by symbol $\sigma_{\bar{x}}$.

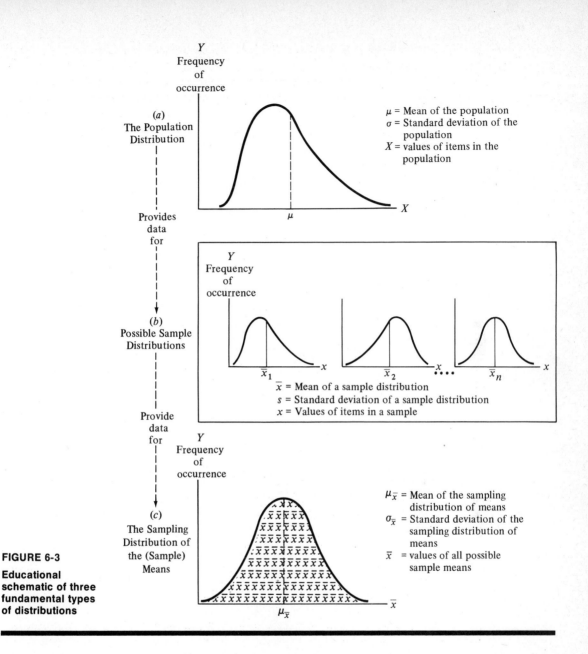

μ = Mean of the population
σ = Standard deviation of the
population
X = values of items in the
population

(a)
The Population
Distribution

Provides
data
for

Y
Frequency
of
occurrence

(b)
Possible Sample
Distributions

$\bar{x}$ = Mean of a sample distribution
s = Standard deviation of a sample distribution
x = Values of items in a sample

Provide
data
for

(c)
The Sampling
Distribution of
the (Sample)
Means

$\mu_{\bar{x}}$ = Mean of the sampling
distribution of means
$\sigma_{\bar{x}}$ = Standard deviation of the
sampling distribution of
means
$\bar{x}$  = values of all possible
sample means

## Mean of the Sampling Distribution of Means

Be alert and pay attention now, because we are about to lay a very important fact on you: *The mean of the sampling distribution of means is equal to the population mean* —i.e., $\mu_{\bar{x}} = \mu$. What's that you said? "I have read some ridiculous statements in the past, and more than a few of them have come from this book, but. . . ." Anticipating just such a skeptical attitude, we have prepared an example to redeem our credibility.

Assume the population is a small group of five students enrolled in a statistics course and an instructor wants to estimate the average amount of time spent by each student preparing for classes each week. Figure 6-4 lists the amount of time each student spends per week preparing for class (but the instructor does not have access

## FIGURE 6-4

Population of students and their weekly preparation time

| Student | Preparation time (hours) |
|---------|--------------------------|
| A | 7 |
| B | 3 |
| C | 6 |
| D | 10 |
| E | 4 |
| | $\Sigma X = 30$ |

$$\mu = \frac{\Sigma X}{N} = \frac{30}{5} = 6$$

to this information). As we can see from Fig. 6-4, the population mean preparation time is 6.

If the instructor takes a sample of three students, what are the possible values of the sample mean? How different from the true mean of 6 might a sample mean be? Figure 6-5 provides the answer; it also provides us with the data needed to compute the mean of the sampling distribution as follows:

$$\mu_{\bar{x}} = \frac{\Sigma(\bar{x}_1 + \bar{x}_2 + \bar{x}_3 + \cdots + \bar{x}_{nCr})}{nCr} \tag{6-1}$$

$$= 60/10$$

$$= 6$$

## FIGURE 6-5

Sampling distribution of means

| Sample combinations | Sample data | Sample means $(\bar{x})$ | $(\bar{x} - \mu_{\bar{x}})$ | $(\bar{x} - \mu_{\bar{x}})^2$ |
|---------------------|-------------|---------------------------|------------------------------|-------------------------------|
| 1. A, B, C | 7, 3, 6 | 5.33 | .67 | .45 |
| 2. A, B, D | 7, 3, 10 | 6.67 | .67 | .45 |
| 3. A, B, E | 7, 3, 4 | 4.67 | 1.33 | 1.77 |
| 4. A, C, D | 7, 6, 10 | 7.67 | 1.67 | 2.79 |
| 5. A, C, E | 7, 6, 4 | 5.67 | .33 | .11 |
| 6. A, D, E | 7, 10, 4 | 7.0 | 1.00 | 1.00 |
| 7. B, C, D | 3, 6, 10 | 6.33 | .33 | .11 |
| 8. B, C, E | 3, 6, 4 | 4.33 | 1.67 | 2.79 |
| 9. B, D, E | 3, 10, 4 | 5.67 | .33 | .11 |
| 10. C, D, E | 6, 10, 4 | 6.67 | .67 | .45 |
| | | 60.0 | | 10.0 |

$$\mu_{\bar{x}} = \frac{\Sigma(\bar{x}_1 + \bar{x}_2 + \bar{x}_3 + \cdots + \bar{x}_{nCr})}{nCr} = \frac{60}{10} = 6$$

where the numerator is the sum of all possible sample means, and the denominator is the number of possible samples.

Thus, as you can see, the mean in Fig. 6-4 equals the mean in Fig. 6-5. That is, $\mu_{\bar{x}} = \mu$.

As the devil's advocate, we may say "So what?" to the fact that $\mu_{\bar{x}} = \mu$. No one in a realistic situation really takes all possible sample combinations and calculates the sample means. In practice only one sample is taken. What benefit is there in discussing the sampling distribution? Shouldn't we really be concerned with the proximity of a single sample mean to the population mean? In essence, the discussion of the sampling distribution *is* concerned with the proximity of a sample mean to the population mean.

You can see from the example in Fig. 6-5 that the possible values of the sample means *tend toward* the population mean. Since these values have frequencies of occurrence, the sampling distribution is essentially a probability distribution. If the sample size is *sufficiently large (n more than 30),* the sampling distribution approximates the *normal distribution whether or not the population is normally distributed.* And the sampling distribution is *normally distributed regardless of sample size if the population is normally distributed.* Figure 6-6 illustrates the sampling distribution as a normal distribution.

You will recall that in a normal probability distribution the likelihood of an outcome is determined by the number of standard deviations from the mean of the distribution. (You do recall all those light bulb problems, don't you?) Therefore, as we see in Fig. 6-6, there is a 68.3 percent chance that a sample selected at random will have a mean that lies within *one* standard deviation $(\sigma_{\bar{x}})$ of the population mean. Also, there is a 95.4 percent chance that the sample mean will lie within *two* standard

**FIGURE 6-6**

$\mu$, $\mu_{\bar{x}}$, and $\sigma_{\bar{x}}$ for the areas under the sampling distribution of means

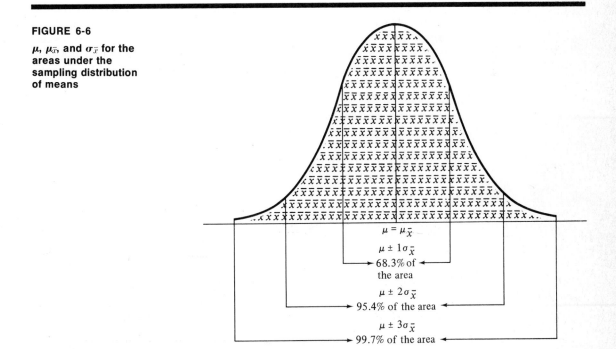

deviations of the population mean. Thus, a knowledge of the properties of the sampling distribution tells us the probable proximity of a sample mean outcome to the value of the population mean. With a knowledge of the sampling distribution, probability statements can be made about the range of possible values a sample mean may assume. This range of possible values can be calculated if a value for the standard deviation of the sampling distribution ($\sigma_{\bar{x}}$) is available. The computation of $\sigma_{\bar{x}}$ is shown in the next section.

**Standard Deviation of the Sampling Distribution of Means**

In order to determine the extent to which a sample mean might differ from the population mean, we need some *measure of dispersion*. In other words, we must be able to compute the likely deviation of a sample mean from the mean of the sampling distribution. The standard deviation of the sampling distribution, in statistical jargon, is given the rather intimidating name of *standard error of the mean* and, as we have seen, is represented by the symbol $\sigma_{\bar{x}}$. For the data given in Fig. 6-5, the calculation of this measure is similar to the calculation of any other standard deviation about a mean:

$$\sigma_{\bar{x}} = \sqrt{\frac{\Sigma(\bar{x} - \mu_{\bar{x}})^2}{N}} \qquad (6\text{-}2)$$

$$= \sqrt{10.00/10}$$

$$= \sqrt{1.00}$$

$$= 1.00$$

where $N$ = total number of possible samples

However, as pointed out in an earlier paragraph, no one (other than an eccentric statistician or a student with an assigned problem) ever deals with all the possible sample combinations. Therefore, an alternate method to compute $\sigma_{\bar{x}}$ must exist.

Since we have seen the relationship that exists between $\mu_{\bar{x}}$ and $\mu$, we might intuitively assume that there is a relationship between the standard error of the mean and the population standard deviation that will produce a shortcut method of computing the standard error. As a matter of fact, we are right, and the *standard error may be computed for a finite population with the following formula:*

$$\sigma_{\bar{x}} = \frac{\sigma}{\sqrt{n}} \sqrt{\frac{N - n}{N - 1}} \qquad (6\text{-}3)$$

where $\sigma$ = the population standard deviation

$N$ = population size

$n$ = sample size

$\sqrt{\dfrac{N - n}{N - 1}}$ = finite population correction factor

From the data in Fig. 6-4, the *standard deviation of the population* is computed as follows:

$$\sigma = \sqrt{\frac{\Sigma(X - \mu)^2}{N}}$$

$$= \sqrt{\frac{1^2 + 3^2 + 4^2 + 2^2}{5}}$$

$$= \sqrt{\frac{1 + 9 + 16 + 4}{5}}$$

$$= 2.45$$

Therefore, the standard error for the data given in Fig. 6-5 is computed as follows:

$$\sigma_{\bar{x}} = \frac{2.45}{\sqrt{3}} \sqrt{\frac{5 - 3}{5 - 1}}$$

$$= 1.4145 \ (.7071)$$

$$= 1.00$$

We can now see that the results of the computations based on formulas (6-2) and (6-3) are equal. Thus, $\sigma_{\bar{x}}$ *may be determined with a knowledge of the population standard deviation, the sample size, and the population size.*

*If the population is infinite in size,* as, for example, in the case of items from an assembly-line operation, the standard error does not require a finite correction factor and may be computed as follows:

$$\sigma_{\bar{x}} = \frac{\sigma}{\sqrt{n}} \qquad \textit{for infinite population} \qquad (6\text{-}4)$$

If the population is infinite, there is no need for the correction factor. *However, a finite population does not necessarily mean that the correction has to be employed.* At this point the last statement may cause you to wince in pain, since there is an apparent contradiction. Let's look at the following example and, we hope, relieve your pain.

Suppose we have a finite population of approximately 200 million, and we have taken a sample of 2,000. If we followed the rules strictly, we would have:

$$\sigma_{\bar{x}} = \frac{\sigma}{\sqrt{n}} \sqrt{\frac{N - n}{N - 1}}$$

$$= \frac{\sigma}{\sqrt{n}} \sqrt{\frac{200,000,000 - 2,000}{200,000,000 - 1}}$$

$$= \frac{\sigma}{\sqrt{n}} (.99999)$$

However, the size of the population is so large that the finite correction factor for all practical purposes is 1. *If the population size is extremely large compared to the sample size, formula (6-4) may be used to calculate the standard error of a finite population.*

**The Relationship between *n* and $\sigma_{\bar{x}}$**

The standard error of the mean is, of course, *a measure of the dispersion* of sample means about the population mean. If the degree of dispersion *decreases*, the range of possible values a sample mean may assume also *decreases*, meaning the value of any single *sample mean* will probably be closer to the value of the *population mean* as the standard error decreases. And with formula (6-3) or (6-4), the value of $\sigma_{\bar{x}}$ obviously must decrease as the size of *n* increases—i.e.,

$$\downarrow \sigma_{\bar{x}} = \frac{\sigma}{\sqrt{n} \uparrow}$$

To add meaning to this mathematical manipulation, let's look at the relationship between *n* and $\sigma_{\bar{x}}$ intuitively. Let's assume we wish to estimate some parameter of a population of 100, and we are initially entertaining the thought of taking a sample of 10 items. Ten items may provide adequate information, but it is clear that more information could be obtained from a larger sample such as 20. More information provides a more precise estimate of the population parameter. As a matter of fact, a sample of 50 or 60 items would provide even more information and thus a more precise estimate. The ultimate option is that we could sample the entire population and obtain complete information, and thus there would be no difference between the sample statistic and the population parameter. In our example, if we were to estimate the population mean, a sample of 100 would have the following standard error:

$$\sigma_{\bar{x}} = \frac{\sigma}{\sqrt{n}} \sqrt{\frac{N-n}{N-1}}$$

$$= \frac{\sigma}{\sqrt{n}} \sqrt{\frac{100-100}{99}}$$

$$= 0$$

*The general principle is that as n increases, $\sigma_{\bar{x}}$ decreases. As the sample size increases, we have more information on which to estimate the population mean, and thus the probable difference between the true value and any sample outcome decreases.* Figure 6-7 summarizes the points made in this section.

**The Central Limit Theorem**

Up to this point we have explained the concept of the sampling distribution of means on a rather intuitive basis. Now, however, we are ready to formalize the concepts developed in previous sections and attribute the properties of the sampling distribution to the *Central Limit Theorem*. The Central Limit Theorem basically states:

*The mean of the sampling distribution of means will be equal to the population mean. The standard deviation of the sampling distribution of means ($\sigma_{\bar{x}}$) will be equal to $\sigma/\sqrt{n}$ for an infinite population, and it will be equal to $\sigma/\sqrt{n} \sqrt{(N-n)/(N-1)}$ for a finite population. If the sample size (n) is sufficiently large, the sampling distribution will approximate the normal probability distribution. If the population is normally distributed, the sampling distribution will be normal regardless of sample size.*

A considerable amount of important theoretical material has been presented in the last few pages. Perhaps it would be appropriate now to consider some example problems to illustrate certain basic concepts.

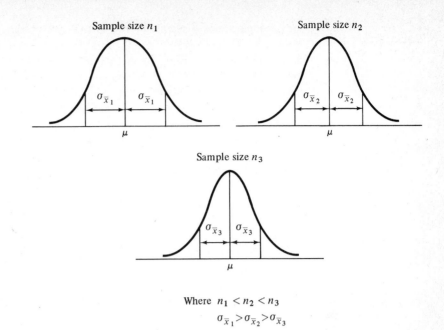

**FIGURE 6-7**

**The relationship between *n* and $\sigma_{\bar{x}}$**

Where $n_1 < n_2 < n_3$

$$\sigma_{\bar{x}_1} > \sigma_{\bar{x}_2} > \sigma_{\bar{x}_3}$$

**Example Problems**

**Example 6-1** The Bigg Truck Company has a fleet of five trucks which have monthly maintenance costs of $200, $175, $185, $210, and $190. An estimate of the average monthly cost for a truck is to be obtained from a sample of 3. What are the mean and the standard deviation of this sampling distribution of size 3?

The population mean is $192 — ($200 + $175 + $185 + $210 + $190)/5. Since the population mean is equal to the mean of the sampling distribution, $\mu_{\bar{x}} = \$192$. The standard deviation of the population is computed as follows:

$$\sigma = \sqrt{\frac{\Sigma(X - \mu)^2}{N}}$$

$$= \sqrt{\frac{8^2 + 17^2 + 7^2 + 18^2 + 2^2}{5}}$$

$$= \sqrt{\frac{64 + 289 + 49 + 324 + 4}{5}}$$

$$= 12$$

Consequently,

$$\sigma_{\bar{x}} = \frac{\sigma}{\sqrt{n}} \sqrt{\frac{N - n}{N - 1}}$$

$$= \frac{12}{\sqrt{3}} \sqrt{\frac{5 - 3}{5 - 1}}$$

$$= 4.9$$

**Example 6-2** The Sam and Janet Evening Catering Company wants to estimate the average amount spent for each order. The company will obtain an estimate by selecting a simple random sample of 100 orders. Let's assume a normal population in which the true mean is $120 and the true standard deviation is $25. What is the standard deviation of the sampling distribution? What is the likelihood that the sample mean will fall between $\mu - 1\ \sigma_{\bar{x}}$ and $\mu + 1\ \sigma_{\bar{x}}$? Within what range of values does $\bar{x}$ have a 95.4 percent chance of falling?

$\sigma_{\bar{x}}$ is $\sigma/\sqrt{n} = \$25/10 = \$2.50$. The likelihood that $\bar{x}$ will fall between $117.50 and $122.50 is 68.3 percent. There is a 95.4 percent chance that $\bar{x}$ will fall between $\mu - 2\sigma_{\bar{x}}$ and $\mu + 2\sigma_{\bar{x}}$, or between $115 and $125.

**Example 6-3** The Write-On Pen Company wants to estimate the average number of pens sold per month on the basis of the mean of a sample of 100 months. Assume that the true mean is 5,650 pens per month and that the standard deviation is 700. What are the chances that the sample mean will have a value within 200 pens of the true mean?

The problem basically asks: "What are the chances that the value of the sample mean will be between $\mu - 200$ and $\mu + 200$?" Remember that the width has the general form $\mu - Z\sigma_{\bar{x}}$ and $\mu + Z\sigma_{\bar{x}}$. (The interval is more commonly expressed as $\mu \pm Z\sigma_{\bar{x}}$.) Consequently, we want $Z\sigma_{\bar{x}}$ to be 200. $\sigma_{\bar{x}}$ can be determined as follows:

$$\sigma_{\bar{x}} = \frac{\sigma}{\sqrt{n}} = \frac{700}{\sqrt{100}} = 70$$

Therefore, $Z$ may be calculated in the following manner:

$$Z\sigma_{\bar{x}} = 200$$
$$Z(70) = 200$$
$$Z = 2.86$$

Since $Z = 2.86$, we can consult Appendix 2 and see that the probability of a value occurring within 2.86 standard deviations to one side of a true mean is 49.79 percent. Since we are concerned with 2.86 standard deviations to both sides of the mean, the total likelihood of a sample mean value between $5,650 \pm 200$ is 99.58 percent.

**Self-testing Review 6-5**

1  A _____ distribution is a distribution of all the possible sample means, while a _____ distribution is the distribution of individual items of a single sample.

2  If Greek letters are to be used only with population parameters, why are they used to denote characteristics of a sampling distribution of sample means?

3  What method of sample selection is assumed when we discuss concepts concerning the sampling distribution of means?

4  How is the mean of the sampling distribution of means related to the population mean?

5  When may we omit the finite correction factor in calculating $\sigma_{\bar{x}}$?

6  When will the sampling distribution approximate the normal probability distribution?

7  Why is it possible for us to make probability statements concerning the value of a sample mean if we know the population mean?

## SAMPLING DISTRIBUTION OF PERCENTAGES

We are frequently interested in estimating population percentages.[2] For example, a company might want to estimate the percentage of defective items produced by a machine, or a company might be interested in the percentage of minority employees in its work force, or we might be interested in the percentage of students who hope to become statisticians. As in the case of estimating a population mean, *we estimate a population percentage on the basis of sample results.* In this section of the chapter we will look at the relationship between a population percentage and the possible values a sample percentage may assume. We will look at the sampling distribution of percentages. *A sampling distribution of percentages is a distribution of the percentages of all possible samples where the samples are simple random samples of fixed size n.*

## Mean of the Sampling Distribution of Percentages

The Greek letter $\pi$ (pi) will be used to denote the *population percentage*, while the lowercase letter $p$ will denote the *sample percentage*. The symbol $\mu_p$ will refer to the mean of the sampling distribution of percentages. The sample percentage is defined as $p = x/n$, where $x$ is the number of items in a sample possessing the characteristic of interest, and $n$ is the sample size.

Suppose we have a population of 5 students and we wish to take a simple random sample of 3 students and estimate the true percentage of students who have made the dean's list. Figure 6-8 lists the population and each student's status concerning the dean's list. What would the sampling distribution look like? Well, in

**FIGURE 6-8**

Population of students and dean's list status

| Student | Dean's list |
|---------|-------------|
| A | Yes |
| B | No |
| C | Yes |
| D | No |
| E | No |

$X = 2$ (the number of students on the dean's list)

$$\pi = \frac{X}{N} = \frac{2}{5} \text{ or 40 percent}$$

where $\pi$ = population percentage
$N$ = population size
$X$ = number of students on the dean's list

---

[2] Many texts deal with the material in the discussion that follows in terms of *proportions* rather than percentages. We prefer to use percentages because many students seem to find the arithmetic easier, and because percentages are more frequently used in every day discussion. Of course, if you prefer to use proportions, you can simply move the decimal two places to the left in all computations. The ultimate results are identical.

**FIGURE 6-9**

Sampling distribution of percentages

| Sample combinations | Sample data | Sample percentage ($p$) |
|---|---|---|
| 1. A, B, C | Yes, no, yes | 0.667 |
| 2. A, B, D | Yes, no, no | 0.333 |
| 3. A, B, E | Yes, no, no | 0.333 |
| 4. A, C, D | Yes, yes, no | 0.667 |
| 5. A, C, E | Yes, yes, no | 0.667 |
| 6. A, D, E | Yes, no, no | 0.333 |
| 7. B, C, D | No, yes, no | 0.333 |
| 8. B, C, E | No, yes, no | 0.333 |
| 9. B, D, E | No, no, no | 0.000 |
| 10. C, D, E | Yes, no, no | 0.333 |
| | | $\Sigma_p = 4.000$ |

$$\mu_p = \frac{\Sigma_p}{N} = \frac{4}{10} \text{ or 40 percent}$$

where $\mu_p$ = mean of the sampling distribution of percentages
$N$ = number of sample combinations – i.e., $nCr$.

Fig. 6-9 the possible percentages a sample might have are listed. And we can see that the mean of the sampling distribution in Fig. 6-9 is equal to the population percentage calculated in Fig. 6-8. Thus, *the mean of a sampling distribution of percentages with simple random samples of size n is equal to the population percentage – i.e., $\mu_p = \pi$.*

**Standard Deviation of the Sampling Distribution of Percentages**

The standard deviation of the sampling distribution of percentages may be computed with knowledge of the population percentage, population size, and sample size. The symbol for this standard deviation (more frequently called the *standard error of percentage*) is $\sigma_p$. The computation of $\sigma_p$ is as follows *for a finite population:*

$$\sigma_p = \sqrt{\frac{\pi(100 - \pi)}{n}} \sqrt{\frac{N - n}{N - 1}} \qquad (6\text{-}5)$$

where $\pi$ = population percentage possessing a particular characteristic
$100 - \pi$ = population percentage not possessing a particular characteristic
$N$ = population size
$n$ = sample size

$\sqrt{\dfrac{N - n}{N - 1}}$ = correction factor for a finite population

As you have probably anticipated, the finite correction factor may be omitted if the population is infinite, or if the size of the population relative to the sample size is extremely large. The computation of the standard deviation of the sampling distribution for *an infinite population* is:

$$\sigma_p = \sqrt{\frac{\pi(100 - \pi)}{n}} \qquad\qquad (6\text{-}6)$$

One last comment should be made about the sampling distribution of percentages. The Central Limit Theorem also has application for sample percentages. *If the sample size is sufficiently large, the sampling distribution will approximate a normal probability distribution.* The implication of such a distribution is that probability statements can be made about the possible value of a sample statistic based on the knowledge of the population percentage. For example, there is a 95.4 percent chance that a sample percentage will fall within $\pm 2\sigma_p$ of $\pi$. And there is approximately a 99.7 percent chance that a sample percentage will assume a value within $\pm 3\sigma_p$ of $\pi$.

A few example problems illustrating some basic concepts of the sampling distribution of means were presented earlier. Let us now do the same thing with some problems related to the material just presented.

**Example Problems**

**Example 6-4** The Kane and Abel Fraternal Organization has a total of eight members whose ages are 27, 32, 33, 26, 43, 52, 28, and 25. The organization has a weird rule which requires a minimum age of 33 for a member to be president. Assume a sample size of 4 is selected to provide an estimate of the population percentage eligible for the presidency. What would be the mean and the standard deviation of the sampling distribution?

The population percentage is computed as follows:

$$\pi = \frac{\text{the number possessing the age qualification}}{\text{the population size}} = \frac{3}{8} \text{ or } 37.5 \text{ percent}$$

Since $\mu_p = \pi$, the mean of the sampling distribution is also 37.5 percent. And since we have a *finite* population, the standard error of percentage is computed as follows:

$$\sigma_p = \sqrt{\frac{\pi(100 - \pi)}{n}} \sqrt{\frac{N - n}{N - 1}} = \sqrt{\frac{37.5(100 - 37.5)}{4}} \sqrt{\frac{8 - 4}{8 - 1}}$$

$$= 18.3 \text{ percent}$$

**Example 6-5** The Tack Nail Company has selected a sample of 100 nails in order to estimate the percentage of nails in a production run which are acceptable. Assume that the population is 90 percent acceptable. What are the chances that the sample percentage will be within 5 percent of the population percentage?

The basic question is: What are the chances that the value of $p$ will be between $\pi - 5$ percent and $\pi + 5$ percent? In this case, $Z\sigma_p$ equals 5 percent because the basic format in determining the width of the interval is $\pi - Z\sigma_p$ and $\pi + Z\sigma_p$. (The interval is commonly expressed in the form $\pi \pm Z\sigma_p$.) We can solve for $\sigma_p$ in the following manner:

$$\sigma_p = \sqrt{\frac{\pi(100 - \pi)}{n}}$$

$$= \sqrt{\frac{(90)(10)}{100}}$$

$$= 3 \text{ percent}$$

And since $\sigma_p = 3$ percent, we can solve for $Z$ as follows:

$$Z\sigma_p \qquad = 5 \text{ percent}$$
$$Z(3 \text{ percent}) = 5 \text{ percent}$$
$$Z \qquad\qquad = 1.67$$

With a $Z$ value of 1.67, we see in Appendix 2 that the area under the normal curve corresponding to a $Z$ of 1.67 is .4515. And with 1.67 standard errors to *each side* of $\pi$, the likelihood of a sample percentage within 5 percent of the population percentage is 90.3 percent.

**Self-testing Review 6-6**

1  How is the mean of the sampling distribution of percentages related to the population percentage?

2  What population parameters need to be known in order to calculate the standard deviation of the sampling distribution of percentages?

3  When may we omit the finite correction factor in calculating $\sigma_p$?

4  When does the sampling distribution of percentages approximate the normal probability distribution?

**SUMMARY**

The purpose of this chapter was to provide you with an understanding of the validity of using a sample statistic to estimate a parameter. If a sample is representative of the population, it is possible to make inferences about the population from the sample results. Although complete information, in the form of a census, may be desirable, advantages in sampling such as reduced cost, reduced time, and accuracy of results often outweigh the disadvantages of sampling error. A variety of sampling methods exists to control sampling error, but there is no one best method that can eliminate all sampling error.

Although sampling will not provide an exact value of a parameter, it is adequate for estimation purposes to know that sample values are governed by population characteristics. When a simple random sample is used, the mean of a sampling distribution is the parameter value to be estimated, while the standard deviation of the sampling distribution of means is determined by the standard deviation of the population. The results of different samples will have different values, but these values tend toward the population value. These properties of the sampling distribution may be attributed to the Central Limit Theorem.

The Central Limit Theorem also states that the shape of the sampling distribution is approximately normal if the sample size is large. Given this normality property, it is possible to make probability statements concerning the possible values a statistic may assume. The value of the sample statistic will tend toward the parameter value, and a probability statement can be made about the proximity of the statistic to the parameter.

The characteristics of two sampling distributions are summarized in Fig. 6-10.

## FIGURE 6-10

Properties of sampling distributions*

| Population | Sampling distribution of | |
| --- | --- | --- |
| | Means ($\bar{x}$) | Percentages ($p$) |
| Finite | $\mu_{\bar{x}} = \mu$ | $\mu_p = \pi$ |
| | $\sigma_{\bar{x}} = \dfrac{\sigma}{\sqrt{n}} \sqrt{\dfrac{N-n}{N-1}}$ | $\sigma_p = \sqrt{\dfrac{\pi(100-\pi)}{n}} \sqrt{\dfrac{N-n}{N-1}}$ |
| Infinite | $\mu_{\bar{x}} = \mu$ | $\mu_p = \pi$ |
| | $\sigma_{\bar{x}} = \dfrac{\sigma}{\sqrt{n}}$ | $\sigma_p = \sqrt{\dfrac{\pi(100-\pi)}{n}}$ |

* Both sampling distributions approximate the normal probability distribution, if the sample size is sufficiently large. As a general rule of thumb "sufficiently large" is over 30.

## Important Terms and Concepts

1 Population
2 Finite population
3 Infinite population
4 Sample
5 Parameter
6 Statistic
7 $\mu = \dfrac{\Sigma X}{N}$
8 $\sigma = \sqrt{\dfrac{\Sigma(X - \mu)^2}{N}}$
9 $\bar{x} = \dfrac{\Sigma x}{n}$
10 $s = \sqrt{\dfrac{\Sigma(x - \bar{x})^2}{n}}$
11 Judgment sample
12 Probability sample
13 Simple random sample

14 Stratified sample
15 Cluster sample
16 Table of random numbers
17 Sampling variation
18 Sampling distribution of means
19 $\sigma_{\bar{x}} = \dfrac{\sigma}{\sqrt{n}}$ or $\sigma_{\bar{x}} = \dfrac{\sigma}{\sqrt{n}} \sqrt{\dfrac{N-n}{N-1}}$
20 Finite population correction factor
21 Central Limit Theorem
22 $\mu = \mu_{\bar{x}}$
23 $\pi = \mu_p$
24 $\sigma_p = \sqrt{\dfrac{\pi(100-\pi)}{n}} \sqrt{\dfrac{N-n}{N-1}}$
 or $\sigma_p = \sqrt{\dfrac{\pi(100-\pi)}{n}}$
25 Sampling distribution of percentages

## Problems

1 Assume that a population consists of 10 items. What is the probability of selection for each possible sample if a simple random sample of 3 is taken?

2 If we have a population of 8, what is the probability of selection for each possible sample if a sample of 5 is taken? What is the probability of selection for each item of the population?

**3** A population consists of 5 students. The number of hours they spend watching television is as follows:

| Student | Hours |
| --- | --- |
| a | 7 |
| b | 16 |
| c | 20 |
| d | 12 |
| e | 22 |

A simple random sample of 3 is to be taken to estimate the population mean, that is, the average number of hours spent watching television.

**a** Calculate the population mean and the population standard deviation.

**b** What is the mean of the sampling distribution?

**c** Calculate the standard deviation of the sampling distribution.

**4** We have a population of 5 motorists. The price they each pay for a gallon of gasoline is as follows:

| Motorist | Price |
| --- | --- |
| a | $0.52 |
| b | 0.48 |
| c | 0.54 |
| d | 0.50 |
| e | 0.53 |

A simple random sample of 3 is to be taken, and an average price per gallon is to be estimated.

**a** Obtain the sampling distribution of $\bar{x}$.

**b** Calculate the mean and the standard deviation of the sampling distribution.

**c** Verify the values of $\mu_{\bar{x}}$ and $\sigma_{\bar{x}}$ with the use of the population parameters.

**5** The Tite Wire Company manufactures wires for circus acts. It has taken a sample of 100 pieces of wire and wants to see if the thickness of a batch of wire meets minimum specifications. Assume that $\mu = 0.45$ inches, with a standard deviation of 0.03 inches.

**a** Calculate the mean and standard deviation of the sampling distribution.

**b** What may be said about the shape of the sampling distribution?

**c** Within what range of values does the sample mean have a 68.3 percent chance of falling?

**d** Within what range of values does the sample mean have a 95.4 percent chance of falling?

**6** Assume we have an infinite population with a mean of 200 and a standard deviation of 15.

**a** Within what range of values will a sample mean have a 95.4 percent chance of falling if we have a sample of 45?

**b** Within what range of values will there be a likelihood of 95.4 percent occurrence for the sample mean if we have a sample of 36? What about a sample of 49? What about a sample of 64?

**c** What relationship can you observe between the sample size and the dispersion of the sampling distribution?

**7** The Keep On Trucking Company wants to estimate the average tonnage of freight handled per month, and it has taken a sample of 36 months. Assume the true average tonnage per month is 225 tons, with a standard deviation of 30 tons. What are the chances that the sample mean will have a value within 7 tons of the true mean?

**8** Dr. D. Zees would like to estimate the average amount charged per patient each visit. He has a sample of 40 patients. Assume that $\mu = \$13$ and $\sigma = \$4$. What are the chances that the sample mean will have a value within $1 of the true mean?

**9** Assume we have a population of 20 high school students and we take a sample of 5 students to estimate the proportion of students who intend to enter college. Assume the true percentage is 60 percent.
  **a** What will be the mean of the sampling distribution of percentages? Why?
  **b** Is the finite correction factor needed to calculate $\sigma_p$? What is the value of the finite correction factor?
  **c** Calculate $\sigma_p$.

**10** The Vanity Press Company wants to estimate the percentage of books printed that are defective and cannot be sold. Assume we have a sample of 100 and the true percentage is 8.5 percent. What are the chances that the sample percentage will be within 1 percent of the population percentage?

**Topics for Discussion**

**1** Explain the difference between a sample distribution and a sampling distribution.

**2** In what situations may your statistics class be a population, and under what conditions may your statistics class be a sample?

**3** "We should never be satisfied with a sample unless it is completely representative of the population." Comment on this statement.

**4** "Since sampling errors will always exist, it is difficult to have any confidence in an estimation." Comment on this statement.

**5** "In practice we take only one sample. It is therefore not necessary to be concerned with the sampling distribution, because no one ever takes all possible samples." Comment on this statement.

**6** "A finite correction factor must always be used in the calculation of $\sigma_{\bar{x}}$ if we have a finite population." Why is this statement incorrect?

**7** What effect does sample size have on the dispersion of sample means about the population mean?

**Answers to Self-testing Review Questions**

6-1

**1 a** Population.
   **b** Sample.
   **c** Sample.
   **d** Population.

**2** Yes: the population may be the major leagues.

**3 a** The parameter is the total membership, 300. It may also be the mean age, 42.

**b** The sample size, 25, and the sample mean, 39, are statistics.

**c** Population mean.

## 6-2

1 The statement is incorrect. Sampling occurs frequently in the course of daily events.

2 The purpose of sampling is to provide sufficient information so that judgments may be made concerning the characteristics of the population.

3 The goal is maximum representation of population characteristics.

4 False. If the errors can be objectively assessed, we have some idea of the precision of the estimate.

5 It can be determined by objectively assessing the amount of sampling error.

## 6-3

1 This statement is true. Representation can be achieved if the right method of sample selection is employed.

2 There is justification for a sample if the sample costs are less than census costs, but costs should never be the sole reason for sampling.

## 6-4

1 There is no one best method for sample selection. The nature of the population and the skills of the researcher determine the appropriate method.

2 It is difficult to say whether a probability sample is more representative than a judgment sample because it is impossible to assess objectively the error in a judgment sample.

3 The probability sample is desirable because it can be objectively assessed.

4 There are 45 possible combinations. Each sample combination will have a 1 in 45 chance of selection. Each individual item will have a 1 in 5 chance of selection.

5 The basic assumption in cluster sampling is that the items within a cluster are representative of the population.

6 We must have some prior knowledge about the structure of the population.

## 6-5

1 Sampling; sample.

2 The sampling distribution is a population of all possible sample means of a fixed sample size.

3 In discussing any sampling distribution, we assume simple random samples.

4 The mean of the sampling distribution is always equal to the population mean.

5 The finite population correction factor may be omitted when there is an infinite

population or when the population is extremely large relative to the sample size.

**6** The sampling distribution will approximate the normal distribution if the sample size is sufficiently large.

**7** The sampling distribution is a probability distribution, and in cases of a large sample size, the sampling distribution approximates a *normal* probability distribution. Since we know the shape and characteristics of a normal distribution, we can make probability statements.

### 6-6

**1** The mean of the sampling distribution of percentages, $\mu_p$, is equal to the population percentage.

**2** We need to know the population percentage, and in the case of a finite population, we need to know the population size.

**3** The finite correction factor may be omitted with an infinite population, or the finite correction factor may be omitted if the population is extremely large relative to the sample size.

**4** The sampling distribution approximates the normal probability distribution when the sample size is large.

# CHAPTER 7

# ESTIMATING MEANS AND PERCENTAGES

**LEARNING OBJECTIVES**

After studying this chapter, working the problems, and answering the discussion questions, you should be able to:

☞ Explain the basic theories and concepts underlying the interval estimation of population means and population percentages.

☞ Compute interval estimates of the population mean at different levels of confidence when the population standard deviation is unknown as well as when it is available.

☞ Understand when and how to use the $t$ distribution rather than the $Z$ distribution for estimation purposes.

☞ Compute interval estimates of the population percentage at different levels of confidence.

☞ Determine the appropriate sample size to use to estimate the population mean or percentage at different levels of confidence.

**CHAPTER OUTLINE**

ESTIMATOR, ESTIMATE, ESTIMATION, ET CETERA
  Point Estimation
  Interval Estimation
  Self-testing Review 7-1

INTERVAL ESTIMATION OF THE POPULATION MEAN: SOME BASIC CONCEPTS
  The Sampling Distribution—Again
  Interval Width Considerations

140

**To guess is cheap,**
**To guess wrongly is expensive.**

**— An old Chinese proverb**

As the proverb implies, a guess or an estimate can be easily made. Anyone can offer an approximation which can be based on a variety of methods. The naïve as well as the expert in an area can produce an estimate if requested to do so. John Q. Public as well as the famous economist Monte Terry Drane can give an estimate of next year's gross national product if asked to do so. However, the task in estimation is *not* to just produce an estimate but to produce an estimate which has some degree of accuracy.

Let us assume that a sales manager who must make a sales forecast for the next period asks you, the trusted assistant, to provide an estimate of the average dollar purchase made by a typical customer. You could provide an estimate based on one of a variety of methods, such as rolling a pair of dice, drawing from a deck of cards, or reading a cup of soggy tea leaves. It is not necessary that your estimate be scientifically calculated, but it is necessary that your estimate have some degree of precision. Consider the consequences of an extremely inaccurate estimate. On the basis of your wild guess, your boss makes a grossly inaccurate forecast which causes the company to lose thousands upon thousands of dollars. This result, in turn, causes the company president to rebuke your boss, and this action could cause you to become a member of the unemployed population.

In this chapter we will cover methods of estimating the population mean and the population percentage with some degree of accuracy. By no means will these methods produce exact population values, and we do not claim to offer any formulas which predict a parameter value exactly. What you will see in this chapter will be methods which permit estimation with some degree of confidence that the estimate approximates the true value. Error will, of course, occur in estimation (as the Roman poet Ovid once wrote, "The judgment of man is fallible"),[1] but the amount of error can be objectively assessed and controlled in the estimation.

The specific topics to be covered in the pages that follow are (1) some *definitions of important terms*, (2) some *basic concepts about the estimation of the population mean*, (3) *procedures for actually estimating the population mean*, and (4) *procedures for estimating the population percentage*. In addition to these topics, a brief explanation of *how to determine sample size* is presented at the end of the chapter.

## ESTIMATOR, ESTIMATE, ESTIMATION, ET CETERA

Although no formal definitions of the words "estimate" and "estimation" were presented prior to this chapter, you probably had an intuitive feeling for their meanings. This section presents formal definitions of words associated with estimation. Since you may wonder why this is necessary when you already have an intuitive feeling for the meanings, let us just say that not everyone's intuition interprets words and concepts in the same way. Although your intuition is undoubtedly correct, the formal definitions are for the benefit of others' intuitions which may have gone astray in interpreting the words "estimate" and "estimation."

Suppose the Adam and Eve Apple Orchards want to estimate the average dollar sales per day, and a sample of days has produced a sample mean of $300. In this case the *statistic* ($\bar{x}$) may be used to estimate the population mean ($\mu$). The sample *value* of $300 is an *estimate* of the population value. *Any statistic used to estimate a parameter is an estimator*. Thus, the sample mean is an estimator of the population mean, and the sample percentage is an estimator of the population percentage. But remember that any *specific value* of a statistic is an *estimate*.

There are several reasons for selecting a particular statistic to be an estimator. A complete discussion of all the reasons is beyond the scope of this book, but we will mention one very important criterion for the selection of a statistic as an estimator. The sample mean is selected as an estimator of $\mu$ and the sample percentage is an estimator of $\pi$ because these statistics are *unbiased*. *An unbiased estimator is one for which the mean of its sampling distribution is equal to the population parameter to be estimated*. From what has been discussed in Chapter 6 about the properties of certain sampling distributions, we know that the possible values of the statistics $\bar{x}$ and $p$ *tend toward* the values of the population parameters $\mu$ and $\pi$, respectively. This tendency of an estimator is, of course, highly desirable.

*The entire process of using an estimator to produce an estimate of the parameter is known as estimation.*

## Point Estimation

*A single number used to estimate a population parameter is known as a point estimate, and the process of estimating with a single number is known as point estimation.* For example, the sample mean of $300 in our previous example is a point esti-

---

[1] *Fasti*, chap. V, line 191. (There is nothing like a little Ovid to mollify the poets who are required to read this text.)

mate because the value is only one point along a scale of possible values. But how likely is a single estimate to be correct?

**Interval Estimation**

Rather than using a single number as an estimate, the parameter is often estimated to be within a *range* of values. Since it is highly unlikely that any particular sample mean will be exactly equal to the population mean, allowances must be made for sampling error. *A range of values used to estimate a population parameter is known as an interval estimate, and the process of estimating with a range of values is known as interval estimation.*

With two methods of estimation, which one should be used? To answer this, let's look at the precision of a point estimate. The mean of a sample taken from a population may assume different values with different samples. It is unlikely, however, that the sample mean will equal $\mu$. Furthermore, we have no way of knowing from just the point estimate what the likely magnitude of difference will be. By looking at just a single number, we have no way of assessing the probable difference between the statistic and the parameter. *A point estimate is not only highly subject to error in estimation; it also does not allow evaluation of the precision of the estimate.*

The precision of an estimate is determined by the degree of sampling error. And although a point estimate will probably be incorrect as a result of sampling error, this does not prevent us from placing *considerable confidence* in the estimate that the parameter will be within a given range of values. For example, we may say that the daily average sales for the orchards is somewhere between \$285 and \$315 instead of simply saying that the true mean is approximately \$300. Thus, the interval estimates the parameter to be within a range of values; and since it makes allowances for sampling error, *the precision of the estimate can be objectively assessed.* Of course, an interval estimate can be wrong, like any other estimate, but in contrast to the point estimate the *probability of error* for the interval *can be objectively determined.*

Do not get the impression that a point estimate is of little value in estimation. As you will see in the following sections of this chapter, the interval estimate is actually based on the point estimate. *In essence, the point estimate is adjusted for sampling error to produce an interval estimate.* The remainder of this chapter discusses interval estimations of $\mu$ and $\pi$, and the accuracy of these estimations can be determined with some degree of confidence.

**Self-testing Review 7-1**

1   $\bar{x}$ is an ___*estimator*___ of $\mu$.

2   A sample of 36 items has produced a sample percentage of 82 percent. Which is the estimator and which is the estimate?

3   What is the difference between a point estimate and an interval estimate?

4   What is the disadvantage of a point estimate?

5   "Since sampling error is considered in an interval estimate, the parameter value will always fall within the range." Comment on this statement.

6   Why is an unbiased estimator desirable?

**7** "Since a point estimate is rarely used as an estimate, it is rarely calculated." Comment on this statement.

## INTERVAL ESTIMATION OF THE POPULATION MEAN: SOME BASIC CONCEPTS

In practice, only one sample of a population is taken, the sample mean is calculated, and an estimate of the population mean is made. In order to estimate the population mean, we must make some kind of assumption about the relationship between the sample mean and the population mean.

### The Sampling Distribution—Again

A quick review of the concepts of the sampling distribution of means will illustrate the theoretical basis for the interval estimation of $\mu$. Suppose we have a sample size which is sufficiently large so that the sampling distribution is approximately normal. Figure 7-1 shows that 95.4 percent of the possible outcomes of $\bar{x}$ are within $2\sigma_{\bar{x}}$ to each side of the mean of the sampling distribution. This means that if a mad statistician took 1,000 samples of the same size from a population, approximately 954 of the sample means would fall within two standard errors to both sides of the population mean.[2]

---

[2] If you are uncertain about the information contained in the last few sentences and in Fig. 7-1, it is suggested that you review Chapter 6.

**FIGURE 7-1**

**An educational schematic of the sampling distribution of the means when the sample size is large**

**There is a 95.4% chance that a $\bar{x}$ will have a value between $\mu \pm 2\sigma_{\bar{x}}$.**

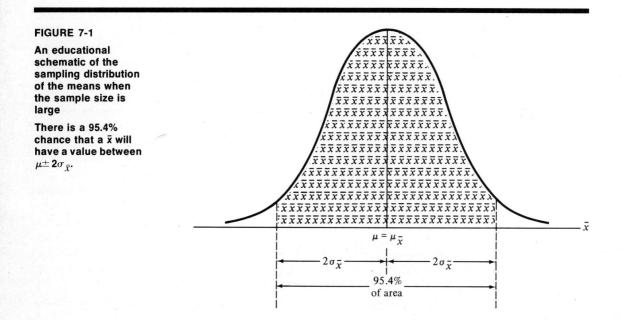

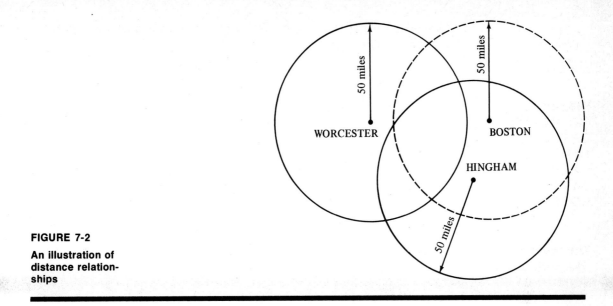

**FIGURE 7-2**

**An illustration of distance relationships**

## Interval Width Considerations

The following statement is logically true (and if it does not seem to be, please read it again carefully): If 95.4 percent of the possible values of $\bar{x}$ fall within $2\sigma_{\bar{x}}$ of the population mean as shown in Fig. 7-1, then obviously $\mu$ will not be farther than $2\sigma_{\bar{x}}$ from 95.4 percent of the possible values of $\bar{x}$.

Now let's illustrate in nonstatistical terms the logic of the preceding statement. Let's assume we have 1,000 towns located *various distances* from the city of Boston, and it happens that 95.4 percent of these towns are within a 50-mile radius of Boston. If 954 towns are within a 50-mile radius of Boston, then logically Boston must fall within a 50-mile radius of each of these 954 towns. If Hingham is within 50 miles of Boston, then Boston will certainly be no farther than 50 miles from Hingham (see Fig. 7-2). If we randomly select a large number of towns from the 1,000, Boston will be within a radius of 50 miles of 95.4 percent of all the towns selected. If all this appears simple and trite, then we have accomplished something.

In returning to the statistical world, substitute the population mean for Boston, let the possible sample means be the towns, and use $2\sigma_{\bar{x}}$ in place of the 50-mile radius. To repeat, then, if 95.4 percent of the sample means are within $2\sigma_{\bar{x}}$ of $\mu$, then certainly $\mu$ must be within $2\sigma_{\bar{x}}$ of 95.4 percent of the sample means. Thus, *if we use the method of $\bar{x} \pm 2\sigma_{\bar{x}}$ to estimate the population mean, and if we construct a large number of intervals, 95.4 percent of the interval estimates will include $\mu$.*

Now let's assume that we have 1,000 possible samples and thus 1,000 sample means, 3 of which are shown in Fig. 7-3. The population mean will be located within 95.4 percent of the 1,000 possible intervals that could be constructed using $\bar{x} \pm 2\sigma_{\bar{x}}$. Any specific single interval *may or may not contain* $\mu$ (note that in Fig. 7-3 the intervals produced using $\bar{x}_1$ and $\bar{x}_2$ *do include* $\mu$, but the interval constructed using $\bar{x}_3$ *fails to reach* $\mu$), but the method employed assures that *if a large number of intervals are constructed, $\mu$ will be included in 95.4 percent of them.*

Unless you plan to use a 95.4 percent probability of estimating the population

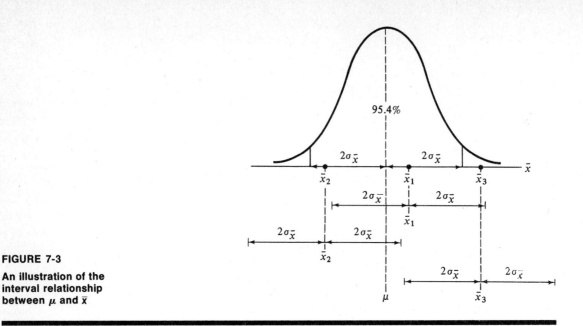

**FIGURE 7-3**

**An illustration of the interval relationship between $\mu$ and $\bar{x}$**

mean for the rest of your life, we need to generalize what has been discussed so far so that we may apply interval estimation to a variety of situations. If the sampling distribution is normal, an interval estimate of $\mu$ may be constructed in the following manner:

$$\bar{x} - Z\sigma_{\bar{x}} \; < \mu < \; \bar{x} + Z\sigma_{\bar{x}}$$

Lower limit      Upper limit

of estimate      of estimate

where $\bar{x}$ = sample mean (and point estimator of $\mu$)

$\sigma_{\bar{x}}$ = standard error of the mean

$Z$ = value determined by the probability associated with the interval estimate —i.e., the value associated with a certain likelihood that $\mu$ will be included in a large number of interval estimates

**The Level of Confidence**

*The level of probability associated with an interval estimate is known as the confidence level, degree of confidence, or confidence coefficient.* The word "confidence" is used because the probability is an indicator of *the degree of certainty that the particular method of interval estimation will produce an estimate which includes $\mu$.* The higher the level of probability associated with an interval estimate, the more certainty there is that the method of estimation will produce an estimate containing the population mean.

In practice, the confidence level is generally specified *prior* to estimation, and the appropriate $Z$ value is then used to construct the interval estimate. For example, 90 percent confidence would mean a *45 percent area on each side of the normal distribution.* In checking Appendix 2 at the back of the book, we see that the $Z$ value corresponding to an area of .45 (or 45 percent) is approximately 1.64. Therefore, the interval estimate of $\mu$ using a 90 percent confidence level is:

$$\bar{x} - 1.64\sigma_{\bar{x}} < \mu < \bar{x} + 1.64\sigma_{\bar{x}}$$

The most frequent confidence levels employed in interval estimation are 90, 95, and 99 percent. The $Z$ values and the general forms of the interval estimates associated with these confidence levels are shown in Fig. 7-4. In short, *the interval estimates based on specified confidence levels are known as confidence intervals,* and the upper and lower limits of the intervals are known as *confidence limits.*

At this point you may wonder why it is necessary to have various confidence levels when it seems logical that the highest level of confidence is desirable in estimating $\mu$. Your thoughts may be characterized in the following words: "If I am required to provide an accurate estimate of the true mean, why shouldn't I always go with a 99 percent confidence level? It makes intuitive sense to have as much confidence as possible in my estimate!"

Unfortunately, intuitive sense is only partially correct in this matter. There is no question that it is highly desirable to have as much confidence as possible in the estimate. However, if more confidence is desired in an estimate, the allowance for sampling error must be increased. In a nutshell, *a higher confidence level produces a wider interval estimate, and thus the precision of the estimate decreases.* Consider the following exchange among three dormitory students anxiously waiting for the mail.

First student: "I have a *feeling* they'll pass out the mail around 2:30 like they usually do."

Second student: "I'm *almost sure* we'll get the mail some time between 2:15 and 2:45."

Third student: "I'm *absolutely sure* we'll get the mail between now and never."

If you want more confidence in your estimate, you must allow for more sampling error. The widths of the intervals will increase, and the estimate will lose some precision. Figure 7-4 shows this relationship between the confidence coefficient and the interval width. *If an interval estimate is too wide, the estimate will have no utility.*

**FIGURE 7-4**

Commonly used confidence coefficients and confidence intervals for a large sample

| Confidence coefficient | $Z$ value | General form of the interval estimate |
|---|---|---|
| 90 | 1.64 | $\bar{x} - 1.64\sigma_{\bar{x}} < \mu < \bar{x} + 1.64\sigma_{\bar{x}}$ |
| 95 | 1.96 | $\bar{x} - 1.96\sigma_{\bar{x}} < \mu < \bar{x} + 1.96\sigma_{\bar{x}}$ |
| 99 | 2.58 | $\bar{x} - 2.58\sigma_{\bar{x}} < \mu < \bar{x} + 2.58\sigma_{\bar{x}}$ |

Assume you must submit a budget request to the finance department for advertising, and the advertising expenditures must be 10 percent of sales for the next period. A sample has produced a mean sales figure of $\bar{x} = \$250,000$ with a $\sigma_{\bar{x}}$ of $\$2,000$. Using the general forms of the interval estimates shown in Fig. 7-4, we can construct confidence intervals for increasing levels of 90 and 99 percent. Basing your advertising budget on the estimate with a coefficient of 90 percent, you would tell the finance department to provide between $21,720 and $28,280; basing your advertising budget on the estimate with a 99 percent confidence level, you would allow more room for error and tell the finance department to provide between $19,840 and $30,160. As the range in your budget request increases, you create more uncertainty in the planning of the finance section. They are then forced to manipulate and tie up more money than necessary. In short, an increase in the confidence level might produce an estimate which cannot be useful.

One very important caution should be mentioned before we proceed to the next section. The confidence coefficient should be stated *prior* to the interval estimation. Many times a novice researcher will calculate a number of interval estimates based on a single sample while varying the confidence level. After having obtained these estimates, he or she then selects the estimate which looks most accurate. Such an approach is really manipulating data so that the results of a sample are the way a researcher would like to see them. Such an approach introduces the bias of the researcher into the study and should be avoided.

**Self-testing Review 7-2**

1 What is the theoretical basis for the interval estimation of $\mu$ with $\bar{x} \pm Z\sigma_{\bar{x}}$?

2 What is the difference between a confidence coefficient and a confidence level?

3 What does a 95 percent confidence level mean?

4 What relationship exists between the confidence level and the interval width?

**ESTIMATING THE POPULATION MEAN: $\sigma$ KNOWN**

Moving on now from the general form of (and the theoretical basis for) the interval estimate, the remainder of this chapter will discuss interval estimation of parameters under specific conditions.

When the *population standard deviation ($\sigma$) is known*, we may directly compute the standard error of the mean. Thus, *the interval estimate may be constructed in the following manner:*

$$\underset{\substack{\text{Lower confi-}\\\text{dence limit}}}{\bar{x} - Z\sigma_{\bar{x}}} < \mu < \underset{\substack{\text{Upper confi-}\\\text{dence limit}}}{\bar{x} + Z\sigma_{\bar{x}}} \qquad (7\text{-}1)$$

And you will recall from Chapter 6 that $\sigma_{\bar{x}}$ may be found by:

$$\sigma_{\bar{x}} = \frac{\sigma}{\sqrt{n}} \qquad \text{for an infinite population}$$

$$\text{or } \sigma_{\bar{x}} = \frac{\sigma}{\sqrt{n}} \sqrt{\frac{N-n}{N-1}} \qquad \text{for a finite population}$$

So let's now use this estimation procedure to consider some example problems.

**Example 7-1** The Papyrus Paper Company wants to estimate the average time required for a new machine to produce a ream of paper. A sample of 36 reams required an average machine time of 1.5 minutes for each ream. Assuming $\sigma = 0.30$ minute, construct an interval estimate with a confidence level of 95 percent.

We have the following data from the problem: $\bar{x} = 1.5$, $\sigma = .30$, $n = 36$, and confidence level = 95 percent. The standard deviation of the sampling distribution $(\sigma_{\bar{x}})$ is computed as follows:

$$\sigma_{\bar{x}} = \frac{\sigma}{\sqrt{n}}$$

$$= .30/\sqrt{36}$$

$$= .05$$

With a 95 percent confidence coefficient, the $Z$ value equals 1.96. Thus, the interval estimate of the true average time $(\mu)$ is constructed as follows:

$$\bar{x} - Z\sigma_{\bar{x}} < \mu < \bar{x} + Z\sigma_{\bar{x}}$$
$$1.5 - 1.96(.05) < \mu < 1.5 + 1.96(.05)$$
$$1.5 - .098 < \mu < 1.5 + .098$$
$$1.402 < \mu < 1.598$$

**Example 7-2** The Ledd Pipe Company has received a shipment of 100 lengths of pipe, and it wants to estimate the average width of the pipes to see if they meet minimum standards. A sample of 50 pipes produced an average diameter of 2.55 inches. In the past, the population standard deviation of the diameter has been 0.07 inches. Construct an interval estimate with a 99 percent degree of confidence.

We have the following data from the problem situation: $\bar{x} = 2.55$, $\sigma = .07$, $n = 50$, $N = 100$, and confidence level = 99 percent. The standard error of the mean is computed as follows:

$$\sigma_{\bar{x}} = \frac{\sigma}{\sqrt{n}} \sqrt{\frac{N - n}{N - 1}}$$

$$= \frac{.07}{\sqrt{50}} \sqrt{\frac{100 - 50}{100 - 1}}$$

$$= .007$$

With a 99 percent confidence coefficient, the $Z$ value is 2.58. Therefore, the interval estimate of $\mu$, the true average diameter of the shipment of pipes, is constructed as follows:

$$\bar{x} - Z\sigma_{\bar{x}} < \mu < \bar{x} + Z\sigma_{\bar{x}}$$
$$2.55 - 2.58(.007) < \mu < 2.55 + 2.58(.007)$$
$$2.55 - .018 < \mu < 2.55 + .018$$
$$2.532 < \mu < 2.568$$

Note that the preceding examples were based on situations in which $\sigma$ *was known (or could be identified), and the sampling distribution was normally distributed.* The general procedure for interval estimation under such conditions is illustrated in Fig. 7-5.

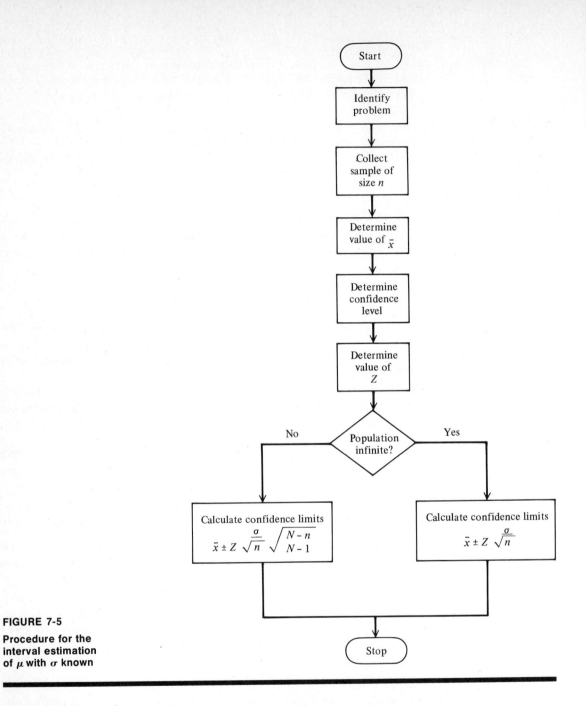

**FIGURE 7-5**

**Procedure for the interval estimation of $\mu$ with $\sigma$ known**

---

1 When does the sampling distribution approximate the normal distribution?

2 Determine the $Z$ value for the following confidence levels:
  a 91 percent
  b 73 percent
  c 86 percent

**3** Construct a confidence interval with the following data: $\bar{x} = 48$, $\sigma = 9$, $n = 36$, and confidence level = 90 percent.

**4** Construct a confidence interval with the following data: $\bar{x} = 104$, $\sigma_{\bar{x}} = 13$, and confidence level = 80 percent.

## ESTIMATING THE POPULATION MEAN: $\sigma$ UNKNOWN

In many situations, not only is the population mean unknown but the population standard deviation is also unknown. In fact, it is only in isolated cases that $\sigma$ *is* known,[3] so it usually must be estimated along with the population mean.

### The Estimator of $\sigma$

Intuitively, it appears that the sample standard deviation is an estimator of the population standard deviation because of their similarity of computation and because the sample mean is an estimator of the population mean. The computations of $s$ and $\sigma$ are as follows:

$$s = \sqrt{\frac{\Sigma (x - \bar{x})^2}{n}} \qquad \sigma = \sqrt{\frac{\Sigma (X - \mu)^2}{N}}$$

But although there is a similarity between $s$ and $\sigma$, we must remember that one of the important criteria for a statistic to qualify as an estimator is the criterion of unbiasedness. *The sample standard deviation is not an unbiased estimator of the population standard deviation.*[4]

The unbiased estimator of the population standard deviation is denoted by $\hat{\sigma}$,[5] and the wonders of mathematical manipulation have shown this unbiased estimator to be computed as follows:

$$\hat{\sigma} = s\sqrt{\frac{n}{n - 1}} \tag{7-2}$$

Of course, $\hat{\sigma}$ is also equal to:

$$\sqrt{\frac{\Sigma (x - \bar{x})^2}{n}} \sqrt{\frac{n}{n - 1}} \quad \text{or} \quad \sqrt{\frac{\Sigma (x - \bar{x})^2}{n} \frac{n}{n - 1}} \quad \text{or} \quad \sqrt{\frac{\Sigma (x - \bar{x})^2}{n - 1}}$$

Obviously, if $\sigma$ is unknown, the standard error of the mean must also be estimated. The estimator of the standard error is denoted by the symbol $\hat{\sigma}_{\bar{x}}$ and may be computed in the following way:

$$\hat{\sigma}_{\bar{x}} = \frac{\hat{\sigma}}{\sqrt{n}} \qquad \text{for an infinite population}$$

$$\hat{\sigma}_{\bar{x}} = \frac{\hat{\sigma}}{\sqrt{n}} \sqrt{\frac{N - n}{N - 1}} \qquad \text{for a finite population}$$

---

[3] The prospect of knowing the population standard deviation has been summed up by a disillusioned statistician with the acronym TANSTAFL (there ain't no such thing as a free lunch).

[4] Would you believe that the sample standard deviation is biased toward *understating* the population standard deviation? It's true. Did you care to know that? Don't tell me your answer; it might depress me.

[5] The little mark (ˆ) over a symbol indicates that the value represented is an *estimated* value.

However, since

$$\hat{\sigma}_{\bar{x}} = \frac{\hat{\sigma}}{\sqrt{n}}$$

$$= \frac{s \sqrt{n/(n-1)}}{\sqrt{n}}$$

$$= \frac{s}{\sqrt{n-1}}$$

it is usually easier to compute the estimator of the standard error in the following way:

$$\hat{\sigma}_{\bar{x}} = \frac{s}{\sqrt{n-1}} \qquad \text{for an infinite population} \qquad (7\text{-}3)$$

or

$$\hat{\sigma}_{\bar{x}} = \frac{s}{\sqrt{n-1}} \sqrt{\frac{N-n}{N-1}} \qquad \text{for a finite population} \qquad (7\text{-}4)$$

Notice that the calculation of the *estimate* of the standard error is the same as the computation for $\sigma_{\bar{x}}$, except that the population standard deviation is replaced by the sample standard deviation $(s)$, and the denominator is $\sqrt{n-1}$ rather than $\sqrt{n}$.

If the population standard deviation is unknown, the sampling distribution of means can be assumed to be approximately normal *only when the sample size is relatively large (over 30)*. With estimated values of $\sigma$ and $\sigma_{\bar{x}}$, the general form of the interval estimate *for large samples* is altered slightly so that it appears as follows:

$$\bar{x} - Z\hat{\sigma}_{\bar{x}} < \mu < \bar{x} + Z\hat{\sigma}_{\bar{x}} \qquad (7\text{-}5)$$

As you can see, we merely substituted the estimated value for the true value of the standard error of the mean. Again, let's look at some examples to illustrate the above points.

**Example 7-3** The Stagger Inn Tavern wants to estimate the average dollar purchase per customer. A sample of 100 customers spent an average of $3.50 each with a *sample* standard deviation of $0.75. Estimate the true average expenditure with a 90 percent confidence level.

We have the following data from the problem situation: $\bar{x} = \$3.50$, $s = .75$, $n = 100$, confidence level $= 90$ percent. Therefore the estimate of $\hat{\sigma}_{\bar{x}}$ is computed as follows:

$$\hat{\sigma}_{\bar{x}} = \frac{s}{\sqrt{n-1}}$$

$$= .75/\sqrt{100-1}$$

$$= .0754 \text{ or } .08$$

With a 90 percent confidence level the $Z$ value is 1.64. The interval estimate is thus

$$\bar{x} - Z\hat{\sigma}_{\bar{x}} < \mu < \bar{x} + Z\hat{\sigma}_{\bar{x}}$$
$$\$3.50 - 1.64(.08) < \mu < \$3.50 + 1.64(.08)$$
$$\$3.50 - .13 < \mu < \$3.50 + .13$$
$$\$3.37 < \mu < \$3.63$$

**Example 7-4** The Pluck M. Cleen Poultry Company has received a shipment of 100 hens, and the manager wants to estimate the true average weight of a hen in order to determine if the hens meet Pluck M. Cleen's standards. A sample of 36 hens has shown an average weight of 3.6 pounds with a *sample* standard deviation of .6. Construct an interval estimate of the true average weight per chicken with a 99 percent confidence level.

We have the following data: $\bar{x} = 3.6$, $s = .6$, $n = 36$, confidence level $= 99$ percent, $N = 100$. The estimate of the standard error is computed as follows:

$$\hat{\sigma}_{\bar{x}} = \frac{s}{\sqrt{n-1}} \sqrt{\frac{N-n}{N-1}}$$

$$= \frac{.6}{\sqrt{36-1}} \sqrt{\frac{100-36}{100-1}}$$

$$= .081$$

With a 99 percent confidence level, the $Z$ value is 2.58. Therefore, the interval estimate is:

$$\bar{x} - Z\hat{\sigma}_{\bar{x}} < \mu < \bar{x} + Z\hat{\sigma}_{\bar{x}}$$
$$3.6 - 2.58(.081) < \mu < 3.6 + 2.58(.081)$$
$$3.6 - .21 < \mu < 3.6 + .21$$
$$3.39 \text{ pounds} < \mu < 2.81 \text{ pounds}$$

When $\sigma$ is unknown, and when the *sample size is large*, the sampling distribution is approximately normally distributed. *However, if $\sigma_{\bar{x}}$ must be estimated, and if the sample size is 30 or less, the sampling distribution will not be normally distributed, and therefore the interval estimate cannot be calculated with the use of the Z distribution.* What distribution should then be used? The following section provides us with the answer.

## Estimation Using the *t* Distribution

Instead of following a normal distribution curve, the sampling distribution of means with a small sample size follows a *t distribution*. The appropriate use of a *t* distribution in estimation computations is summarized in Fig. 7-6. A *t* distribution is similar to a $Z$ distribution with a zero mean and a symmetrical shape. However, unlike the shape of the $Z$ distribution, the *t* distribution's shape depends on the *sample size.* (There is a different distribution for each sample size.) In general, the shape of a *t* distribution is *flatter* than that of a $Z$ distribution. As the sample size increases and approaches 30, however, the shapes of the *t* distributions lose their flatness and approximate the shape of the $Z$ distribution (see Fig. 7-7).

*If $\sigma$ is unknown, and if the sample size is small, the interval estimate of the population mean has the following form:*

$$\bar{x} - t_{\alpha/2}\hat{\sigma}_{\bar{x}} < \mu < \bar{x} + t_{\alpha/2}\hat{\sigma}_{\bar{x}}$$

Lower confi-  Upper confi-  (7-6)
dence limit   dence limit

Like the $Z$ value, the value of $t$ depends on the confidence level.

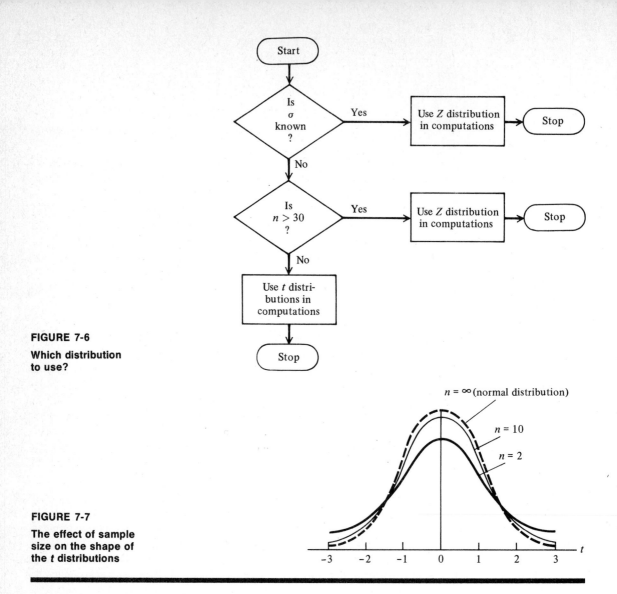

**FIGURE 7-6**

**Which distribution to use?**

**FIGURE 7-7**

**The effect of sample size on the shape of the *t* distributions**

Appendix 4 at the back of the book is a table of *t* distribution values. Unfortunately, the *t* table is constructed in a different manner of presentation than the *Z* table. If, for example, we are interested in making an estimate at the 95 percent confidence level, the *t*-table format is not designed to emphasize the 95 percent chance of including $\mu$ in the estimate; rather, the presentation focuses attention on the 5 percent chance of *not including $\mu$. This chance of error is labeled $\alpha$ (alpha) and in decimal form equals 1.00 minus the confidence coefficient* — e.g., if the confidence coefficient is .95 (or 95 percent), then $\alpha$ will be $1.00 - .95$ or .05. Since $\alpha$ represents the *total* chance of error — i.e., the chance of not including $\mu$ — and since the particular $t$ distribution being used is symmetrical, the total error is divided evenly between the chance of overestimation and the chance of underestimation. As indicated by the shaded portion in the figure at the top of Appendix 4, however, the $t$ table *only deals with areas to one side of the distribution*. Consequently, the subscript $\alpha/2$ follows $t$ in the interval formula (7-6) shown above. With a 95 percent confidence level, we want only a 5 percent chance of error, and therefore we look under the *column* designated by $t_{.025}$ in Appendix 4.[6]

Another factor that must be known before an appropriate $t$ value can be determined is the degrees of freedom (*df*). For our purposes here, the *degrees of freedom* identify the *appropriate row* in the table, and they *are simply equal to* $n - 1$. (Each *df* row refers to a *different* $t$ distribution in the family of curves.) Let us assume that we have a sample size of 17 and we want a 95 percent confidence level for an interval estimate. The $\alpha$ value would be .05, and thus we would refer to *column* $t_{.025}$ in Appendix 4. And since $df = 16$ $(17 - 1)$, the necessary $t$ value of 2.120 is found at the intersection of the 16 *df* row and the .025 column.

Let us now look at some examples of the use of the $t$ distribution in estimating the population mean.

**Example 7-5** The Dew Drop Inn wants to estimate the average number of gallons of a product sold per day. Twenty business days were monitored, and an average of 32 gallons was sold daily. The sample standard deviation was 12 gallons. Calculate the confidence limits at the 95 percent confidence level.

We have the following information: $\bar{x} = 32$, $s = 12$, $n = 20$, and confidence level = 95 percent. With a confidence level of 95 percent, and with a sample size of 20, $\alpha$ is 5 percent and the degrees of freedom are 19. Thus, from Appendix 4 under the $t_{.025}$ column, we see that the $t$ value is 2.093. The estimate of $\sigma_{\bar{x}}$ is computed as follows:

$$\hat{\sigma}_{\bar{x}} = \frac{s}{\sqrt{n - 1}}$$

$$= 12.0/\sqrt{20 - 1}$$

$$= 2.75$$

The interval estimate is:

$$\bar{x} - t_{\alpha/2}\hat{\sigma}_{\bar{x}} < \mu < \bar{x} + t_{\alpha/2}\hat{\sigma}_{\bar{x}}$$
$$32 - t_{.025}(2.75) < \mu < 32 + t_{.025}(2.75)$$
$$32 - 2.093(2.75) < \mu < 32 + 2.093(2.75)$$
$$26.24 < \mu < 37.76$$

Finally, the confidence limits are 26.24 and 37.76 gallons.

---

[6] The use of a $t$ table is admittedly rather confusing to beginning students. Some find it convenient to locate the correct column by looking at the last row in the table (where *df* is $\infty$). This row presents $Z$ values, since the last distribution in the family of $t$ distributions is the normal distribution. Thus, when you know that a $Z$ value of, say, 1.96 would have been used if the sample size had been large, *you can locate the proper column* by finding 1.96 in the last row.

**Example 7-6** The Moe Doe Bread Company wants to estimate its average daily usage of flour. With a sample of 14 days, the point estimate of $\mu$ is 173 pounds, with $s = 45$ pounds. Construct a confidence interval with a 99 percent confidence coefficient.

The data from the problem are as follows: $\bar{x} = 173$, $s = 45$, $n = 14$, and confidence coefficient $= 99$ percent. The estimate of $\sigma_{\bar{x}}$ is computed as follows:

$$\hat{\sigma}_{\bar{x}} = \frac{s}{\sqrt{n-1}}$$

$$= 45/\sqrt{14-1}$$

$$= 12.48$$

**FIGURE 7-8**

**Procedure for the interval estimation of $\mu$ with $\sigma$ unknown**

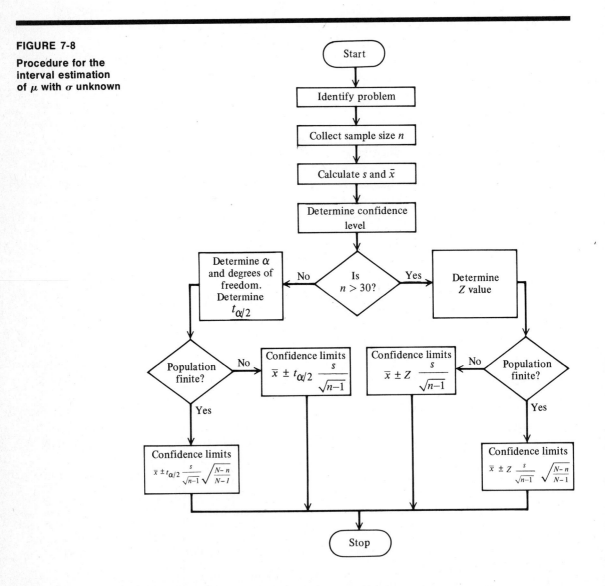

With a 99 percent confidence coefficient, $\alpha$ is 1 percent. With a sample size of 14, there are 13 degrees of freedom. Therefore, consulting Appendix 4 under $t_{.005}$ we see that the $t$ value is 3.012. The interval estimate is:

$$\bar{x} - t_{.005}\hat{\sigma}_{\bar{x}} < \mu < \bar{x} + t_{.005}\hat{\sigma}_{\bar{x}}$$
$$173 - 3.012(12.48) < \mu < 173 + 3.012(12.48)$$
$$173 - 37.59 < \mu < 173 + 37.59$$
$$135.41 < \mu < 210.59$$

The general approach to constructing an interval estimate with an unknown $\sigma$ is summarized in Fig. 7-8. Note that the procedures for large and small sample sizes are basically the same. The only difference is the use of a $t$ or a $Z$ distribution.

**Self-Testing Review 7-4**

1 Suppose we have a sample of 24 items and $\sigma$ is unknown. Would you use a $t$ or a $Z$ distribution in your interval estimation?

2 What is the $t$ value in interval estimation if the sample size is 27 and the desired confidence level is 95 percent?

3 Assume that a sample of 100 customer charge accounts showed an average balance of $42 with a sample standard deviation of $16. Construct a confidence interval with a coefficient of 95 percent.

4 If you have been given a sample of 20 candles from a shipment of 1,000 candles and are asked to provide an interval estimate of their average burning life, how would you proceed and what information would be needed?

5 How does the shape of a $t$ distribution differ from the shape of a $Z$ distribution?

6 Identify the factor that affects the shape of a $t$ distribution.

7 The $Z$ table is constructed so that we are concerned with the probability of including the population mean in the interval estimate. How could the format of the $t$ table in Appendix 4 be characterized?

8 Why isn't the sample standard deviation used as an estimator of the population standard deviation?

**INTERVAL ESTIMATION OF THE POPULATION PERCENTAGE**

Since the mean of the sampling distribution of percentages is equal to the population percentage, the *sample* percentage ($p$) is an unbiased estimator of the *population* percentage ($\pi$). If the sample size is sufficiently large, the sampling distribution approximates the normal distribution, and thus we are able to make probability statements about the interval estimates of $\pi$ that are based on sample percentages. In this section we will discuss only the large-sample case in the interval estimation of $\pi$, since the small-sample approach is beyond the scope of this book.

As indicated below, the *sample percentage* serves as the basis for constructing the interval estimate of the population percentage:

$$\underset{\substack{\text{Lower confi-}\\\text{dence limit}}}{p - Z\hat{\sigma}_p} < \pi < \underset{\substack{\text{Upper confi-}\\\text{dence limit}}}{p + Z\hat{\sigma}_p} \qquad (7\text{-}7)$$

A $Z$ value is used here in exactly the same way it was used in estimating a population mean. The symbol $\hat{\sigma}_p$ is the *estimate* of the standard deviation of the sampling distribution of percentages—i.e., an estimate of the standard error of percentage. An unbiased estimate of the standard error of percentage may be computed in the following manner:

$$\hat{\sigma}_p = \sqrt{\frac{p(100-p)}{n-1}} \; \sqrt{\frac{N-n}{N-1}} \qquad \text{for a finite population} \qquad (7\text{-}8)$$

or $\qquad \hat{\sigma}_p = \sqrt{\frac{p(100-p)}{n-1}} \qquad$ for an infinite population $\qquad (7\text{-}9)$

The *estimate* of the standard error is *always used* in the construction of an interval estimate. Why? Because the true standard error cannot be computed for an interval estimate of $\pi$. This fact is obvious from the following formula:

$$\sigma_p = \sqrt{\frac{\pi(100-\pi)}{n}}$$

where, as you can see, the calculation of $\sigma_p$ requires knowledge of $\pi$. Yet $\pi$ is what we are trying to estimate! In order to resolve this dilemma, we must therefore use formulas (7-8) and (7-9).

We are now in a position to illustrate the similarity of the procedures used to estimate population means and percentages by considering the following example problems.

**Example 7-7** The Highland Fling Scottish Boomerang Company wants to estimate the percentage of credit customers who have submitted checks in payment for boomerangs and whose checks have bounced. A sample of 150 accounts showed that 15 customers have passed bad checks. Estimate at the 95 percent confidence level the true percentage of credit customers who have passed bad checks.

We have the following data: $p = 15/150 = 10$ percent, $n = 150$, and confidence coefficient = 95 percent. The estimate of $\sigma_p$ is computed as follows:

$$\hat{\sigma}_p = \sqrt{\frac{p(100-p)}{n-1}}$$

$$= \sqrt{\frac{10(90)}{149}}$$

$$= 2.46 \text{ percent}$$

With a confidence coefficient of 95 percent, the $Z$ value is 1.96. Therefore, the interval estimate of the true percentage of credit customers who pass bad checks is:

$p - Z\hat{\sigma}_p < \pi < p + Z\hat{\sigma}_p$
10 percent $- 1.96(2.46$ percent$) < \pi < 10$ percent $+ 1.96(2.46$ percent$)$
10 percent $- 4.82$ percent $< \pi < 10$ percent $+ 4.82$ percent
5.18 percent $< \pi < 14.82$ percent

**Example 7-8** A high school student counselor, Ms. Kerr Reer, was interested in the proportion of male students who would volunteer for military service. Out of 600 male students, she sampled 50 and found that 15 of them expressed a favorable

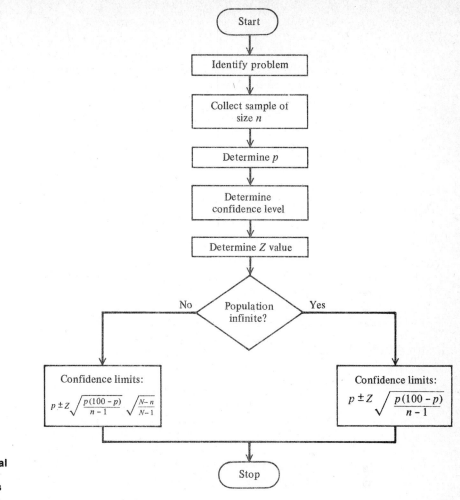

**FIGURE 7-9**

**Procedure for interval estimation of $\pi$ using large samples**

opinion toward enlistment. Use a 99 percent confidence coefficient to estimate the true percentage.

The data are: $p = 15/50 = 30$ percent, $n = 50$, confidence coefficient $= 99$ percent. The estimate of the standard error is computed as follows:

$$\hat{\sigma}_p = \sqrt{\frac{p(100 - p)}{n - 1}} \sqrt{\frac{N - n}{N - 1}}$$

$$= \sqrt{\frac{30(70)}{49}} \sqrt{\frac{600 - 50}{600 - 1}}$$

$$= 6.28 \text{ percent}$$

With a confidence coefficient of 99 percent, the $Z$ value is 2.58. Therefore, the interval estimate is:

$$p - Z\hat{\sigma}_p < \pi < p + Z\hat{\sigma}_p$$

30 percent $-$ 2.58(6.28 percent) $< \pi <$ 30 percent $+$ 2.58(6.28 percent)

30 percent $-$ 16.2 percent $< \pi <$ 30 percent $+$ 16.2 percent

13.8 percent $< \pi <$ 46.2 percent

(Notice the large width of this interval estimate as a result of the high level of confidence specified and the relatively small sample size.)

The general procedure for constructing an interval estimate of $\pi$ in the large-sample case is summarized in Fig. 7-9.

**Self-testing Review 7-5**

1 What assumption is made about the sample size when the $Z$ distribution is used in calculating an interval estimate of $\pi$?

2 Compare the formulas for $\hat{\sigma}_p$ and $\sigma_p$. Why is there a difference in the denominators?

3 Why must we always use $\hat{\sigma}_p$ rather than $\sigma_p$ in the interval estimation of $\pi$?

**DETERMINATION OF SAMPLE SIZE**

In this chapter, the basic problem situation in the estimation of population means (or percentages) can be summarized as follows: "A sample of size $n$ has been collected. We have the calculated values for the sample mean and standard deviation (or for the sample percentage). Compute an interval estimate with a confidence level of _____ percent."

The sample was assumed to have been collected, and our task was to calculate an estimate based on the sample data given. We had to live with whatever confidence interval resulted from a specified confidence level, no matter how wide it might have been. Of course, one method of controlling the interval width is to change the confidence level, but it is improper to manipulate the confidence level so that the interval range comes out the way we want it to appear.

**General Considerations about Sample Size**

It is very often the case that the precision of the estimate must be specified before a sample is ever taken. For example, you may be checking the average diameter of machined parts which should not have too much error or else they cannot be used in finely machined equipment. In such a case, you would take a sample of parts, but you would want an estimate with as little sampling error as possible in the interval estimate. You would want a precise estimate. Too much sampling error produces an interval width too large to be of any use.

*We can control sampling error by selecting a sample of adequate size.* Remember that sampling error arises because the entire population is not studied; someone or something is always left out of the investigation. Whenever sampling is performed, we always miss some bit of information about the population which would be helpful in our estimation. If we want a high level of precision, we must sample enough of the population to provide the necessary and sufficient information. The following sections will discuss the methods of determining the sample size necessary to achieve a specified level of precision.

**Determining the Sample Size for the Estimation of $\mu$**

Consider the following situation: A hardware wholesaler receives a shipment of 10,000 widgets, and before he pays for the shipment he would like to know if the widgets were properly made and if they meet tolerance specifications. He'd like to estimate the average width or diameter of the widgets. He would also like the estimate to be within ±0.01 inches of the true average diameter, and he'd like a 95 percent confidence level associated with his estimate. How does he determine the sample size?

First, look at what is actually desired by the hardware wholesaler. With a sample mean of $\bar{x}$, he wants the interval estimate to have *limits* which are no more than 0.01 inches *above* the point estimate and no more than 0.01 inches *below* the point estimate. He wants this interval estimate to have a 95 percent confidence level of containing the true average. Thus, the desired confidence limits have been specified to be

$$\bar{x} \pm 0.01 \text{ inches}$$

Since the general form of the confidence limits is

$$\bar{x} \pm Z\sigma_{\bar{x}}$$

the hardware wholesaler is saying that he wants $Z\sigma_{\bar{x}}$ to equal .01.

We are now in a position to determine the necessary sample size by solving the equation $Z\sigma_{\bar{x}} = .01$. Since a confidence level of 95 percent is desired in our estimation of $\mu$, the $Z$ value is 1.96. Therefore,

$$Z\sigma_{\bar{x}} = .01$$
$$1.96\sigma_{\bar{x}} = .01$$
$$\sigma_{\bar{x}} = .01/1.96$$
$$\sigma_{\bar{x}} = .005$$

and the standard error should be .005. Assuming the finite correction factor is not applicable, the formula for $\sigma_{\bar{x}}$ is

$$\sigma_{\bar{x}} = \frac{\sigma}{\sqrt{n}}$$

With some mathematical manipulation which we shall omit, the sample size is then

$$n = \frac{\sigma^2}{\sigma_{\bar{x}}^2}$$

We know that $\sigma_{\bar{x}}$ should be .005, but what about the value of $\sigma$? At this point it is necessary that we make an assumption concerning the value of the population standard deviation. *In determining a sample size for an interval estimation, it is always necessary to make an assumption about the value of $\sigma$.*

On the basis of previous shipments, it might be possible to assume that the population standard deviation of the diameter of widgets is 0.05 inches. The necessary sample size for the wholesaler's desired level of precision is then computed as follows:

$$n = \frac{\sigma^2}{\sigma_{\bar{x}}^2}$$

$$= \frac{(.05)^2}{(.005)^2}$$

$$= 100$$

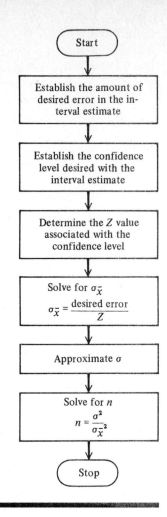

**FIGURE 7-10**

**General procedure for determining the sample size for the estimation of $\mu$**

Figure 7-10 summarizes the general procedure for determining the sample size in the estimation of $\mu$ with a specified amount of precision.

**Determining the Sample Size for the Estimation of $\pi$**

The procedure for determining the sample size for the estimation of $\pi$ is very similar to the procedure for determining the sample size for the estimation of $\mu$. Consider the general form of the interval estimate, which is:

$$p - Z\sigma_p < \pi < p + Z\sigma_p$$

If it is specified that $\pi$ must be estimated within a certain amount of *desired error*, it is essentially required that the confidence limits be:

$$p \pm Z\sigma_p = p \pm \text{desired error}$$

where $Z\sigma_p$ must, of course, equal *desired error*.

Following through with some mathematical manipulation which we will not illustrate, we see that $\sigma_p$ is computed as follows:

$$\sigma_p = \frac{\text{desired error}}{Z}$$

And since the formula for $\sigma_p$ is:

$$\sigma_p = \sqrt{\frac{\pi(100 - \pi)}{n}}$$

then

$$\sigma_p{}^2 = \frac{\pi(100 - \pi)}{n}$$

Therefore, you may now verify for yourself that

$$n = \frac{\pi(100 - \pi)}{\left(\dfrac{\text{desired error}}{Z}\right)^2}$$

where $\sigma_p = $ desired error$/Z$

*It is necessary at this point to approximate the value of $\pi$.* The skeptics among you may wonder how it is possible to approximate $\pi$ when $\pi$ is what we want to estimate. Well, many times you have a rough idea of the true population percentage. For example, you may not know the true percentage of United States citizens who are black, but you know that the percentage is more than 10 percent but less than 20 percent. In many cases an experienced researcher has enough knowledge about the population to approximate the true percentage. Judgment provides an approximation of the parameter, but the sample results provide an estimate of the parameter which can be objectively assessed.

What if we are completely ignorant concerning characteristics of the population and are unable to make any approximation of the parameter? Since one picture is worth a thousand words, let's look at Fig. 7-11 and assume that we have specified a desired amount of error and confidence level and our computations show that $\sigma_p$ should be 2 percent. Figure 7-11 shows the necessary sample size under various assumptions about the population percentage. You can see *the symmetry in results*.

**FIGURE 7-11**

Illustration of the relationship between the assumed value of $\pi$ and the sample size*

| Assumed value of $\pi$ (%) (1) | $\pi(100 - \pi)$ (2) | $n = \dfrac{\pi(100 - \pi)}{\sigma_p{}^2}$ (3) |
|---|---|---|
| 20 | (20)(80) = 1600 | 400 |
| 40 | (40)(60) = 2400 | 600 |
| 50 | (50)(50) = 2500 | 625 |
| 60 | (60)(40) = 2400 | 600 |
| 80 | (80)(20) = 1600 | 400 |

* Given: $\sigma_p{}^2 = (2 \text{ percent})^2 = 4$ percent

The necessary sample size for an assumption of $\pi = 20$ percent is the same as that for an assumption of $\pi = 80$ percent. (A glance at column 2 of Fig. 7-11 will show you why this is the case.)

As you can see in Fig. 7-11, the largest sample size arises when the population percentage is assumed to be 50 percent. *When you have absolutely no idea about the true population percentage, you should assume that $\pi = 50$ percent and obtain the largest sample size possible which also gives you as much information as possible to make an estimate of the population parameter.*

Perhaps a problem example would be helpful. Suppose you wished to estimate the percentage of students at a university who would be willing to donate a pint of blood. Since the Red Cross is planning its schedule for the coming months, it would like you to provide an estimate that would be within ±5 percent of the true percentage. A confidence level of 95 percent is desired. How big should the sample be? Assume you have no idea of the true percentage.

With the data available, you can say the $Z\sigma_p$ should equal 5 percent, since the

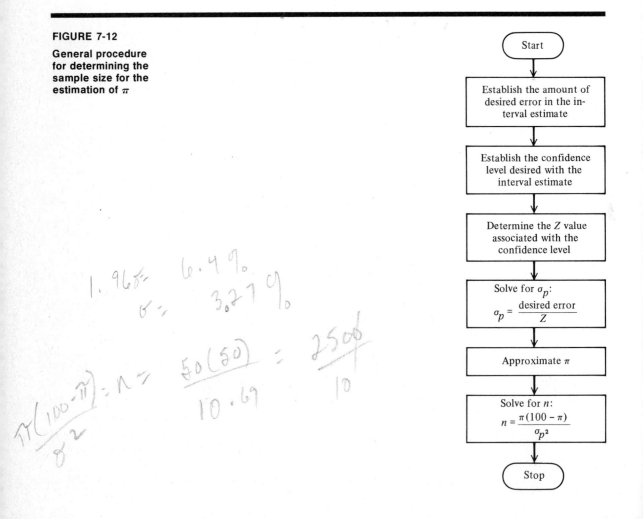

**FIGURE 7-12**

**General procedure for determining the sample size for the estimation of $\pi$**

Start

Establish the amount of desired error in the interval estimate

Establish the confidence level desired with the interval estimate

Determine the $Z$ value associated with the confidence level

Solve for $\sigma_p$:
$$\sigma_p = \frac{\text{desired error}}{Z}$$

Approximate $\pi$

Solve for $n$:
$$n = \frac{\pi(100 - \pi)}{\sigma_p{}^2}$$

Stop

confidence limits are computed as $p \pm Z\sigma_p$. With a confidence level of 95 percent you know that $Z$ is 1.96, and therefore

$$Z\sigma_p = 5 \text{ percent}$$
$$1.96\sigma_p = 5 \text{ percent}$$
$$\sigma_p = 5 \text{ percent}/1.96$$
$$\sigma_p = 2.55 \text{ percent}$$

Since we have no knowledge of the true percentage at all, we must obtain the largest sample size possible by assuming that $\pi = 50$ percent. Therefore the necessary sample size is computed as follows:

$$n = \frac{\pi(100 - \pi)}{\sigma_p{}^2}$$

$$= \frac{50(50)}{2.55^2}$$

$$= \frac{2,500}{6.50}$$

$$= 385$$

Figure 7-12 summarizes the general procedure for determining the sample size for the estimation of $\pi$.

**Self-testing Review 7-6**

1  A Chamber of Commerce wishes to determine the mean price of new single-family residences built in the Chamber's city in the last 12 months. Data are available from local builders, realtors, and the building permit office of the city. A recent survey in a nearby city showed that the $\sigma$ amount for single-family housing in that city was $5,000. The Chamber manager wants to be 90 percent confident that the results of a study will yield an estimate that is within $1,000 of the true mean price. Since you are assigned to conduct the study, what sample size would you use?

2  Executives of the Surface Transit Company are considering a new policy of reducing bus fares for senior citizens (over 65 years of age) during specified periods of the year. Before making a final decision, however, they would like to estimate what percentage of their passengers are senior citizens. The executives want to be 95 percent confident that the estimate obtained is within 3 percent of the true figure. What size sample of passengers should be taken?

**SUMMARY**

An unbiased estimator is desirable for estimation because the possible outcomes of the sample statistic tend toward the value of the population parameter. Although the statistic possesses this tendency, it is very unlikely that the value of a point estimate will be exactly equal to the value of the parameter. Consequently, an interval estimate is desired over a point estimate because allowances are made for sampling error.

On the basis of the properties of the sampling distribution, it is possible to

**FIGURE 7-13**

Summary of interval estimation under various conditions

| Population | Estimating $\mu$ | | | Estimating $\pi$ |
|---|---|---|---|---|
| | $\sigma$ Known | $\sigma$ Unknown | | $n > 30$ |
| | | $n \leq 30$ | $n > 30$ | |
| Finite | $\bar{x} \pm Z \dfrac{\sigma}{\sqrt{n}} \sqrt{\dfrac{N-n}{N-1}}$ | $\bar{x} \pm t_{\alpha/2} \dfrac{s}{\sqrt{n-1}} \sqrt{\dfrac{N-n}{N-1}}$ | $\bar{x} \pm Z \dfrac{s}{\sqrt{n-1}} \sqrt{\dfrac{N-n}{N-1}}$ | $p \pm Z \sqrt{\dfrac{p(100-p)}{n-1}} \sqrt{\dfrac{N-n}{N-1}}$ |
| Infinite | $\bar{x} \pm Z \dfrac{\sigma}{\sqrt{n}}$ | $\bar{x} \pm t_{\alpha/2} \dfrac{s}{\sqrt{n-1}}$ | $\bar{x} \pm Z \dfrac{s}{\sqrt{n-1}}$ | $p \pm Z \sqrt{\dfrac{p(100-p)}{n-1}}$ |

\* Note: $\hat{\sigma} = s \sqrt{\dfrac{n}{n-1}}$.

construct an interval estimate of $\mu$ or $\pi$ with some degree of certainty. The width of the interval estimate increases as the level of confidence increases, since allowance must be made for more sampling error.

The sample size must be considered in estimation, since it affects the width of the confidence interval. If the sample size is sufficiently large, the sampling distribution approximates the normal distribution. If we are estimating $\mu$ and the sample size is 30 or less, the sampling distribution approximates a $t$ distribution. The shape of a $t$ distribution is determined by the sample size.

The procedure for the interval estimation of $\pi$ is similar to the procedure for estimating $\mu$ except for the small-sample case.

Figure 7-13 summarizes the various forms of interval estimation covered in this chapter. A discussion of how to determine the appropriate sample size when estimating the population mean or population percentage has also been covered in this chapter.

**Important Terms and Concepts**

1   Estimator
2   Unbiased estimator
3   Estimate
4   Estimation
5   Point estimate
6   Interval estimate
7   $Z$ distribution
8   Confidence level
9   Confidence coefficient
10   Confidence interval
11   Confidence limits
12   Precision of the estimate
13   $t$ distribution
14   Degrees of freedom
15   $\alpha$
16   $\hat{\sigma} = s \sqrt{\dfrac{n}{n-1}}$
17   $s = \sqrt{\dfrac{\Sigma (x - \bar{x})^2}{n}}$
18   $\hat{\sigma}_{\bar{x}} = \dfrac{s}{\sqrt{n-1}}$
19   $\hat{\sigma}_p = \sqrt{\dfrac{p(100-p)}{n-1}}$

**20**  $n = \dfrac{\sigma^2}{\sigma_{\bar{x}}^2}$

**21**  $n = \dfrac{\pi(100 - \pi)}{\sigma_p^2}$

**Problems**

**1** The following set of data represents a simple random sample of temperatures from 24 cities in a southern state. These temperatures were observed at the same hour and on the same day.

| 105 | 97 | 101 | 88 | 96 | 100 | 87 | 110 |
|-----|-----|-----|-----|-----|-----|-----|-----|
| 99 | 92 | 99 | 93 | 93 | 87 | 101 | 101 |
| 103 | 107 | 95 | 92 | 89 | 95 | 102 | 100 |

**a** What estimator would you use to estimate the true average temperature within the state?
**b** What is the point estimate of the true average temperature?
**c** Calculate the sample standard deviation.

**2** The following set of data represents a simple random sample of IQ scores of 32 students at an eastern university:

| 137 | 141 | 128 | 132 | 129 | 122 | 140 | 119 |
|-----|-----|-----|-----|-----|-----|-----|-----|
| 126 | 133 | 121 | 138 | 111 | 124 | 121 | 116 |
| 120 | 127 | 129 | 122 | 113 | 125 | 126 | 118 |
| 117 | 132 | 124 | 116 | 135 | 123 | 126 | 131 |

**a** Calculate a point estimate of the true average IQ of the students at the eastern university.
**b** Calculate the sample standard deviation.

**3** Using the data in problem **1**, construct a confidence interval of $\mu$ with a confidence coefficient of 99 percent.

**4** Using the data in problem **2**, construct a confidence interval of $\mu$ with a confidence coefficient of 95 percent.

**5** Assume you are constructing an interval estimate with a large sample size. Determine the $Z$ value for each of the following confidence coefficients:
**a** 75 percent
**b** 93 percent
**c** 88 percent
**d** 95 percent

**6** Determine the $t$ value for the following:
**a** $n = 15$, 99 percent confidence level
**b** $n = 23$, 90 percent confidence level
**c** $n = 28$, 95 percent confidence level
**d** $n = 27$, 95 percent confidence level
**e** $n = 25$, 95 percent confidence level
**f** $n = 20$, 95 percent confidence level

**7** Look at the results of parts **c** through **f** in problem **6**. What relationship do you see between the interval width and the sample size?

**8** Assume a student is interested in determining the average amount of money she spends per day in the month of September. A sample of 10 days shows an average of $6.24 per day with a standard deviation of $1.20. Calculate an interval estimate of $\mu$ with a 90 percent confidence level.

**9** The Howe, Doo, Yoo, and Dew Answering Service wants to estimate the average number of calls handled daily. A sample of 50 days produced a mean of 326 calls per day with $s = 48$.

    **a** Compute an estimate of the standard deviation of the sampling distribution of means.

    **b** Construct a confidence interval with a confidence coefficient of 90 percent.

**10** The Automated Automaton Assembly Plant has had a large number of employees quit their jobs soon after they start employment. The personnel manager, Mr. Rowe Botts, would like to estimate the average stay with the company of each employee who quits. A sample of 15 former employees' records indicated that the average stay with the company was 54 days. The sample standard deviation was 16 days. Construct an interval estimate of the true average with a 99 percent confidence level.

**11** Suppose the average height of 20 students in a peculiar high school is 5 feet 8 inches with a sample standard deviation of 6 inches. Construct a confidence interval of the true average with only a 5 percent chance of error.

**12** The Bill Fold Wallet Company wants to know the average number of wallets sold each day. A sample of 36 days produced an average of 114 wallets sold daily. Assume $\sigma = 17$. Construct an interval estimate of $\mu$ with a 95 percent confidence coefficient.

**13** The Russell Caddle Beef Company wants to estimate the average tonnage of beef processed daily. A simple random sample of 50 days showed that the average daily tonnage was 100 tons. The population standard deviation is 22 tons. Construct an interval estimate of the true average with a 90 percent confidence coefficient.

**14** A sample of 50 farm workers contained 36 workers who indicated that they would like to unionize.

    **a** What is the estimator of $\pi$?

    **b** What is the point estimate?

    **c** What further information is needed for an interval estimate?

**15** The Just-for-the-Halibut Fish Store received a shipment of fish, and the store would like to estimate the percentage of fish which meet its weight standards. Of a sample of 37 fish, 6 were too small. Estimate the true percentage of acceptable fish, using a 99 percent confidence level.

**16** The Fish Friar and Chip Monk of a monastery kitchen would like to know what percentage of the 200 monks in residence actually like fish and chips. A sample of 40 friars contained 30 who like fish and chips. Estimate at the 95 percent confidence level the true percentage of monks who actually like fish and chips.

**17** Bjorn Talooz, a student at a Norwegian university, wants to determine the feasibility of campaigning for the presidency of the students' association. A sample of 50 students showed that 22 percent of the students would vote for him. Estimate the true percentage with a confidence level of 99 percent.

**18** In order to respond to a federal government survey, the director of a state office of education needs to estimate the percentage of women teachers in the state public school system. The director wants to be 99 percent confident that the estimate of the percentage of women teachers is within 3 percent of the true population percentage. What size sample should be selected from the state files?

19 In problem **10** above, what size sample would be needed if Rowe Botts wanted to be 95 percent confident that an estimate of the average employment time would be within 3 days of the true mean time? (Use the sample standard deviation of 16 in problem **10** as an approximation of the population standard deviation.)

## Topics for Discussion

1 Why is an unbiased estimator desirable?

2 "If a point estimate is rarely used in estimation, it is rarely computed." Discuss the fallacy of this statement.

3 "A 95 percent confidence level means that there is a 95 percent chance that $\mu$ will fall within the computed estimate." Comment on this statement.

4 What effect does an increase in the confidence level have on the width of the confidence interval?

5 "We should always use the highest confidence coefficient in interval estimation." Discuss this statement.

6 When does the sampling distribution approximate the normal distribution?

7 What can be said about the sample size when $\sigma$ is unknown and when the $Z$ distribution applies to the interval estimation?

8 If a $t$ distribution is applicable, what must be known in order to determine the $t$ value for an interval estimation?

9 Why isn't the sample standard deviation used as an estimator of $\sigma$?

10 If the parameter $\sigma_{\bar{x}}$ may sometimes be used in an interval estimation of $\mu$, why isn't it possible to sometimes use the parameter $\sigma_p$ in an interval estimation of $\pi$?

11 "If an interval estimate is too wide, we should lower the specified confidence level so that the interval range narrows." Why is this statement incorrect?

## Answers to Self-testing Review Questions

### 7-1

1 Unbiased estimator.

2 $p$ is an unbiased estimator of $\pi$, while 82 percent is a point estimate of $\pi$.

3 An interval estimate is actually a point estimate with an allowance for sampling error. An interval estimate estimates the parameter to be within a range of values, while a point estimate is a single value.

4 A point estimate by itself gives no indication of the precision of the estimate.

5 The statement is false. Although allowance is made for sampling error, an interval estimate, like any other estimate, can be wrong.

**6** The mean of the sampling distribution of an unbiased estimator is equal to the population parameter to be estimated. This means that the possible outcomes of samples tend toward the value of the parameter.

**7** The statement is false. The point estimate serves as a basis for construction of the interval estimate.

## 7-2

**1** The theoretical basis for estimating $\mu$ with the interval lies in the properties of the sampling distribution of means.

**2** There is no difference between a confidence coefficient and a confidence level.

**3** A 95 percent confidence level means that the particular method being used for interval estimation can produce a large number of intervals of which approximately 95 percent will contain the parameter.

**4** A positive relationship exists. A higher confidence level will produce a wider confidence interval.

## 7-3

**1** The sampling distribution approximates the normal distribution when the population is normally distributed or when the sample size is more than 30.

**2 a** $Z = 1.70$
  **b** $Z = 1.10$
  **c** $Z = 1.48$

**3** With $\sigma = 9$ and $n = 36$,

$$\sigma_{\bar{x}} = \frac{\sigma}{\sqrt{n}} = \frac{9}{\sqrt{36}} = 1.5$$

With the confidence coefficient equal to 90 percent, $Z = 1.64$. The confidence interval is $\bar{x} \pm Z\sigma_{\bar{x}} = 48 \pm 1.64(1.5)$. The confidence limits are 45.54 and 50.46.

**4** With the confidence coefficient equal to 80 percent, $Z = 1.28$. The confidence interval is $\bar{x} \pm Z\sigma_{\bar{x}} = 104 \pm 1.28(13)$. The confidence limits are 87.36 and 120.64.

## 7-4

**1** A $t$ distribution should apply because the sample size is less than 30.

**2** With 26 degrees of freedom, $t_{.025} = 2.056$.

**3** $\$38.85 < \mu < \$45.15$.

**4** You should calculate the sample mean and the sample standard deviation. You should then use the sample standard deviation to calculate $\hat{\sigma}_{\bar{x}}$. In order to construct the interval estimate you must specify a confidence level.

**5** A $t$ distribution is generally flatter than a $Z$ distribution.

**6** The sample size affects the shape of a $t$ distribution.

**7** The format of the $t$ table is based on the total chance of error in estimation.

**8** The sample standard deviation is not an unbiased estimator of the population standard deviation.

### 7-5

**1** When we use the $Z$ value in interval estimation, we assume a large sample size — i.e., $n > 30$.

**2** The denominator $n - 1$ in $\hat{\sigma}_p$ makes $\hat{\sigma}_p$ an unbiased estimator of $\sigma_p$.

**3** The formula for $\sigma_p$ requires knowledge of $\pi$. However, $\pi$ is what is to be estimated.

### 7-6

**1** Since $Z\sigma_{\bar{x}} = \$1,000$, and since $Z = 1.64$, $\sigma_{\bar{x}} = \$1,000/1.64$ or $\$609.76$. Therefore, since $\sigma$ may also be estimated to be $\$5,000$ in the Chamber's city, the necessary sample size is computed as follows:

$$n = \$5,000^2/\$609.76^2 = 67.24 \text{ or } 68$$

**2** Since $Z\sigma_p = 3$ percent, and since $Z = 1.96$, $\sigma_p = 3$ percent/1.96 or 1.53 percent. Therefore, since we have no idea what the value of the population percentage is, the necessary sample size is computed as follows:

$$n = \frac{(50)(50)}{1.53^2} = \frac{2,500}{2.34} = 1,068$$

# CHAPTER 8

# TESTING HYPOTHESES AND MAKING DECISIONS: ONE-SAMPLE PROCEDURES

**LEARNING OBJECTIVES**

After studying this chapter, working the problems, and answering the discussion questions, you should be able to:

☞ Explain the necessary steps in the general hypothesis-testing procedure.

☞ Compute one-sample hypothesis tests of means (both one- and two-tailed versions) when the population standard deviation is unknown as well as when it is available.

☞ Compute one-sample hypothesis tests of percentages for both one- and two-tailed testing situations.

**CHAPTER OUTLINE**

In the previous chapter on estimation, the value of the population parameter was unknown, and the results of a sample were manipulated to provide some insight into the true value. In this chapter the sample results will be used for a different purpose. Although the exact value of a parameter may be unknown, there is often some hunch or hypothesis about the true value. Sample results may bolster the hypothesis, or the sample results may indicate that the assumption is untenable. For example, Dean I. V. Leeg may state that the average IQ of the students at his university is 130. This statement is an assumption on his part, and there should be some way of testing the dean's claim. One possible method of validation (assuming that the dean really knows) is to strap him in a chair and administer a lie detector test.[1] Another method, which is more feasible and attractive to the dean and the researcher, is sampling. If a random sample of these students had an average IQ of 104, it would be easy to reject the assumption that the true average is 130 because of the large discrepancy between the sample mean and the assumed value of the population mean. Similarly, if a sample mean were 131, it would be reasonable to accept the dean's statement. Unfortunately, life's decisions are not always as easy as this. Many times decisions fall under the general category of ulcer inducers. In the case of testing a hypothesis, the difference between the values of the sample statistic and the assumed parameter is usually neither too large nor too small, and thus obvious and clear-cut decisions are often rare. Suppose, for example, the average IQ of a sample was 134; or suppose it was 127. Does either value warrant rejection of the statement that $\mu = 130$? Obviously such a decision cannot be eyeballed. There must be some sort of criterion on which the decision process can be focused.

This chapter discusses procedures for objectively determining, under various conditions, whether sample results support a hypothesis about a parameter value, or whether the sample results indicate that a hypothesis should be rejected. More specifically, in this chapter we will consider (1) a *general hypothesis-testing procedure*, (2) *one-sample hypothesis tests of means*, and (3) *one-sample hypothesis tests of percentages*. In the next chapter we will deal with *two-sample hypothesis tests of both means and percentages*, and in Chapters 9 and 11 we will look at hypothesis-testing procedures that involve *more than two* sample means or sample percentages.

---

[1] This method further assumes that you can locate the dean. Father Damian Fandal, formerly academic dean at the University of Dallas, has *facetiously* formulated two rules for deans: Rule 1 – Hide!!!; Rule 2 – If they find you, lie!!! See Thomas L. Martin, Jr., *Malice in Blunderland,* McGraw-Hill Book Company, New York, 1973, p. 90.

## THE HYPOTHESIS-TESTING PROCEDURE IN GENERAL

Before we present a formal enumeration of the steps in the hypothesis-testing procedure, let us consider an example. Suppose the mayor of a rural town has stated that the average per-capita income of the town's citizens is $5,000, and you have an emerging statistician friend—Stan D. Viate—who has been assigned by the town council to verify or discredit the mayor's claim. Obviously, Stan's knowledge about sampling variation tells him that even if the true mean were $5,000 as stated, a sample mean would *most likely not equal* the parameter value. As an educated person versed in the rudiments of statistics, Stan realizes there will probably be a difference between the sample mean and the assumed value. The immediate problem confronting him is how large or *significant* should the difference between $\bar{x}$ and the assumed value be in order to provide sufficient reason to dismiss the mayor's claims? Is a difference in values of $100 significant? Is a difference of $1,000 significant? Well, the significant differences can be determined through statistical techniques.

## Still Another Look at the Sampling Distribution of Means

Let's look at Fig. 8-1 and assume that we have a sampling distribution of means where (1) the true mean ($\mu$) is actually equal to the hypothesized value ($\mu_{H0}$) of $5,000 and (2) the standard error is equal to $100. *In other words, we are assuming that the mayor was actually correct and $\mu$ is indeed $5,000.* (Of course, Stan and the town council are not aware of this fact.) Suppose further that Stan takes a sample of townspeople, with the result that the sample mean per-capita income is equal to $5,100. Is it reasonable for Stan to expect this result with a $\mu_{H0}$ of $5,000 and a $\sigma_{\bar{x}}$ of $100? How likely is it that a $\bar{x}$ of $5,100 will occur in this situation? As a more general question, what are the chances of Stan's getting a sample mean which differs from the $\mu_{H0}$ of $5,000 *by $100*?

Since the sampling distribution in Fig. 8-1 is approximately normal, Stan can determine the likelihood that a sample mean will equal $5,100 or $4,900 by determining how many standard errors from the $\mu$ of $5,000 a difference of $100 represents. How many $Z$ values does $5,100 or $4,900 lie from the true and assumed population mean of $5,000—i.e., what is the *standardized difference* or the number of *standard units*? Stan can calculate the standard units in the following way:

## FIGURE 8-1

Educational schematic of a normally shaped sampling distribution where the assumed mean and the true mean happen to be of equal value

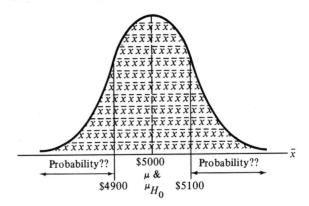

Where: $\sigma_{\bar{x}} = \$100$

$$Z = \frac{\bar{x} - \mu_{H0}}{\sigma_{\bar{x}}}$$

$$= \frac{\$5,100 - \$5,000}{\$100} \quad \text{and} \quad \frac{\$4,900 - \$5,000}{\$100}$$

$$= 1.00 \qquad\qquad\qquad = -1.00$$

Thus, we can see that if a sample mean in our example differs from the assumed value by $100, it differs by one standard unit or one standard error. Consulting the Z table in Appendix 2, we see that the area under one side of the distribution up to $Z = 1$ is .3413, and the total area between $Z = \pm 1$ is .6826. This means that there is a .1587 chance that a sample mean could be *larger* than the true and assumed population mean by *one or more standard errors,* and there is also a .1587 chance that $\bar{x}$ may be *less than* the population mean by *one or more standard errors.* All this is demonstrated in Fig. 8-2, where it is shown that there is a total chance of 31.74 percent that $\bar{x}$ will differ from $\mu$ by one standard unit or more. Consequently, Stan could report to the town council that a sample mean of $5,100 is likely to occur in this example and that the $100 difference is not sufficiently significant for him to reject the mayor's claim.

Suppose Stan's sample mean had been $5,200 instead of $5,100. Would he reject the mayor's claim with this sample result? (Remember, he doesn't know the true value of the population mean.) Converting this $200 difference between $\bar{x}$ and $\mu_{H0}$ into standard units, we get:

$$Z = \frac{\bar{x} - \mu_{H0}}{\sigma_{\bar{x}}}$$

$$= \frac{\$5,200 - \$5,000}{\$100}$$

$$= 2.00$$

**FIGURE 8-2**

**Illustration of the likelihood of obtaining a $\bar{x}$ which differs from the true mean by one standard error or more**

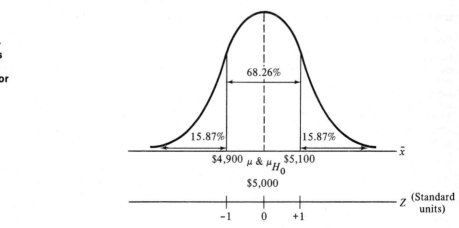

Where: $\sigma_{\bar{x}} = \$100$

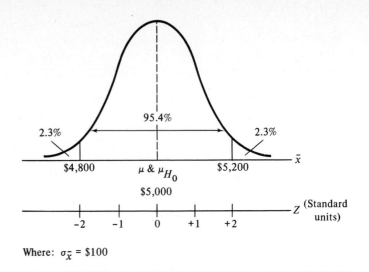

**FIGURE 8-3**

**Illustration of the likelihood of obtaining a x̄ which differs from the true mean by two or more standard errors**

95.4%

2.3%                                    2.3%

$4,800          μ & μ$_{H_0}$          $5,200          x̄

$5,000

Z (Standard units)

−2      −1      0      +1      +2

Where: $\sigma_{\bar{x}} = \$100$

Thus, the total chance that a x̄ will differ from our true mean of $5,000 by two or more standard errors is only approximately 4.6 percent, as shown in Fig. 8-3. Given such a low chance of occurrence, Stan would probably be justified in *rejecting* the mayor's claim. There is sufficient statistical evidence for him to conclude that the mayor's claim is incorrect.

As you can now see, the difference between the value of an obtained sample mean and an assumed value of a hypothetical population mean is considered significantly large to warrant rejection of the hypothesis if the likelihood of the value of a sample mean is too low. The criteria of "too low" will vary with the standards of researchers. At this point it is sufficient to state that all hypothesis tests must have some established rule which rejects a hypothesis if the likelihood of a value of x̄ falls below a minimum acceptable probability level.

Unfortunately, if Stan were unaware in the above example that the true population mean was indeed $5,000, he might have justifiably but erroneously rejected the mayor's claim if he had obtained a sample mean of $5,200. As a matter of fact, if he had established a rule that any sample mean value which differed from the assumed mean of $5,000 by two or more standard errors in either direction of the sampling distribution would cause rejection of the hypothesis, and *if* indeed the true mean were $5,000, he would erroneously reject the mayor's claim 4.6 percent of the time in a large number of tests. In other words, a particular sample mean may be a part of a sampling distribution in which the value of the true mean happens to be equal to the assumed value, but the likelihood of that particular sample mean occurring may be so low that there is sufficient reason not to accept the hypothesized value as the true value. In short, the minimum acceptable likelihood of a sample mean is also the *risk of rejecting a statement which is actually true.*

With this basic example in mind, we are now ready to study the formal steps in a hypothesis-testing procedure.

| **Steps in the Hypothesis-testing Procedure** | **Stating the null and alternative hypotheses** The *first* step in hypothesis testing is to state specifically the assumed value of the parameter *prior* to sampling. *This assumption to be tested is known as the null hypothesis.* Suppose we want to test the hypothesis that the population mean is equal to 100. The format of this hypothesis would be: |
|---|---|

$$H_0: \mu = 100$$

As we have noted earlier, the hypothesized value of the population mean when used in calculations is identified by the symbol $\mu_{H0}$.

If the sample results do not support the null hypothesis, we must obviously conclude something else. *The conclusion which is accepted contingent on the rejection of the null hypothesis is known as the alternative hypothesis.* There are three possible alternative hypotheses to the null hypothesis stated above:

$$H_1: \mu \neq 100$$
$$H_1: \mu > 100$$
$$H_1: \mu < 100$$

The selection of an alternative hypothesis depends on the nature of the problem at hand, and later sections of this chapter will discuss these alternative hypotheses. As with the null hypothesis, the alternative hypothesis should be stated *prior* to actual sampling.

**Selecting the level of significance** Having stated the null and alternative hypotheses, the *second* step is to establish a criterion for rejection or acceptance of the null hypothesis. If the true mean is actually the assumed value, we know that the probability of the differences between sample means and $\mu_{H0}$ diminishes as the size of the difference increases—i.e., extremely large differences are unlikely. We must state, *prior* to sampling, the minimum acceptable probability of occurrence for a difference between $\bar{x}$ and $\mu_{H0}$. In our previous example involving the mayor's claim, a difference between $\bar{x}$ and $\mu_{H0}$ with a likelihood of only 4.6 percent or less was considered unlikely, and thus Stan felt there was sufficient reason to reject the hypothesis. In such a case, a 4.6 percent chance of occurrence would have been the minimum acceptable probability level. As indicated earlier, if the true mean is indeed the assumed value, *the minimum acceptable probability level is also the risk of erroneously rejecting the null hypothesis when the null hypothesis is true.* Therefore, this next step in the hypothesis-testing procedure is to state the level of risk you desire in rejecting a null hypothesis when the null hypothesis is true. *This risk of erroneous rejection is known as the level of significance, which is denoted by the Greek letter $\alpha$ (alpha).*[2]

---

[2] Technically, $\alpha$ is known as the risk of a Type I error—i.e., the risk that a true hypothesis will be rejected. When a *false* hypothesis is erroneously *accepted* as true, it is known as a Type II error. (Some students have been unkind enough to suggest to one of the authors that registering for his course is known on campus as a Type III error.)

**Determining the test distribution to use** Once the level of significance has been selected, it is then necessary to determine the appropriate probability distribution to use for the particular test. In this chapter and the one which follows, we will be concerned only with the normal distribution (using the $Z$ table) and the $t$ distributions; in Chapters 10 and 11, other test distributions are introduced and used. For our purposes in this chapter, then, Fig. 7-6, page 154, may be used to identify the appropriate distribution to employ for testing purposes.

**Defining the rejection or critical regions** Once the appropriate test distribution has been determined, it is then possible to specify in standard units what a significant difference is. *If the sampling distribution is normally distributed, the level of significance can be expressed in standard units using the $Z$ table.* Suppose in a test it is stated that the desirable risk of erroneous rejection of the null hypothesis is $\alpha = .05$. This means that the hypothesis will not be acceptable if the difference expressed in standard units between $\bar{x}$ and $\mu_{H0}$ has only a 5 percent or less chance of occurring. Since the hypothesis can be rejected if $\bar{x}$ is too high or if $\bar{x}$ is too low, we may want a .025 chance of erroneous rejection in each tail of the sampling distribution if the true mean were equal to the assumed value. In this case, an $\alpha$ value of .05 represents the *total* risk of error. Figure 8-4 indicates how the normal curve is partitioned. With .025 in each tail, the remaining area *in each half* of the sampling distribution is .4750 (.500 − .025). Appendix 2 indicates that the appropriate $Z$ value corresponding to an area figure of .4750 is 1.96.

What does the partitioning of the normal curve in Fig. 8-4 mean? Figure 8-5 shows that if a sample mean differs from the hypothetical mean by 1.96 or more standard errors in either direction, there is sufficient reason to reject the null hypothesis at the .05 level of significance. Thus, a $Z$ value of 1.96 represents the level in standard units at which the difference between $\bar{x}$ and $\mu_{H0}$ becomes significant enough to raise doubt that $\mu_{H0} = \mu$. *The significant difference is the degree of difference between $\bar{x}$ and $\mu_{H0}$ that leads to the rejection of the null hypothesis.*

Therefore, after the level of significance has been stated and the appropriate test distribution has been determined, the *fourth step* in our procedure is to deter-

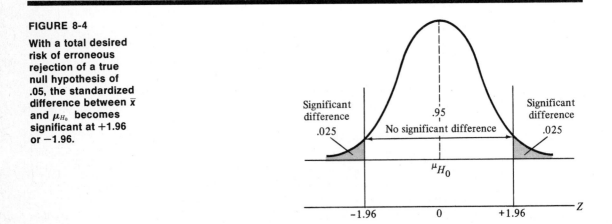

**FIGURE 8-4**

**With a total desired risk of erroneous rejection of a true null hypothesis of .05, the standardized difference between $\bar{x}$ and $\mu_{H_0}$ becomes significant at +1.96 or −1.96.**

Significant difference .025

.95

No significant difference

Significant difference .025

$\mu_{H_0}$

−1.96     0     +1.96     Z

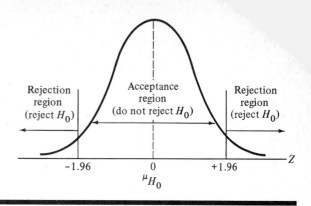

**FIGURE 8-5**

Construction of
acceptance and
rejection regions
with a significant
level of .05

mine the rejection or critical regions of the sampling distribution, which are represented in standard units. *If the difference between an obtained x̄ and the assumed mean has a value which falls into a critical region, the null hypothesis will be rejected.* (If the difference does not fall into a rejection region, of course, there is no statistical reason to doubt the hypothesis.) However, a word of caution concerning conclusions about the validity of a null hypothesis is appropriate here. Theoretically, a test *never proves* that a hypothesis is true. Rather, a test merely provides statistical evidence for not rejecting a hypothesis. The only standard of truth is the population mean, and since the true value of the mean is unknown, the assumption can never be proved. Thus, when it is said that a hypothesis is accepted, it essentially means that there is no statistically sufficient reason for rejection of the assumption.

**Stating the decision rule** After we have stated the hypotheses, selected the level of significance, determined the test distribution to use, and defined the rejection regions, the *fifth step* is a formal statement of the rules on which a conclusion will be made about the null hypothesis. *A decision rule should clearly state the appropriate conclusion based on sample results.* The general format of a decision rule is:

Accept $H_0$ if the standardized difference between $\bar{x}$ and $\mu_{H0}$ falls into the acceptance region.

*or*

Reject $H_0$ if the standardized difference between $\bar{x}$ and $\mu_{H0}$ falls into a rejection region.

**Making the necessary computations** After all the ground rules have been laid out for the test, the *next step* is the actual data analysis. A sample of items must be collected, and an estimate of the parameter must be calculated. Assuming that we are testing a hypothesis about the value of the population mean, we *first* calculate the value of a sample mean. In order to convert the difference between $\bar{x}$ and $\mu_{H0}$ into a standardized value, it is *then necessary* to determine the standard

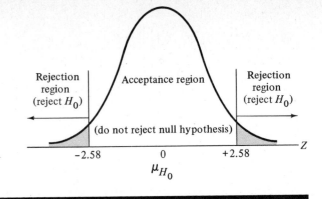

**FIGURE 8-6**

**Acceptance and rejection regions with $\alpha = .01$**

Rejection region (reject $H_0$)    Acceptance region    Rejection region (reject $H_0$)

(do not reject null hypothesis)

$-2.58$     0     $+2.58$ — $Z$

$\mu_{H_0}$

---

error of the mean. *The standardized difference between the statistic and the assumed parameter is called the critical ratio, because this value is critical in determining the acceptance or rejection of the null hypothesis.* The critical ratio (CR) for a hypothesis test of a population mean is determined as follows:

$$CR = \frac{\bar{x} - \mu_{H0}}{\sigma_{\bar{x}}}$$

**Making a statistical decision** *If the value of the critical ratio falls into a rejection region, the null hypothesis is rejected.* For example, Fig. 8-6 shows the rejection regions of a normal curve with $\alpha = .01$. Referring to the Z table, a *total* risk of 1 percent corresponds to Z values of $-2.58$ and $+2.58$. Suppose a sample produced a critical ratio of 2.60. Since the CR falls into a rejection region, there is sufficient reason to reject the null hypothesis, and the risk of erroneous rejection is only 1 percent.

At this point, your head may be dizzy with definitions and procedural steps. In order to help you sort out your head, Fig. 8-7 summarizes the general procedure for a hypothesis test.

**Managerial Decisions and Statistical Decisions: A Caution**

Let's conclude this section on a nonstatistical note. Although statistical laws may provide convenient and objective methods for assessing hypotheses, a statistical conclusion by no means represents a final decision in decision making. Consumers of statistical reports use quantitative results merely as one form of input in a complex network of factors that affect an ultimate decision. Undoubtedly decision making is full of uncertainty, and statistical results serve to reduce and control some of the uncertainty; but they do not completely eliminate uncertainty. Problems may be quantified and a result obtained, but the solution is only as good as the input that has gone into structuring the problem. Statistical results, although objectively determined, should not be accepted with blind faith. Other situational factors must be considered. For example, a statistical test may tell a production manager that a machine is not producing as much as the manager had assumed. This

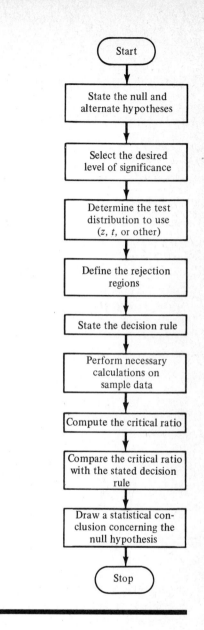

**FIGURE 8-7**

**The general hypothesis-testing procedure**

result, however, does not tell the manager the appropriate course of action to be taken. The manager may replace the machine, fix the machine, or leave the machine in its present condition. The ultimate decision is made by consideration of the available money for replacement, the repair record of the machine, the availability of new machines, and so forth. Thus, *the statistical conclusion is not necessarily the managerial conclusion;* it is simply one factor which must be considered in the context of the whole problem.

1  What is a null hypothesis?

2  What is an alternative hypothesis?

3  What is a significance level?

4  When is a $Z$ distribution applicable in hypothesis testing?

5  What is meant by a significant difference in hypothesis testing?

6  What is a rejection region?

7  How is the difference between $\bar{x}$ and $\mu_{H0}$ standardized?

8  "If a critical ratio is not within a rejection region, there is proof that the hypothesis is true." Comment on this statement.

9  "Since a hypothesis test is based on statistical laws, and since the conclusions have been objectively determined, decision making becomes easier." Discuss this statement.


## ONE-SAMPLE HYPOTHESIS TESTS OF MEANS

This section considers the one-sample testing procedure for means under various circumstances. You will recall that there are three possible alternative hypotheses, and the selection of an appropriate alternative conclusion if the null hypothesis is rejected depends on the nature of the problem. This section discusses circumstances under which the possible alternative hypotheses are appropriate.


## Two-tailed Tests When $\sigma$ Is Known

When the null and alternative hypotheses are in the format:

$H_0$: $\mu =$ assumed value

$H_1$: $\mu \neq$ assumed value

if the null hypothesis cannot be accepted, it is merely concluded that the population mean does not equal the assumed value. It *doesn't matter* if the true value might be more or less than the assumed value. The only conclusion is that the true value and the hypothesized value are not the same.

The nature of the above hypotheses requires a *two-tailed test*. A two-tailed test rejects the null hypothesis if the sample mean is significantly *higher or lower than* the assumed value of the population mean. With a two-tailed test, therefore, there are *two* rejection regions, as shown in Fig. 8-8. Since the hypothesis may be rejected with a sample mean that is too high or too low, the total risk of error in rejecting $\mu_{H0}$ is evenly distributed for each tail—i.e., the area in *each* rejection region is $\alpha/2$.

*If $\sigma$ is known or if the sample size is large, the boundaries of the rejection regions are determined through the use of the Z table.* The boundaries of the rejection regions are determined by the $Z$ value corresponding to the probability $.5000 - \alpha/2$. For example, with a two-tailed test and $\alpha = .05$, the area in each tail is $.025$ and the boundaries of the rejection region are $Z = -1.96$ and $Z = +1.96$ (see Fig. 8-8).

The appropriate *decision rule* in this example for a two-tailed test using the $Z$ distribution is:

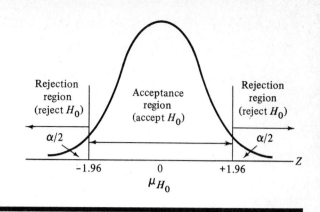

**FIGURE 8-8**

Illustration of
acceptance and
rejection regions for
a two-tailed test
with $\alpha = .05$

Accept $H_0$ if CR falls between[3] $\pm 1.96$.

*or*

Reject $H_0$ and accept $H_1$ if CR $< -1.96$ or CR $> +1.96$.

The following examples show the use of two-tailed tests when $\sigma$ is known.

**Example 8-1** The owner of the Kate and Edith Cake Company stated that the average number of buns sold daily was 1,500. A worker in the store wants to test the accuracy of the boss's statement. A random sample of 36 days showed that the average daily sales were 1,450 buns. Using a level of significance of $\alpha = .01$ and assuming $\sigma = 120$, what should the worker conclude?

The *hypotheses are:*

$H_0$: $\mu = 1,500$ buns
$H_1$: $\mu \neq 1,500$ buns

and this is a two-tailed test because a sample mean which is significantly too high or too low is sufficient to reject the null hypothesis. The interest of this test is only whether or not $\mu = 1,500$; no other conclusion is to be drawn.

The Z distribution is applicable here because $\sigma$ is known, and with $\alpha = .01$, the risk of erroneous rejection is .005 in each tail. This means that the chance of correct acceptance of $H_0$ on one side of the normal curve is .4950. Consulting the Z table, the Z value corresponding to an area of .4950 is approximately 2.58. (The rejection regions for this problem are illustrated in Fig. 8-9.)

---

[3] What if the CR is exactly 1.96? In this and other similar situations throughout the book, we will interpret the decision rule to include 1.96 in the acceptance region. A value of 1.97, of course, would be in the rejection region.

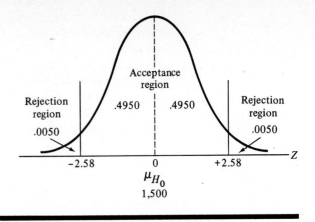

FIGURE 8-9

Under these circumstances, the *decision rule* may be stated as follows:

Accept $H_0$ if CR falls between $\pm 2.58$.

*or*

Reject $H_0$ and accept $H_1$ if CR $< -2.58$ or CR $> +2.58$.

*The computation of the critical ratio (CR) is:*

$$CR = \frac{\bar{x} - \mu_{H0}}{\sigma_{\bar{x}}}$$

$$= \frac{1,450 - 1,500}{\sigma/\sqrt{n}}$$

$$= \frac{-50}{120/\sqrt{36}}$$

$$= -2.5$$

*Conclusion:* Since CR $= -2.5$, which is between $\pm 2.58$, the null hypothesis should not be rejected at the .01 level of significance.

**Example 8-2** A local insurance agent, Mr. Hy R. Fee, claimed that the average amount paid for personal injury in an automobile accident is $4,500. A policyholder, Mr. Kent Difault, wanted to check the accuracy of the claim and was allowed to sample randomly 16 cases involving personal injury. The sample mean was $4,715. Assuming that $\sigma = \$800$, test the claim of the agent with $\alpha = .05$.

The *hypotheses are:*

$H_0$: $\mu = \$4,500$
$H_1$: $\mu \neq \$4,500$

This is a two-tailed test because the policyholder is only interested in concluding nonequality between the assumed value and the true value if the null hypothesis is rejected.

*Since $\sigma$ is given, the Z distribution applies regardless of sample size.* With $\alpha = .05$, there is a risk of 0.25 in each tail. The corresponding Z value for an area of $.5000 - .0250$ or .4750 is 1.96. Therefore, the *decision rule should be:*

Accept $H_0$ if CR falls between $\pm 1.96$.

*or*

Reject $H_0$ and accept $H_1$ if CR $< -1.96$ or CR $> +1.96$.

The *critical ratio* is calculated as follows:

$$CR = \frac{\bar{x} - \mu_{H0}}{\sigma_{\bar{x}}}$$

$$= \frac{\$4,715 - \$4,500}{\$800/\sqrt{16}}$$

$$= 1.075$$

*Conclusion:* Since CR falls between $\pm 1.96$, there is not enough evidence to reject the null hypothesis at the .05 level of significance. The sample mean of \$4,715 lies only 1.075 standard errors to the right of the assumed mean, and sampling variation could easily account for this fact.

**Self-testing Review 8-2**

1  What is a two-tailed test?

2  Determine the appropriate Z values for a two-tailed test for the following:
   a  $\alpha = .01$
   b  $\alpha = .08$
   c  $\alpha = .05$
   d  $\alpha = .03$

3  The population mean is assumed to be 500, with $\sigma = 50$. A sample of 36 had $\bar{x} = 475$. Conduct a two-tailed test with $\alpha = .01$.

**One-tailed Tests When $\sigma$ Is Known**

Many times it is unsatisfactory simply to conclude that the true value is *not equal* to the assumed value. If the null hypothesis is not tenable, we often want to know if the rejection occurs because the true value is *probably higher* or *probably lower* than the assumed value. In other words, was the hypothesis rejected because the true value is likely to be greater than or less than the assumed value? In such situations, the null hypothesis is still of the form:

$H_0$: $\mu$ = assumed value

But the *alternative hypothesis* may be one of the following:

$H_1$: $\mu >$ assumed value

*or*

$H_1$: $\mu <$ assumed value

The nature of either hypothesis indicates a one-tailed test. *In a one-tailed test, there is only one rejection region and the null hypothesis is rejected only if the value of a sample mean falls into this single rejection region.*

**Right-tailed tests** If the alternative hypothesis is:

$H_1$: $\mu >$ assumed value

the rejection region is in the right tail of the sampling distribution, and this is known as a *right-tailed test*. The null hypothesis will be rejected in this case *only* if the value of a sample mean is *significantly high*. If the value of a sample mean is extremely low compared to the assumed value, the null hypothesis will not be rejected. The *major pitfall with a right-tailed test* is that the true value may be less than the assumed value, but because of the structure of the right-tailed test the null hypothesis will not be rejected. In such a test, the attention is focused on rejecting $H_0$ *solely* on the basis that the true value might be greater than the assumed value.

If you are confused by the above paragraph, consider this analogy. Suppose you and a friend are guessing a third person's age. Your friend hypothesizes that the third person is 23 years old, while you believe that he is older. Finally, you ask the third person, "Are you more than 23 years old?" He says, "No." As a result, you cannot reject your friend's assertion, but, of course, the assertion may be incorrect because you didn't ask the third person if he was less than 23 years of age. The nonrejection of $H_0$ in a right-tailed test is similar to this analogy.

**Left-tailed tests** If the alternative hypothesis is:

$H_1$: $\mu <$ assumed value

we are interested in determining if the true value is *less* than the assumed value. Such a hypothesis is indicative of a *left-tailed test*. In a left-tailed test the single rejection region is on the left side of the sampling distribution. In this case, the null hypothesis will be rejected only if the value of a sample mean is *significantly low*. In a left-tailed test, the null hypothesis will not be rejected if the true value is likely to be more than the assumed value.

The distinctions between a left-tailed test and a right-tailed test are illustrated in Fig. 8-10.

**Level of significance considerations** The level of significance ($\alpha$) is the *total risk* of erroneously rejecting $H_0$ when it is actually true. In a two-tailed test, the total risk is evenly divided between each tail. However, in a one-tailed test (since there is only one rejection region) *an area in the single tail is assigned the total risk or $\alpha$.* If the $Z$ distribution is applicable, the appropriate $Z$ value is thus determined by the one-tailed probability of $.5000 - \alpha$. For example, if the level of sig-

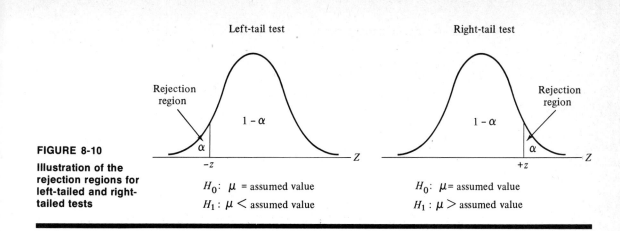

**FIGURE 8-10**

**Illustration of the rejection regions for left-tailed and right-tailed tests**

Left-tail test

Rejection region

$1 - \alpha$

$\alpha$

$-z$

$Z$

$H_0$: $\mu$ = assumed value

$H_1$: $\mu <$ assumed value

Right-tail test

Rejection region

$1 - \alpha$

$\alpha$

$+z$

$Z$

$H_0$: $\mu$ = assumed value

$H_1$: $\mu >$ assumed value

nificance were .05 for a left-tailed test, the boundary of the rejection region would be a $Z$ value of $-1.64$ (see Fig. 8-11).

**Decision rule statements**  If it is assumed that the $Z$ distribution is applicable, the decision rule for a *left-tailed test* is of the following form (remember the elementary rules of first-year algebra here):

> Accept $H_0$ if CR $\geqq -Z$ value.
>
> *or*
>
> Reject $H_0$ and accept $H_1$ if CR $< -Z$ value.

The decision rule for a *right-tailed test* is:

> Accept $H_0$ if CR $\leqq Z$ value.
>
> *or*
>
> Reject $H_0$ and accept $H_1$ if CR $> Z$ value.

**FIGURE 8-11**

**Rejection region for a left-tailed test at the .05 level of significance**

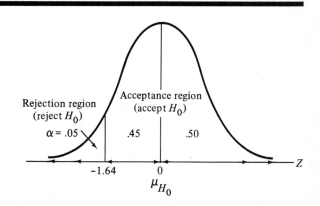

Rejection region (reject $H_0$)

$\alpha = .05$

Acceptance region (accept $H_0$)

.45          .50

$-1.64$          0          $Z$

$\mu_{H_0}$

The computation of the critical ratio is the same for a one-tailed test as for a two-tailed test.

Let's now look at the following examples, in which a one-tailed test is applicable and in which $\sigma$ is known.

**Example 8-3** Mr. Tyrone Hops, the supervisor of the local brewery, wants to make sure that the average volume of the Super-Duper can is 16 ounces. If the average volume is significantly less than 16 ounces, customers (and various regulatory agencies) would likely complain, prompting undesirable publicity. The physical size of the can does not allow an average volume significantly above 16 ounces. A random sample of 36 cans showed a sample mean of 15.7 ounces. Assuming $\sigma$ is 0.2 ounces, conduct a hypothesis test with $\alpha = .01$.

The *hypotheses are:*

$H_0$: $\mu = 16$ ounces
$H_1$: $\mu < 16$ ounces

The nature of the problem indicates that if the null hypothesis is rejected, Mr. Hops wants to conclude that the hypothesis was rejected because the sample mean was significantly low.

With $\sigma$ known, the $Z$ distribution is applicable. Thus, with $\alpha = .01$, and with a left-tailed test, the rejection region begins at a $Z$ value to the left of $-2.33$. Therefore, *the decision rule is:*

Accept $H_0$ if CR $\geq -2.33$. (Remember, $-2.32 > -2.33$.)

*or*

Reject $H_0$ and accept $H_1$ if CR $< -2.33$. (Also, remember that $-2.34 < -2.33$.)

The *critical ratio* is computed as follows:

$$CR = \frac{\bar{x} - \mu_{H0}}{\sigma_{\bar{x}}}$$

$$= \frac{15.7 - 16}{.2/\sqrt{36}}$$

$$= -9.00$$

*Conclusion:* Since CR $< -2.33$, the brewery must reject $H_0$ and improve its filling process. It is *very* unlikely that a sample selected from a sampling distribution that had a true mean of 16 ounces would have a mean located 9.00 standard errors to the left of the true mean!

**Example 8-4** Mr. Hal I. Tosis, a mouthwash distributor, has stated that the average cost to process a sales order is $13.25. Ms. Minnie Mize, cost controller, fears that the average cost of processing is more than $13.25. She is interested in taking action if costs are high, but she does not care if the actual average is below the assumed value. A random sample of 100 orders had a sample mean of $13.35. Assuming the $\sigma$ is $0.50, conduct a test at the .01 level of significance.

The *hypotheses are:*

$H_0$: $\mu =$ $13.25 cost

$H_1$: $\mu >$ $13.25 cost

This is a right-tailed test because only a significantly high sample mean value will reject the null hypothesis. In the problem, there has been no concern expressed for a sample mean value which might be too low. With $\sigma$ known, the $Z$ distribution is applicable. With $\alpha = .01$, the appropriate $Z$ value is 2.33. The appropriate *decision rule is:*

Accept $H_0$ if CR $\leqslant$ 2.33.

*or*

Reject $H_0$ and accept $H_1$ if CR $>$ 2.33.

The *critical ratio* is computed as follows:

$$CR = \frac{\bar{x} - \mu_{H0}}{\sigma_{\bar{x}}} = \frac{\$13.35 - \$13.25}{\$0.50/\sqrt{100}} = 2.00$$

*Conclusion:* Since CR $<$ 2.33, Ms. Mize has no evidence to reject Mr. Tosis' statement at the .01 level of significance.

**Self-testing Review 8-3**

1 What is a one-tailed test?

2 How do the alternative hypotheses of a left-tailed and a right-tailed test differ?

3 Determine the appropriate $Z$ values for the following:
   a  Two-tailed test, $\alpha = .01$
   b  Two-tailed test, $\alpha = .05$
   c  Left-tailed test, $\alpha = .05$
   d  Right-tailed test, $\alpha = .01$

4 Assume that you have the following null hypothesis: $H_0$: $\mu = 100$. Conduct a left-tailed test with $\alpha = .05$, $\sigma = 15$, $n = 36$, and $\bar{x} = 88$.

5 Assume that you have the following data: $H_0$: $\mu = 24$, $\sigma = 3$, $n = 16$, $\bar{x} = 26$, and $\alpha = .01$. Conduct a right-tailed test.

6 Mr. X stated that the true mean was 500. Mr. Y disagreed, saying that Mr. X was overstating the value. What are the null and alternative hypotheses?

7 A widget machine must produce widgets with a width of 2.5 inches. A mean width of a batch of widgets that is less than 2.5 inches requires the batch to be destroyed. If the mean width is larger than 2.5 inches, the widgets can still be sold for the same price but for different uses. What are the hypotheses?

8 "If $H_0$ is not rejected in a right-tailed test, we can say that $\mu$ equals the assumed value ($\mu_{H0}$)." Discuss this statement.

## Two-tailed tests When σ Is Unknown

Up to now, we have been working hypothesis tests with $\sigma$ known. However, as indicated in the previous chapter, knowledge of $\sigma$ is rare. More often than not, the sample standard deviation ($s$) must be used in the testing procedure.

With $\sigma$ unknown, *the following aspects of the hypothesis-testing procedure are affected:*

1 The appropriate sampling distribution can no longer be assumed to be approximately normally shaped.

2 In the computation of the critical ratio (CR), $\hat{\sigma}\bar{x}$ must be used instead of $\sigma_{\bar{x}}$.

In other words, when $\sigma$ is unknown, the $Z$ distribution (and Appendix 2) can only be used to determine the rejection regions when the sample size is more than 30. When the sample size is 30 or less in this situation, the sampling distribution takes the shape of a $t$ distribution. The $t$ value used to determine the boundary of a rejection region is based on the level of significance and the degrees of freedom (which are $n - 1$). For example, suppose you are making a *two-tailed test* at the .05 level of significance with a sample size of 16. In the $t$ table in Appendix 4, the $t$ value with 15 degrees of freedom is 2.131.[4]

As we shall see in the following examples, the testing procedure is the same with an unknown $\sigma$ as with a given $\sigma$, with the exceptions (1) that the appropriate test distribution ($Z$ or $t$) must be determined and (2) that the correct method of calculating the critical ratio must be used.

**Example 8-5** Mr. Drinkwater of the D-T's Liquor Shoppe thinks that his business sells an average of 17 pints of Derelict's Delight daily. His partner, Mr. Teetotaler, thinks this estimate is wrong. A random sample of 36 days showed a mean of 15 pints and a sample standard deviation ($s$) of 4 pints. Test the accuracy of Mr. Drinkwater's statement at the .10 level of significance.

The *hypotheses are:*

$H_0$: $\mu = 17$ pints
$H_1$: $\mu \neq 17$ pints

This is a two-tailed test because we only wish to determine the validity of Mr. Drinkwater's statement. An extremely high value of $\bar{x}$ or an extremely low value of $\bar{x}$ will reject the null hypothesis. With $n = 36$, the sampling distribution approximates the normal distribution, and thus the $Z$ distribution applies.

Since this is a two-tailed test and $\alpha = .10$, the risk of error in each tail is .05. The $Z$ value corresponding to $.5000 - .05 = .45$ is approximately 1.64. The appropriate *decision rule is:*

Accept $H_0$ if CR falls between ±1.64.

*or*

Reject $H_0$ and accept $H_1$ if CR $< -1.64$ or CR $> +1.64$.

---

[4] That is, $t_{\alpha/2 \text{ or } .025} = 2.131$. The $t$ table is set up to show the rejection region in one tail.

With $s = 4$ and $n = 36$, $\hat{\sigma}_{\bar{x}}$ is estimated in the following manner:

$$\hat{\sigma}_{\bar{x}} = \frac{s}{\sqrt{n-1}} = \frac{4}{\sqrt{36-1}} = .676$$

Therefore, the *critical ratio* is computed as follows:

$$CR = \frac{\bar{x} - \mu_{H0}}{\hat{\sigma}_{\bar{x}}}$$

$$= \frac{15 - 17}{.676}$$

$$= -2.96$$

*Conclusion:* Since CR $< -1.64$, it is necessary to reject Mr. Drinkwater's claim at the .10 level of significance.

**Example 8-6** Mr. R. U. High claims that the average height of citizens in Biglandia is 64 inches. A sociologist, Mr. Kerr E. Uss, took a random sample of 16 Biglandians and found that the mean was 62.9 inches and the standard deviation was 2.5 inches. Can it be concluded that Mr. High is correct at the .05 level of significance?

The *hypotheses are:*

$H_0$: $\mu = 64$ inches
$H_1$: $\mu \neq 64$ inches

The nature of the problem indicates a *two-tailed test,* with a risk of erroneous rejection of .025 in each tail. The *t distribution* is applicable in this case because the sample size is only 16. With 15 degrees of freedom and a .025 risk in each tail, $t_{.025} = 2.131$.

The *decision rule for this problem is:*

Accept $H_0$ if CR falls between $\pm 2.131$.

*or*

Reject $H_0$ and accept $H_1$ if CR $< -2.131$ or CR $> +2.131$.

With $s = 2.5$ and $n = 16$,

$$\hat{\sigma}_{\bar{x}} = \frac{s}{\sqrt{n-1}} = \frac{2.5}{\sqrt{16-1}} = .645$$

And the *critical ratio* is computed as follows:

$$CR = \frac{\bar{x} - \mu_{H0}}{\hat{\sigma}_{\bar{x}}} = \frac{62.9 - 64.0}{.645} = -1.71$$

*Conclusion:* Since CR falls between $\pm 2.131$, there is no reason to reject Mr. High's statement at the .05 level of significance.

1   How do the hypothesis-testing procedures with $\sigma$ known and $\sigma$ unknown differ?

2   Assume that you have the following data: $H_0$: $\mu = 612$, $\bar{x} = 608$, $s = 5$, $n = 13$, and $\alpha = .05$. Conduct a two-tailed test.

3   Assume that you have the following data: $H_0$: $\mu = 243$, $\bar{x} = 269$, $s = 15$, $n = 36$, and $\alpha = .01$. Conduct a two-tailed test.

**One-tailed Tests
When $\sigma$ Is
Unknown**

The two preceding examples were two-tailed tests. The following examples are one-tailed tests made when $\sigma$ is unknown. You will notice that the testing procedure is essentially unchanged.

**Example 8-7** The manager of the Take-It-for-Granite Rock Company, Seymour Pebbles (isn't that awful!), is under the impression that the average truckload delivered is 4,500 pounds. A stockholder, Mr. Chip Stone, contends that this is an overinflated figure to lure new investors. Mr. Stone randomly sampled the records of 25 trucks and found the mean load to be 4,460 pounds with a standard deviation of 250 pounds. Can Mr. Stone reject Pebbles's claim using a significance level of .05?

The *hypotheses are:*

$H_0$: $\mu = 4{,}500$ pounds
$H_1$: $\mu < 4{,}500$ pounds

This is a *left-tailed test* because only a sample mean that is significantly low will cause rejection of the null hypothesis. And since this is a one-tailed test, the area in the single rejection region is equal to the significance level of .05.

With $n = 25$, the $t$ distribution applies, there are 24 degrees of freedom, and $t_{.05} = 1.711$. The *decision rule then is:*

Accept $H_0$ if CR $\geq -1.711$.

*or*

Reject $H_0$ and accept $H_1$ if CR $< -1.711$.

With $s = 250$ and $n = 25$,

$$\hat{\sigma}_{\bar{x}} = \frac{s}{\sqrt{n-1}} = \frac{250}{\sqrt{25-1}} = 51.03$$

The *critical ratio* is computed as follows:

$$\text{CR} = \frac{\bar{x} - \mu_{H0}}{\hat{\sigma}_{\bar{x}}} = \frac{4{,}460 - 4{,}500}{51.03} = -.784$$

*Conclusion:* Since CR $> -1.711$, there is no significant reason for Mr. Stone to doubt Pebbles's claim. It is quite possible that a sample could be selected with a mean located only $-0.784$ standard error from a true mean.

**Example 8-8**   Mr. Hiram N. Fyrem of the H & F Employment Agency believes that the agency receives an average of 16 complaints per week from companies which hire the agency's clients. Mr. B. S. DeGree, an interviewer, is concerned that the true mean is higher than Mr. Fyrem believes. If Fyrem's hypothesis is an understatement, something must be done about the agency's screening procedures. A sample of 10 weeks yielded an average of 18 complaints with a standard deviation of 3 complaints. Conduct a test at the .01 level.

The *hypotheses are:*

$H_0$: $\mu = 16$ complaints
$H_1$: $\mu > 16$ complaints

If the statistical evidence cannot support the null hypothesis, Mr. DeGree would want to conclude that the parameter value is more than assumed. Thus, this is a *right-tailed* test.

With $n = 10$ and $\alpha = .01$, the *t value* with 9 degrees of freedom is $t_{.01} = 2.821$. The *decision rule* therefore is:

Accept $H_0$ if CR $\lessgtr 2.821$.
*or*
Reject $H_0$ and accept $H_1$ if CR $> 2.821$.

With $s = 3$ and $n = 10$,

$$\hat{\sigma}_{\bar{x}} = \frac{s}{\sqrt{n-1}} = \frac{3}{\sqrt{10-1}} = 1.00$$

and the *critical ratio* is computed as follows:

$$CR = \frac{\bar{x} - \mu_{H0}}{\hat{\sigma}_{\bar{x}}} = \frac{18 - 16}{1.00} = 2.00$$

*Conclusion:* Since CR $< 2.821$, there is no sufficient reason at the .01 level of significance to doubt Mr. Fyrem's hypothesis.

So far in this chapter, we have discussed one-sample hypothesis tests of means under various conditions. The general testing procedure was essentially the same under all conditions. The testing differences that exist are reflected in the differences that appear in the decision rules. These differences are summarized in Fig. 8-12.

**FIGURE 8-12**

Decision rules under various conditions of hypothesis testing of means

|  | $\sigma$ Known or $n > 30$ | $\sigma$ Unknown and $n \leq 30$ |
|---|---|---|
| Two-tailed test | Accept $H_0$ if CR is between $\pm Z$ value<br>Reject $H_0$ and accept $H_1$ if CR $> +Z$ value<br>or CR $< -Z$ value | Accept $H_0$ if CR is between $\pm t_{\alpha/2}$ value<br>Reject $H_0$ and accept $H_1$ if CR $> +t_{\alpha/2}$ value or<br>CR $< -t_{\alpha/2}$ value |
| Left-tailed test | Accept $H_0$ if CR $\geq -Z$ value<br>Reject $H_0$ and accept $H_1$ if CR $< -Z$ value | Accept $H_0$ if CR $\geq -t_a$ value<br>Reject $H_0$ and accept $H_1$ if CR $< -t_a$ value |
| Right-tailed test | Accept $H_0$ if CR $\leq +Z$ value<br>Reject $H_0$ and accept $H_1$ if CR $> +Z$ value | Accept $H_0$ if CR $\leq +t_a$ value<br>Reject $H_0$ and accept $H_1$ if CR $< +t_a$ value |

**1** Assume that the null hypothesis is $H_0$: $\mu$ = assumed value. Determine the appropriate $t$ value for the following:

    **a**   $n = 23$, $\alpha = .01$, $H_1$: $\mu <$ assumed value
    **b**   $n = 16$, $\alpha = .05$, $H_1$: $\mu \neq$ assumed value
    **c**   $n = 26$, $\alpha = .01$, $H_1$: $\mu >$ assumed value
    **d**   $n = 27$, $\alpha = .05$, $H_1$: $\mu >$ assumed value

**2** Assume that you have the following data: $H_0$: $\mu = 400$, $\bar{x} = 389$, $\hat{\sigma}_{\bar{x}} = 8$, $n = 23$, and $\alpha = .01$. Conduct a left-tailed test.

**3** Assume that you have the following data: $H_0$: $\mu = \$6{,}425$, $\bar{x} = \$6{,}535$, $\hat{\sigma}_{\bar{x}} = \$55$, $n = 27$, and $\alpha = .05$. Conduct a right-tailed test.

## ONE-SAMPLE HYPOTHESIS TESTS OF PERCENTAGES

The hypothesis-testing procedure for percentages in the large-sample case is essentially the same procedure employed for testing means with a large sample size. (This section will discuss only the large-sample case for percentage testing because the complexity of the small-sample case is beyond the scope of this book.)

*The only significant change in conducting a test of percentages rather than a test of means is in the computation of the critical ratio.* The critical ratio for percentages is computed as follows:

$$\text{CR} = \frac{p - \pi_{H0}}{\sigma_p}$$

where $\pi_{H0}$ = hypothesized value of the population percentage

The standard error of percentage is computed using the hypothesized value of $\pi$. Thus, $\sigma_p$ is computed as follows:

$$\sigma_p = \sqrt{\frac{\pi_{H0}(100 - \pi_{H0})}{n}}$$

To demonstrate the testing procedure for percentages, let us first examine a two-tailed test situation and then look at one-tailed test examples.

## Two-tailed Testing

**Example 8-9** A local newspaper, *The Weekly Daily*, has stated that only 25 percent of all college students read newspapers daily. A random sample of 200 college students showed that 45 of them were daily readers of newspapers. Test the accuracy of the newspaper's statement, and use a significance level of .05.

The *hypotheses are:*

$H_0$: $\pi = 25$ percent readership
$H_1$: $\pi \neq 25$ percent readership

This is a two-tailed test because $H_0$ may be rejected if the sample percentage is too high or too low. With a large sample size, the *Z distribution* is applicable for determining the rejection regions. With $\alpha = .05$, there is an area of .025 in each rejection region, and the appropriate $Z$ value is 1.96. Thus, the *decision rule* is:

Accept $H_0$ if CR falls between $\pm 1.96$.

*or*

Reject $H_0$ and accept $H_1$ if CR $< -1.96$ or CR $> +1.96$.

With an assumed value of 25 percent, the standard error of percentage is computed as follows:

$$\sigma_p = \sqrt{\frac{\pi_{H0}(100 - \pi_{H0})}{n}} = \sqrt{\frac{(25)(75)}{200}} = 3.1 \text{ percent}$$

And the *critical ratio* is computed as follows:

$$CR = \frac{p - \pi_{H0}}{\sigma_p} = \frac{22.5 \text{ percent} - 25 \text{ percent}}{3.1 \text{ percent}} = -.806$$

*Conclusion:* Since the CR falls between $\pm 1.96$, there is not enough evidence to reject the newspaper's assertion.

**One-tailed Testing**

**Example 8-10** The manager of the Big-Wig Executive Hair Stylists, Hugo Bald, has advertised that 90 percent of the firm's customers are satisfied with the company's services. Ms. Polly Tician, a political activist, feels that this is an exaggerated statement which might require some legal action. In a random sample of 150 of the company's clients, 132 said they were satisfied. What should be concluded if a test were conducted at the .05 level of significance?

The *hypotheses are:*

$H_0$: $\pi = 90$ percent satisfied
$H_1$: $\pi < 90$ percent satisfied

This is a *left-tailed test* because Bald's claim will be discredited only if the value of the sample percentage is significantly low. With the .05 level of significance, the rejection region is bounded by $Z = -1.64$. Thus, the *decision rule* is:

Accept $H_0$ if CR $\geq -1.64$.

*or*

Reject $H_0$ and accept $H_1$ if CR $< -1.64$.

With $\pi_{H0} = 90$ percent and $n = 150$, the standard error of percentage is computed as follows:

$$\sigma_p = \sqrt{\frac{\pi_{H0}(100 - \pi_{H0})}{n}} = \sqrt{\frac{(90)(10)}{150}} = 2.4 \text{ percent}$$

The *critical ratio* is computed as follows:

$$CR = \frac{p - \pi_{H0}}{\sigma_p} = \frac{88 \text{ percent} - 90 \text{ percent}}{2.4 \text{ percent}} = -.833$$

*Conclusion:* Since the CR is greater than $-1.64$, there is no sufficient reason to doubt Big Wig's claim. Polly should look for another cause.

**Example 8-11** The Howard Hurtz Patent Medicine Company assumes that the bottling machine is operating properly if only 5 percent of the processed bottles are not full. A random sample of 100 bottles had 7 bottles which were not full. Using a significance level of .01, conduct a statistical test to determine if the machine is operating properly.

The *hypotheses are:*

$H_0$: $\pi = 5$ percent not full
$H_1$: $\pi > 5$ percent not full

This is a right-tailed test because the company is concerned that the true percentage might be more than anticipated. With a .01 level, the *Z value* is 2.33, and therefore the *decision rule* is:

Accept $H_0$ if CR $\leqslant$ 2.33.
*or*
Reject $H_0$ and accept $H_1$ if CR > 2.33.

With $\pi_{H0} = 5$ percent and $n = 100$, the standard error of percentage is computed as follows:

$$\sigma_p = \sqrt{\frac{\pi_{H0}(100 - \pi_{H0})}{n}} = \sqrt{\frac{(5)(95)}{100}} = 2.18 \text{ percent}$$

The *critical ratio* is computed as follows:

$$CR = \frac{p - \pi_{H0}}{\sigma_p} = \frac{7 \text{ percent} - 5 \text{ percent}}{2.18 \text{ percent}} = .917$$

*Conclusion:* Since the CR is less than 2.33, the machine appears to be operating properly.

**Self-testing Review 8-6**

1  What is assumed in this section concerning the sample size in hypothesis tests of percentages?

2  What values are used in the calculation of $\sigma_p$?

3  Other than the different standard errors, is there any significant difference between the test procedure for means and the test procedure for percentages?

**SUMMARY**

Rather than simply estimating a parameter value, a sample is often taken to confirm or reject some hypothesis made about the value of a parameter. The stated hypothesis, subject to statistical test, is known as the null hypothesis, and the alternative to the null hypothesis is known as the alternative hypothesis.

In conducting a hypothesis test, the level of significance must be stated prior to sampling. The level of significance is the risk of rejecting the null hypothesis when the null hypothesis is actually true. On the basis of the specified level of significance, a rejection region may be determined. A rejection region indicates, in standard units, what values of a statistic should be considered significantly different from the assumed value. If the standardized difference between the statistic and the assumed parameter falls into a rejection region, the null hypothesis will be rejected.

Depending on the nature of the problem, an alternative hypothesis may have one of three possible forms. If the null hypothesis may be rejected by a value of a statistic which is judged to be either too high or too low, it is a two-tailed test. If there is only one rejection region in the test, it is a one-tailed test. Both one- and two-tailed tests may be employed in one-sample tests of hypotheses about population means and percentages. When testing hypotheses about population means, it is often necessary to determine the appropriate test distribution to use.

## Important Terms and Concepts

1  Null hypothesis

2  Alternative hypothesis

3  Significance level

4  $\alpha$ (alpha)

5  Rejection or critical region

6  Decision rule

7  $CR = \dfrac{\bar{x} - \mu_{H0}}{\sigma_{\bar{x}}}$

8  $CR = \dfrac{p - \pi_{H0}}{\sigma_p}$

9  Two-tailed test

10  One-tailed test

11  Left-tailed test

12  Right-tailed test

13  $\sigma_p = \sqrt{\dfrac{\pi_{H0}(100 - \pi_{H0})}{n}}$

## Problems

1  A Chamber of Commerce has stated that the average price per acre of land in Suburbanville is $3,125. A real estate salesperson, Mr. Sel N. Aker, would like to determine the veracity of the statement. A random sample of 36 acres on sale was priced at $\bar{x} = \$3,250$. Assume $\sigma = \$310$.
   a  What are the null and alternative hypotheses?
   b  Is this a two-tailed or a one-tailed test?
   c  Is it possible to conduct a hypothesis test with the data given? Why or why not?

2  Refer to problem 1. Conduct a hypothesis test with $\alpha = .05$.

3  Determine the Z value for the following:
   a  Left-tailed test, $\alpha = .05$
   b  Two-tailed test, $\alpha = .10$
   c  Right-tailed test, $\alpha = .01$
   d  $H_1: \mu <$ assumed value, $\alpha = .01$
   e  $H_1: \pi >$ assumed value, $\alpha = .05$

4  Establish the decision rules for hypothesis tests with the following data:
   a  $n = 36$, $\alpha = .05$, $H_1: \pi <$ assumed value
   b  $n = 14$, $\alpha = .01$, $H_1: \mu \neq$ assumed value
   c  $n = 23$, $\alpha = .05$, $H_1: \mu >$ assumed value
   d  $n = 46$, $\alpha = .05$, $H_1: \pi >$ assumed value

**5** The population mean has been assumed to be 600. Given the following data, conduct a left-tailed test: $\bar{x} = 592$, $n = 36$, $s = 10$, and $\alpha = .05$.

**6** It has been stated that $\mu = 69$. Conduct a right-tailed test with the following data: $\bar{x} = 75$, $n = 19$, $s = 6$, and $\alpha = .10$.

**7** The population percentage has been assumed to be 82 percent. Conduct a right-tailed test with the following data: $p = .85$, $n = 81$, and $\alpha = .05$.

**8** Know-It-All Consultants, Ltd., has stated in its promotional brochure that the average cost for its advice is $5,600 per client. Assume a random sample of 36 clients had a $\bar{x} = \$5,750$ with $s = \$175$. Conduct a test of the consultants' statement with $\alpha = .05$.

**9** Professor O. D. Statt declared that only 33 percent of a college's students have a job while attending school. A student, Fuller Doutt, thinks the professor has underestimated the zeal of his peers. A random sample of 49 students showed that 17 of them worked after school. Using $\alpha = .01$, determine who is correct.

**10** A physical education instructor, Mr. Wate Lifter, claims that his method of exercising permits an individual to do an average of 60 consecutive sit-ups after 1 week of training. Ms. Mussel, a fellow instructor, does not think Lifter is correct. A random sample of 25 individuals who have undergone the course of training had $\bar{x} = 69$ with $s = 7$. Conduct a test with $\alpha = .01$.

**11** The Big Cluck Chicken Farm states that the average weight of its chickens is 3.6 pounds. A wholesaler believes the stated value is too high. A sample of 24 chickens had $\bar{x} = 3.3$ pounds with $s = .25$. Conduct a test with $\alpha = .05$.

**12** On the average, a well-functioning machine should produce items which are 90 percent acceptable. If the percentage is less than 90 percent, the machine must be repaired. A sample of 100 items has $p = 87$ percent. Determine if this machine should be repaired, using a significance level of .05.

**13** Wild Guess Surveys, Inc., reported in a preelection poll that 68 percent of the voters favor Mr. M. T. Hedd for public office. A sample of 36 voters showed that 26 of them would vote for Mr. Hedd. What can be concluded about the accuracy of the report, using a significance level of .05?

**Topics for Discussion**

**1** What is meant by a significant difference in hypothesis testing?

**2** What does a significance level of .05 actually mean?

**3** Why should the term "accept" be taken lightly in hypothesis testing?

**4** What is a rejection region?

**5** When should a one-tailed test be employed?

**6** Indicate whether the $Z$ or the $t$ distribution applies in each of the following:
  **a** $n = 16$, $s = 24$
  **b** $n = 19$, $\sigma = 15$
  **c** $n = 43$, $\sigma = 98$
  **d** $n = 102$, $s = 48$

**7** How can a hypothesis be proved?

**8-1**

1 A null hypothesis is the assumption of a parameter value which is subject to a statistical test.

2 An alternative hypothesis is the conclusion to be drawn contingent on the rejection of the null hypothesis.

3 A significance level is the level at which the null hypothesis would be rejected when it is actually true.

4 The $Z$ distribution is applicable when $\sigma$ is known or when the sample size is large.

5 A significant difference is the size of the difference between an obtained statistic and the assumed value which warrants rejection of the null hypothesis.

6 A rejection region is the area of the sampling distribution which leads to rejection of a null hypothesis.

7 The difference is standardized through the use of the critical ratio.

8 The statement is false because a hypothesis can never be proved. Since knowledge of the parameter value is the means of proving the hypothesis, a statistical test merely provides evidence not to reject a hypothesis.

9 This statement is only partially correct. Statistical results merely serve as one form of information input in decision making.

**8-2**

1 A two-tailed test rejects the null hypothesis if $\bar{x}$ is significantly too low or significantly too high compared to the assumed value. The conclusion on rejection of the null hypothesis is simply that the assumed value is not the true value.

2 a $Z = 2.58$
 b $Z = 1.75$
 c $Z = 1.96$
 d $Z = 2.17$

3 The hypotheses are:
$H_0$: $\mu = 500$
$H_1$: $\mu \neq 500$
and with $\alpha = .01$, $Z = 2.58$.
The decision rule is:
Accept $H_0$ if CR falls between $\pm 2.58$.
*or*
Reject $H_0$ and accept $H_1$ if CR $> 2.58$ or CR $< -2.58$.
The critical ratio is computed as follows:

$$CR = \frac{\bar{x} - \mu_{H0}}{\sigma/\sqrt{n}}$$

$$= \frac{475 - 500}{50/\sqrt{36}}$$

*Conclusion:* Reject $H_0$.　　　　$= -3.00$

**8-3**

1 In a one-tailed test, there is only one rejection region. The alternative hypothesis states that the true value is likely to be higher or lower than the assumed value.

2 For a left-tailed test, $H_1$: $\mu$ < assumed value. For a right-tailed test, $H_1$: $\mu$ > assumed value.

3   **a**   2.58
    **b**   1.96
    **c**   $-1.64$
    **d**   2.33

4 The hypotheses are:
$H_0$: $\mu = 100$
$H_1$: $\mu < 100$
and with $\alpha = .05$, $Z = -1.64$.
The decision rule for this left-tailed test is:
Accept $H_0$ if CR $\geqslant -1.64$.
*or*
Reject $H_0$ and accept $H_1$ if CR $< -1.64$.
The critical ratio is computed as follows:

$$CR = \frac{88 - 100}{15/\sqrt{36}} = -4.8$$

*Conclusion:* Reject $H_0$.

5 The hypotheses are:

$H_0$: $\mu = 24$
$H_1$: $\mu > 24$

and with $\alpha = .01$, $Z = +2.33$.
The decision rule for this right-tailed test is:

Accept $H_0$ if CR $\lesseqgtr 2.33$.
*or*
Reject $H_0$ and accept $H_1$ if CR $> 2.33$.

The critical ratio is computed as follows:

$$CR = \frac{26 - 24}{3/\sqrt{16}} = 2.67$$

*Conclusion:* Reject $H_0$.

6   $H_0$: $\mu = 500$
    $H_1$: $\mu < 500$

7   $H_0$: $\mu = 2.5$
    $H_1$: $\mu < 2.5$

8 This statement is false. With the structure of a right-tailed test, the true value may be less than the assumed value, but the null hypothesis will not be rejected.

**8-4**

**1** They differ in the following manner:

**a**  The normal (Z) distribution is used regardless of sample size if $\sigma$ is known; with $\sigma$ unknown, normality cannot be assumed and sample size determines the test distribution to use.

**b**  The critical ratio must be computed with $\hat{\sigma}_{\bar{x}}$ instead of $\sigma_{\bar{x}}$.

**2**  The hypotheses are:

$H_0$: $\mu = 612$
$H_1$: $\mu \neq 612$

and with $\alpha = .05$ and $n = 13$, the $t$ value with 12 degrees of freedom is $t_{025} = 2.179$.

The appropriate decision rule is:

Accept $H_0$ if CR falls between $\pm 2.179$.

*or*

Reject $H_0$ and accept $H_1$ if CR $> 2.179$ or CR $< -2.179$.

The estimate of $\sigma_{\bar{x}}$ is computed as follows:

$$\hat{\sigma}_{\bar{x}} = \frac{s}{\sqrt{n-1}} = \frac{5}{\sqrt{13-1}} = 1.44$$

The critical ratio is computed as follows:

$$CR = \frac{\bar{x} - \mu_{H0}}{\hat{\sigma}_{\bar{x}}} = \frac{608 - 612}{1.44} = -2.77$$

*Conclusion:* Reject $H_0$.

**3**  The hypotheses are:

$H_0$: $\mu = 243$
$H_1$: $\mu \neq 243$

and with $\alpha = .01$ and $n = 36$, the $Z$ value is 2.58.

The decision rule is:

Accept $H_0$ if CR falls between $\pm 2.58$.

*or*

Reject $H_0$ and accept $H_1$ if CR $> 2.58$ or CR $< -2.58$.

The estimate of $\sigma_{\bar{x}}$ is computed as follows:

$$\hat{\sigma}_{\bar{x}} = \frac{s}{\sqrt{n-1}} = \frac{15}{\sqrt{36-1}} = 2.54$$

The critical ratio is computed as follows:

$$CR = \frac{\bar{x} - \mu_{H0}}{\hat{\sigma}_{\bar{x}}} = \frac{269 - 243}{2.54} = 10.24$$

*Conclusion:* Reject $H_0$.

**8-5**

**1 a** $-2,508$

   **b** 2.131

   **c** 2.485

   **d** 1.706

**2** The hypotheses are

$H_0$: $\mu = 400$
$H_1$: $\mu < 400$

and with $\alpha = .01$, $n = 23$, and a left-tailed test, the $t$ value with 22 degrees of freedom is $t_{01} = -2.508$.
The decision rule is:

Accept $H_0$ if CR $\geqslant -2.508$.
*or*
Reject $H_0$ and accept $H_1$ if CR $< -2.508$.

The critical ratio is computed as follows:

$$CR = \frac{\bar{x} - \mu_{H0}}{\hat{\sigma}_{\bar{x}}} = \frac{389 - 400}{8} = -1.38$$

*Conclusion:* Accept $H_0$.

**3** The hypotheses are:

$H_0$: $\mu = \$6,425$
$H_1$: $\mu > \$6,425$

and with $\alpha = .05$, $n = 27$, and a right-tailed test, the $t$ value with 26 degrees of freedom is $t_{05} = 1.706$.
The decision rule is:

Accept $H_0$ if CR $\leqslant 1.706$.
*or*
Reject $H_0$ and accept $H_1$ if CR $> 1.706$.

The critical ratio is computed as follows:

$$CR = \frac{\bar{x} - \mu_{H0}}{\hat{\sigma}_{\bar{x}}} = \frac{\$6,535 - \$6,425}{\$55} = 2.00$$

*Conclusion:* Reject $H_0$.

**8-6**

**1** In this book we deal with only the large-sample case in the testing of percentages.

**2** The values of $\pi_{H0}$ and $n$ must be used in the calculation of $\sigma_p$.

**3** Essentially there is no difference.

# CHAPTER 9

# TESTING HYPOTHESES AND MAKING DECISIONS: TWO-SAMPLE PROCEDURES

**LEARNING OBJECTIVES**

After studying this chapter, working the problems, and answering the discussion questions, you should be able to:

☞ Explain *(a)* the purpose of two-sample hypothesis tests of means and percentages and *(b)* the procedures to be followed in conducting these tests.

☞ Understand the concepts associated with both the sampling distribution of the differences between sample means and the sampling distribution of the differences between sample percentages. You should also be able to explain how such sampling distributions may be created.

☞ Perform the necessary computations and make the appropriate statistical decisions in two-sample hypothesis-testing situations.

**CHAPTER OUTLINE**

203

Decision makers are often interested in determining whether two populations are similar or different with respect to some characteristic. For example, an instructor may be curious as to whether male professors receive higher salaries than female professors for the same teaching load. Or a psychologist may want to determine objectively whether one group responds differently to an experimental stimulus than another group. Or the purchasing agent of a firm that manufactures cooling towers may be interested in learning whether the cooling fan motors of one supplier are more durable than those of another vendor. In short, there are many situations which require groups to be compared on the basis of a given trait or characteristic.

Chapter 8 discussed procedures for testing the validity of an assumed value of a parameter. The assumed value was a single, absolute, specific value which was sub-jected to statistical testing. In this chapter, however, we are concerned with the parameters of *two populations,* but we are not primarily interested in estimating the *absolute values* of the parameters. Rather, the topic of interest is the *relative values* of the parameters. That is, does one population appear to possess more or less of a trait than the other?

The purpose of this chapter, then, is to use the hypothesis-testing procedures introduced in the last chapter along with the data obtained from two *independent samples*[1] to make relative comparisons between (1) *two population means* and (2) *two population percentages.* In order to maintain simplicity of presentation, how-ever, we will only consider situations where the samples are relatively large (over 30) and where it is thus possible (by the use of the Central Limit Theorem) to assume that the appropriate sampling distributions are approximated by normal curves.

---

[1] The requirement that the samples be *independent*—i.e., that they be taken from different groups and that the sample selected from the first group not be related to the sample selected from the second group—is an important prerequisite for the use of the procedures discussed in this chapter. Thus, a study designed to measure the effects of an intensive statistics review course by first testing a sample group of students before they take the review course and by then retesting the same students after the course is completed could not utilize the procedures presented in this chapter. You will be delighted (?) to know, however, that such a "before" and "after" study can be made if test procedures described in Chapter 15 are used.

## TWO-SAMPLE HYPOTHESIS TESTS OF MEANS

Our purpose here in conducting two-sample hypothesis tests of means is to see, through the use of sample data, if there is likely to be a statistically significant difference between the means of two populations. The null hypothesis in such tests of the differences between two means is:

$$H_0:\ \mu_1 = \mu_2$$

The null hypothesis states that the true mean of the first population is equal to the true mean of the second population. *In essence, $H_0$ states that the populations are the same on the basis of a given characteristic.*

If the null hypothesis cannot be supported, there are three possible alternative hypotheses which may be concluded:

$H_1:\ \mu_1 \neq \mu_2$      two-tailed alternative
$H_1:\ \mu_1 > \mu_2$      right-tailed alternative
$H_1:\ \mu_1 < \mu_2$      left-tailed alternative

If the nature of the problem simply indicates that it is sufficient to reject the null hypothesis without further inferences about the differences between $\mu_1$ and $\mu_2$, a two-tailed test is used. If the nature of the problem indicates that $H_0$ should be rejected only on the basis that $\mu_1$ is significantly higher than $\mu_2$, a right-tailed test is employed. And if there are indications that the first population might possibly be less than the second population on a given characteristic, a left-tailed test is applied.

## The Sampling Distribution of the Differences between Sample Means

Now, let's discuss the underlying concepts in testing for equality between groups. When the null hypothesis states that the true mean of group 1 is equal to the true mean of group 2 ($\mu_1 = \mu_2$), it is essentially asserting that the *difference between the parameters of the two groups is zero*—i.e., $\mu_1 - \mu_2 = 0$. This idea makes it possible for us to visualize yet another type of sampling distribution—*the sampling distribution of the differences between means.*

The conceptual schematic of this new sampling distribution in Fig. 9-1 may help your understanding. Distribution A in Fig. 9-1 is the sampling distribution of the means for population 1, and distribution B is the corresponding sampling distribution for population 2. As we have seen, each of these theoretical distributions is developed from the means of all the possible samples of a given size that could be drawn from a population. Now if we were to select a single sample mean from distribution A and another sample mean from distribution B, we could subtract the value of the second mean from the value of the first mean and get a difference—i.e., $\bar{x}_1 - \bar{x}_2 =$ difference. This difference could be either a negative or positive value, as shown in the examples between distributions A and B in Fig. 9-1. We could theoretically continue to select sample means from each population and continue to compute differences until we reached an advanced stage of senility. If we then constructed a frequency distribution of *all the sample differences,* we could have distribution C in Fig. 9-1, which is the *sampling distribution of the differences be-*

**FIGURE 9-1**

Conceptual
schematic of the
sampling distribution
of the differences
between sample
means

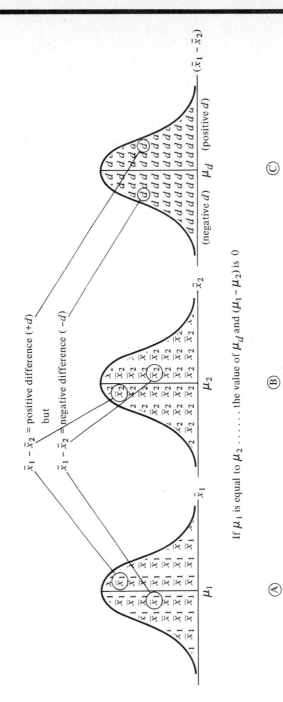

$\bar{x}_1 - \bar{x}_2 =$ positive difference $(+d)$

but

$\bar{x}_1 - \bar{x}_2 =$ negative difference $(-d)$

(negative $d$)   $\mu_d$ (positive $d$)

$(\bar{x}_1 - \bar{x}_2)$

$\bar{x}_1$   $\mu_1$

$\bar{x}_2$   $\mu_2$

If $\mu_1$ is equal to $\mu_2$ . . . . . . the value of $\mu_d$ and $(\mu_1 - \mu_2)$ is 0

(A)

(B)

(C)

Ⓐ = Sampling distribution of the means — population 1
Ⓑ = Sampling distribution of the means — population 2
Ⓒ = Sampling distribution of the differences between sample means

*tween sample means.* And, as noted in Fig. 9-1, *if $H_0$ is true, and if $\mu_1$ is equal to $\mu_2$, then* the value of the mean of the sampling distribution of the differences ($\mu_d$) will be equal to $\mu_1 - \mu_2$—i.e., $\mu_d$ will have a value of zero. In short, the negative differences and the positive differences would cancel, and the mean would be zero.

If it is assumed that the mean of the sampling distribution of differences is zero, the characteristics of the sampling distribution of differences allow for variation in the value of $\bar{x}_1 - \bar{x}_2$ from zero. If the parameters were truly equal, and if samples were taken from the two populations, it would be highly unlikely that the difference of $\bar{x}_1 - \bar{x}_2$ would exactly equal zero; there would be some sampling variation. However, if the true means were equal, the likelihood of an extremely *large* difference between $\bar{x}_1$ and $\bar{x}_2$ would be small. *Thus, if an extremely large difference were to occur between $\bar{x}_1$ and $\bar{x}_2$, it would be justifiable to conclude that the true means are not equal.* The immediate problem, of course, is to determine when the difference between samples becomes significant so that the null hypothesis of equality of $\mu_1$ and $\mu_2$ can be rejected.

If you have absorbed the discussion in Chapter 8, you will notice some familiarity between what is presented here and what has been presented previously. In a test for differences between means, we look for a significant difference between $\bar{x}_1$ and $\bar{x}_2$, which leads to rejection of the null hypothesis.

In the case of a large sample from both populations, the shape of the sampling distribution of the differences between means is approximately normal. Thus, the middle 68.26 percent of the differences in the sampling distribution will be found within one standard deviation of the mean, and two standard deviations to either side of the mean would account for 95.4 percent of the differences. The standard deviation of the sampling distribution of differences goes by the horrendous name *standard error of the difference between means* and is identified by the symbol $\sigma_{\bar{x}_1 - \bar{x}_2}$ (see Fig. 9-2).

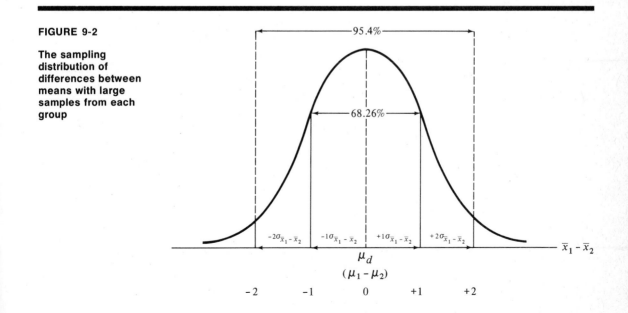

**FIGURE 9-2**

**The sampling distribution of differences between means with large samples from each group**

With the use of the $Z$ table, it is possible to establish the significant difference level once the risk of erroneously rejecting the null hypothesis has been stated. And once a level of significance ($\alpha$) has been stated, the boundaries of the rejection region of the distribution may be established. The process of establishing rejection regions and the construction of decision rules are exactly the same as discussed in Chapter 8, when the $Z$ distribution was applicable. For example, with a two-tailed test and a .05 level of significance, the $Z$ values for the boundaries of the rejection regions are $\pm 1.96$. For a left-tailed test and a .05 level, the $Z$ value is $-1.64$.

As in the case of other hypothesis tests in which the rejection regions are determined by the $Z$ values, the test between means must use the standardized difference between $\bar{x}_1$ and $\bar{x}_2$; that is, a *critical ratio must be calculated. The critical ratio for the difference between means is computed as follows:*

$$CR = \frac{(\bar{x}_1 - \bar{x}_2) - (\mu_1 - \mu_2)}{\sigma_{\bar{x}_1 - \bar{x}_2}} \quad \text{or} \quad CR = \frac{\bar{x}_1 - \bar{x}_2}{\sigma_{\bar{x}_1 - \bar{x}_2}} \quad (9\text{-}1)$$

since $\mu_1 - \mu_2$ is assumed to be zero if $H_0$ is true. And (if it is assumed that we have two large independently drawn random samples) the standard error of the difference is computed as follows:

$$\sigma_{\bar{x}_1 - \bar{x}_2} = \sqrt{\frac{\sigma_1^2}{n_1} + \frac{\sigma_2^2}{n_2}} \quad (9\text{-}2)$$

Let's now look at the following examples, which illustrate two-tailed and one-tailed tests.

**Two-tailed Testing When $\sigma_1$ and $\sigma_2$ Are Known**

As noted above, the steps that are followed when making two-sample hypothesis tests are the same as the ones introduced for the general hypothesis-testing procedure in Chapter 8. We can illustrate this fact with the following two-tailed testing situation.

The Russ Trate Traffic Signal Company has decided to install microcomputers in its traffic light assemblies in order to more efficiently monitor and control traffic flows. Microcomputers from two suppliers are judged to be suitable for the application. In order to have more than one source of supply, the Trate Company would prefer to buy microcomputers from both suppliers, provided there is no significant difference in durability. A random sample of 35 computer assemblies of brand A and a sample of 32 computers of brand B are tested. The mean time between failure (MTBF) for the brand A computers is found to be 2,800 hours, and the MTBF for the brand B units is found to be 2,750 hours. Information from industry sources indicates that the population standard deviation is 200 hours for brand A and 180 hours for brand B. At the .05 level of significance, is there a difference in durability?

The *hypotheses are:*

$H_0: \mu_1 = \mu_2$
$H_1: \mu_1 \neq \mu_2$

Since the Trate Company is only interested in testing for a significant difference, this is a two-tailed test. Also, the level of significance is specified at the .05 level, and the large sample sizes enable us to use the $Z$ distribution. Thus, the rejection regions are bounded by $Z = \pm 1.96$, and the *decision rule is:*

> Accept $H_0$ if CR falls between $\pm 1.96$.
>
> *or*
>
> Reject $H_0$ and accept $H_1$ if CR $< -1.96$ or CR $> +1.96$.

With $\sigma_1 = 200$ hours, $n_1 = 35$, $\sigma_2 = 180$ hours, and $n_2 = 32$,

$$\sigma_{\bar{x}_1 - \bar{x}_2} = \sqrt{\frac{\sigma_1^2}{n_1} + \frac{\sigma_2^2}{n_2}}$$

$$= \sqrt{\frac{200^2}{35} + \frac{180^2}{32}}$$

$$= 46.43 \text{ hours}$$

Therefore, the *critical ratio is* computed as follows:

$$\text{CR} = \frac{\bar{x}_1 - \bar{x}_2}{\sigma_{\bar{x}_1 - \bar{x}_2}} = \frac{2,800 - 2,750}{46.43} = 1.08$$

*Conclusion:* Since the critical ratio falls within the acceptance region of $\pm 1.96$, we can conclude that there is no significant difference in the durability of the two microcomputer brands.

**One-tailed Testing When $\sigma_1$ and $\sigma_2$ Are Known**

Discount Stores Corporation owns outlet A and outlet B. For the past year, outlet A has spent more dollars on advertising widgets than outlet B. The corporation wants to determine if the advertising has resulted in more sales for outlet A. A sample of 36 days at outlet A had a mean of 170 widgets sold daily. A sample of 36 days at outlet B had a mean of 165. Assuming $\sigma_1^2 = 36$ and $\sigma_2^2 = 25$, what would be concluded if a test were conducted at the .05 level of significance?

The *hypotheses are:*

> $H_0$: $\mu_1 = \mu_2$
>
> $H_1$: $\mu_1 > \mu_2$

This is a *right-tailed test* because the corporation wants to find out if the performance of group 1 (outlet A) is better than the performance of group 2. With a .05 level, the $Z$ value that bounds the rejection region is 1.64. Thus, the *decision rule is:*

> Accept $H_0$ if CR $\leq 1.64$.
>
> *or*
>
> Reject $H_0$ and accept $H_1$ if CR $> 1.64$.

The *critical ratio* is computed as follows:

$$\text{CR} = \frac{\bar{x}_1 - \bar{x}_2}{\sigma_{\bar{x}_1 - \bar{x}_2}} = \frac{170 - 165}{\sqrt{36/36 + 25/36}} = \frac{5}{1.3017} = 3.84$$

*Conclusion:* Since CR is more than 1.64, there is sufficient reason to believe that outlet A has sold more widgets than outlet B.

1 "The purpose of the two-sample hypothesis-testing procedures presented in this chapter is to determine the absolute values of the parameters." Comment on this statement.

2 What conditions must be met by the samples if the procedures described in this chapter are to remain valid?

3 Two brands of truck tires are being compared by a trucking firm. A sample of tires of brand X (group 1) had an average life of 45,000 miles, while the average life of a sample of 40 brand Y tires (group 2) was 46,500 miles. Assuming that $\sigma_1$ was 2,000 miles and $\sigma_2$ was 1,500 miles, is there any significant difference in quality at the .01 level?

4 Two brands of batteries are being compared by the same trucking company. A sample of 36 Die Dead batteries had an average life of 42 months, while the average life of a sample of 35 Crank-Up batteries was 45 months. Assuming that the $\sigma$ of Die Dead batteries is 6 months and the $\sigma$ of Crank-Up batteries is 3 months, are the Die Dead batteries likely to be inferior to the Crank-Up brand at the .05 level?

**Two-tailed
Testing When
$\sigma_1$ and $\sigma_2$ Are
Unknown**

When the population standard deviations are unknown—the usual situation—the sample standard deviations may be used to estimate the population values in the following way:[2]

$$\hat{\sigma}_1 = s_1 \sqrt{\frac{n_1}{n_1 - 1}} \quad \text{and} \quad \hat{\sigma}_2 = s_2 \sqrt{\frac{n_2}{n_2 - 1}} \tag{9-3}$$

And an estimated standard error of the difference between means is then computed as follows:

$$\hat{\sigma}_{\bar{x}_1 - \bar{x}_2} = \sqrt{\frac{\hat{\sigma}_1{}^2}{n_1} + \frac{\hat{\sigma}_2{}^2}{n_2}} \tag{9-4}$$

Let's now consider the following example.

Dr. I. M. Sain, a psychologist, administered IQ tests to determine if female college students were as smart as male students. The sample of 40 females had a mean score of 131 with a standard deviation of 15. The sample of 36 males had a mean of 126 and a standard deviation of 17. At the .01 level of significance, is there a difference?

The *hypotheses are:*

---

[2] It is assumed that the sample standard deviations have been computed using this formula:

$$s = \sqrt{\frac{\Sigma (x - \bar{x})^2}{n}}$$

$$H_0: \mu_1 = \mu_2$$
$$H_1: \mu_1 \neq \mu_2$$

Since Dr. Sain is only interested in concluding equality or nonequality between groups, this is a two-tailed test. And since our sample sizes are large, the $Z$ distribution may be used. Thus, the rejection regions are bounded by $Z = -2.58$ and $+2.58$, and the *decision rule is:*

Accept $H_0$ if CR falls between $\pm 2.58$.

*or*

Reject $H_0$ and accept $H_1$ if CR $< -2.58$ or CR $> +2.58$.

With $s_1 = 15$, $n_1 = 40$, $s_2 = 17$, and $n_2 = 36$,

$$\hat{\sigma}_1 = s_1 \sqrt{\frac{n_1}{n_1 - 1}} \qquad \hat{\sigma}_2 = s_2 \sqrt{\frac{n_2}{n_2 - 1}}$$
$$= 15\sqrt{40/39} \qquad = 17\sqrt{36/35}$$
$$= 15.19 \qquad = 17.24$$

Therefore, the *critical ratio* is computed as follows:

$$\text{CR} = \frac{\bar{x}_1 - \bar{x}_2}{\sqrt{\dfrac{\hat{\sigma}_1{}^2}{n_1} + \dfrac{\hat{\sigma}_2{}^2}{n_2}}} = \frac{131 - 126}{\sqrt{\dfrac{15.19^2}{40} + \dfrac{17.24^2}{36}}} = \frac{5}{3.745} = 1.34$$

*Conclusion:* Since the critical ratio falls between $\pm 2.58$, we can conclude that there is no significant difference. (In IQ, that is.)

**One-tailed Testing When $\sigma_1$ and $\sigma_2$ Are Unknown**

A Chamber of Commerce is seeking to attract new industry to its area. One argument it has been using is that average wages paid for a particular type of job are lower than in other parts of the nation. A rather skeptical company president assigns his brother-in-law the task of testing this claim. A sample of 60 workers (group 1) performing the particular job in the Chamber's area is taken, and the sample mean is found to be $7.75 per hour with a sample standard deviation of $2.00 per hour. Another sample of 50 workers (group 2) taken in region NE produced a sample mean of $8.25 per hour with a sample standard deviation of $1.25 per hour. At the .01 level, what report should the brother-in-law give to the president?

The *hypotheses are:*

$$H_0: \mu_1 = \mu_2$$
$$H_1: \mu_1 < \mu_2$$

This is a one-tailed test because the validity of the Chamber's claim is being evaluated—i.e., that wages paid are less than in other areas. At the .01 level, the $Z$ value that bounds the rejection region is $-2.33$. Thus, the *decision rule is:*

Accept $H_0$ if CR $\geq -2.33$.

*or*

Reject $H_0$ and accept $H_1$ if CR $< -2.33$.

With $s_1 = \$2.00$, $n_1 = 60$, $s_2 = \$1.25$, and $n_2 = 50$,

$$\hat{\sigma}_1 = \$2.00\sqrt{60/59} \qquad \hat{\sigma}_2 = \$1.25\sqrt{50/49}$$

$$= \$2.02 \qquad\qquad\qquad = \$1.26$$

Therefore, the *critical ratio* is computed as follows:

$$\text{CR} = \frac{\bar{x}_1 - \bar{x}_2}{\sqrt{\dfrac{\hat{\sigma}_1^2}{n_1} + \dfrac{\hat{\sigma}_2^2}{n_2}}} = \frac{\$7.75 - \$8.25}{\sqrt{\dfrac{\$2.02^2}{60} + \dfrac{\$1.26^2}{50}}} = \frac{-\$0.50}{\$0.316} = -1.58$$

*Conclusion:* Since CR falls into the area of acceptance, the Chamber's claim is not supported by the sample results at the .01 level. The brother-in-law is relieved to report that the test results support the president's doubts.

**Self-testing Review 9-2**

1 Assume the following data are available:

| | Group | |
|---|---|---|
| | 1 | 2 |
| $\bar{x}$ | $5,600 | $5,300 |
| $s$ | $1,120 | $ 725 |
| $n$ | 38 | 36 |

Conduct a two-tailed test at the .10 level.

2 What would be the hypotheses and the decision rule if a right-tailed test had been made at the .10 level using the data in problem 1 above?

**TWO-SAMPLE HYPOTHESIS TESTS OF PERCENTAGES**

The purpose of conducting two-sample hypothesis tests of percentages is to see, through the use of sample data, if there is likely to be a statistically significant difference between the percentages of two populations. The null hypothesis in such tests of the differences between percentages is:

$H_0: \pi_1 = \pi_2$

If the null hypothesis cannot be supported, there are three possible alternative hypotheses which may be accepted:

$H_1: \pi_1 \neq \pi_2$      two-tailed alternative

*or*

$H_1: \pi_1 > \pi_2$      right-tailed alternative

*or*

$H_1: \pi_1 < \pi_2$      left-tailed alternative

| **The Sampling Distribution of the Differences between Sample Percentages** | The *sampling distribution of the differences between sample percentages* is theoretically analogous to the sampling distribution of the differences between sample means. The mean of the sampling distribution of the differences between percentages is zero *if the null hypothesis is true*—i.e., if $\pi_1 = \pi_2$. If the sample size of each group is large, the shape of the sampling distribution is approximately normal, and the standard deviation of the sampling distribution of the differences between percentages is calculated as follows:[3] |
|---|---|

$$\sigma_{p_1 - p_2} = \sqrt{\frac{\pi_1(100 - \pi_1)}{n_1} + \frac{\pi_2(100 - \pi_2)}{n_2}}$$

Unfortunately, the computation of $\sigma_{p_1 - p_2}$ requires a knowledge of the parameters. If these values were known in the first place, there would be no need to conduct a test! Therefore, *in a test of differences between percentages the estimate $\hat{\sigma}_{p_1 - p_2}$ must always be used:*

$$\hat{\sigma}_{p_1 - p_2} = \sqrt{\frac{p_1(100 - p_1)}{n_1 - 1} + \frac{p_2(100 - p_2)}{n_2 - 1}} \qquad (9\text{-}5)$$

And the *critical ratio* is computed as follows:

$$\mathrm{CR} = \frac{(p_1 - p_2) - (\pi_1 - \pi_2)}{\hat{\sigma}_{p_1 - p_2}} \quad \text{or} \quad \mathrm{CR} = \frac{p_1 - p_2}{\hat{\sigma}_{p_1 - p_2}} \qquad (9\text{-}6)$$

since $\pi_1 - \pi_2$ will equal zero if the null hypothesis is true.

| **Testing the Differences between Percentages** | The procedure for testing the differences between percentages is not any different than the test for differences between means. Consequently, there is little need here for further discussion of this matter. The following is an example of a two-tailed test. (You should by now be able to determine for yourself the procedure for a one-tailed test.) |
|---|---|

Ken Kharisma, candidate for public office, feels that male voters as well as female voters have the same opinion of him. A sample of 36 male voters had 12 persons favoring his election. A sample of 49 female voters had a $p_2$ of 36 percent. Test the validity of Ken's assumption, using a significance level of .05.

The *hypotheses are:*

$H_0$: $\pi_1 = \pi_2$

$H_1$: $\pi_1 \neq \pi_2$

This is a two-tailed test because Ken is only interested in the equality or non-equality of opinions between the two groups. The $Z$ distribution is applicable, and the rejection regions are bounded by $Z = \pm 1.96$.

---

[3] You guessed it—the formal name of this standard deviation is the *standard error of the difference between percentages.*

The *decision rule is:*

Accept $H_0$ if CR falls between $\pm 1.96$.

*or*

Reject $H_0$ and accept $H_1$ if CR $< -1.96$ or CR $> +1.96$.

And the *critical ratio* is calculated as follows:

$$CR = \frac{p_1 - p_2}{\sqrt{\dfrac{p_1(100 - p_1)}{n_1 - 1} + \dfrac{p_2(100 - p_2)}{n_2 - 1}}}$$

$$= \frac{33 \text{ percent} - 36 \text{ percent}}{\sqrt{\dfrac{(33)(67)}{35} + \dfrac{(36)(64)}{48}}}$$

$$= \frac{-3.00 \text{ percent}}{10.54}$$

$$= -.28$$

*Conclusion:* Since the critical ratio is between $\pm 1.96$, there is no reason to reject Ken Kharisma's claim. Apparently, both sexes have about the same low opinion of Ken!

**Self-testing Review 9-3**

1 Voters in two cities are polled to determine if they are in favor of a proposal to limit the amount of property tax revenue that the state can collect. A sample of 200 is taken in each city, and 120 voters in city A are in favor of the proposal, while 109 voters in city B favor it. At the .05 level, is there a significant difference in voter opinion between the two cities?

2 A television producer believes her new program will be more popular with urban viewers than with rural viewers. To test this claim, a network shows the program to 300 urban viewers and 100 rural viewers. It is noted that 65 of the urban viewers and 18 of the rural viewers enjoy the program. At the .05 level, should the producer's claim be accepted?

**SUMMARY**

Our purpose in this chapter has been to use hypothesis-testing procedures to make relative comparisons between two population means or two population percentages. The null hypothesis to be tested in both situations is that the two parameters are equal—i.e., there is no significant difference between the two parameters. The alternative hypothesis may be either one-tailed or two-tailed depending on the logic of the situation. Examples of both one- and two-tailed situations have been presented in the chapter.

The sampling distributions of the differences between sample means and the differences between sample percentages are used in analyzing whether the differences between two sample values (and thus two parameter values) are statistically significant or whether these differences can logically be explained by sampling variation. Although the computation of the critical ratio in the two-sample examples presented in this chapter requires the use of standard deviation measures taken from the sampling distributions of the differences between means and between percentages, the other steps in the testing procedure are similar to those presented in Chapter 8.

**Important Terms and Concepts**

1 Independent samples

2 Sampling distribution of the differences between sample means

3 Standard error of the difference between means

4 $CR = \dfrac{\bar{x}_1 - \bar{x}_2}{\sigma_{\bar{x}_1 - \bar{x}_2}}$

5 $\sigma_{\bar{x}_1 - \bar{x}_2} = \sqrt{\dfrac{\sigma_1^2}{n_1} + \dfrac{\sigma_2^2}{n_2}}$

6 $\hat{\sigma}_{\bar{x}_1 - \bar{x}_2} = \sqrt{\dfrac{\hat{\sigma}_1^2}{n_1} + \dfrac{\hat{\sigma}_2^2}{n_2}}$

7 $CR = \dfrac{p_1 - p_2}{\hat{\sigma}_{p_1 - p_2}}$

8 $\hat{\sigma}_{p_1 - p_2} = \sqrt{\dfrac{p_1(100 - p_1)}{n_1 - 1} + \dfrac{p_2(100 - p_2)}{n_2 - 1}}$

9 Sampling distribution of the differences between sample percentages

10 Standard error of the difference between percentages

**Problems**

1 The following sample data have been gathered by a researcher employed by a department store chain:

|  | Downtown store | Ritzy Mall store | Discount City store | Rural Retreat store | Ski Texas store | Gunn Mall store |
|---|---|---|---|---|---|---|
| Average purchase amount ($\bar{x}$) | $36.00 | $40.00 | $33.50 | $28.25 | $22.80 | $26.00 |
| Population standard deviations | $ 6.00 | $ 8.20 | $ 9.50 | — | — | — |
| Sample standard deviations | — | — | — | $10.15 | $10.50 | $ 8.75 |
| Sample size | 40 | 38 | 32 | 42 | 50 | 58 |

a At the .05 level, is there a significant difference in average purchase amounts between the downtown store and the Ritzy Mall store?

b The manager of the downtown store is convinced that the average amount of purchases made at her store is higher than at the Discount City store. Is this belief supported at the .01 level of significance?

c At the .01 level, is there a significant difference in average purchase amounts between the Ski Texas store and the Gunn Mall store?

d Is the average purchase amount significantly greater at the Rural Retreat store than at the Gunn Mall store? Use the .10 level of significance.

**2** The public relations manager of Tailspin Airlines (TA) is concerned about a recent increase in the number of customer claims of damage to luggage. A sampling of the records at three TA terminals yields the following data:

|  | Bayburg terminal | Pitt City terminal | Beantown terminal |
|---|---|---|---|
| Sample count of luggage handled | 760 | 610 | 830 |
| Number of items damaged | 44 | 53 | 60 |

**a** At the .05 level, is there a significant difference in damage claims between Bayburg and Pitt City?

**b** The terminal manager at Bayburg believes that the baggage handling at Beantown is sloppy and that his terminal experiences lower damage claims. At the .01 level, would you agree?

**Topics for Discussion**

**1** "The purpose of this chapter is not to estimate the absolute values of parameters." Discuss this statement.

**2** What assumptions must be made about the samples if the procedures discussed in this chapter are to be properly employed?

**3 a** What is a sampling distribution of the differences between sample means?
**b** Explain how such a sampling distribution is created.

**4** When will the mean of the sampling distribution of the differences between sample means equal zero?

**5 a** What is a sampling distribution of the differences between sample percentages?
**b** Explain how such a distribution would be created.
**c** If the null hypothesis is true, what would be the value of the mean of this distribution?

**6** Why must an estimated standard error of the difference between percentages be used in hypothesis testing?

**Answers to Self-testing Review Questions**

9-1

**1** This statement is false. The purpose here is to determine the *relative* values of the parameters.

**2** The samples must be independent samples selected from different groups, and they must be large—i.e., over 30.

**3** The hypotheses are:

$H_0: \mu_1 = \mu_2$
$H_1: \mu_1 \neq \mu_2$

And at the .01 level, $Z = 2.58$.
The decision rule is:

Accept $H_0$ if CR falls between $\pm 2.58$.
*or*
Reject $H_0$ and accept $H_1$ if CR $> +2.58$ or CR $< -2.58$.

The critical ratio is computed as follows:

$$CR = \frac{\bar{x}_1 - \bar{x}_2}{\sqrt{\dfrac{\sigma_1{}^2}{n_1} + \dfrac{\sigma_2{}^2}{n_2}}} = \frac{45,000 - 46,500}{\sqrt{\dfrac{2,000^2}{50} + \dfrac{1,500^2}{40}}} = \frac{-1,500}{369.12} = -4.06$$

Conclusion: Reject $H_0$; there is a significant difference in tire quality at the .01 level.

**4** The hypotheses are:

$H_0$: $\mu_1 = \mu_2$
$H_1$: $\mu_1 < \mu_2$

And at the .05 level, $Z = 1.64$.
The decision rule is:

Accept $H_0$ if CR $\geq -1.64$.
*or*
Reject $H_0$ and accept $H_1$ if CR $< -1.64$.

The critical ratio is computed as follows:

$$CR = \frac{\bar{x}_1 - \bar{x}_2}{\sqrt{\dfrac{\sigma_1{}^2}{n_1} + \dfrac{\sigma_2{}^2}{n_2}}} = \frac{42 - 45}{\sqrt{\dfrac{6^2}{36} + \dfrac{3^2}{35}}} = \frac{-3.00}{1.12} = -2.68$$

Conclusion: Reject $H_0$; it is likely that the Die Dead batteries have a shorter life expectancy than the Crank-Up brand.

**9-2**

**1** The hypotheses are:

$H_0$: $\mu_1 = \mu_2$
$H_1$: $\mu_1 \neq \mu_2$

At the .10 level, $Z = 1.64$.
The decision rule is:

Accept $H_0$ if CR falls between $\pm 1.64$.
*or*
Reject $H_0$ and accept $H_1$ if CR $> +1.64$ or CR $< -1.64$.

To compute CR:

$$\hat{\sigma}_1 = s_1 \sqrt{\frac{n_1}{n_1 - 1}} = \$1{,}120 \ \sqrt{38/37} = \$1{,}135$$

$$\hat{\sigma}_2 = s_2 \sqrt{\frac{n_2}{n_2 - 1}} = \$725 \quad \sqrt{36/35} = \$735$$

And

$$CR = \frac{\$5{,}600 - \$5{,}300}{\sqrt{\frac{\$1{,}135^2}{38} + \frac{\$735^2}{36}}} = \frac{\$300}{\$221} = 1.36$$

Conclusion: Since $1.36 < 1.64$, the null hypothesis would be accepted.

**2** $H_0\colon \mu_1 = \mu_2$
$H_1\colon \mu_1 > \mu_2$

The decision rule is:

Accept $H_0$ if CR $\lessgtr 1.28$.
*or*
Reject $H_0$ and accept $H_1$ if CR $> 1.28$.

**9-3**

**1** The hypotheses are:

$H_0\colon \pi_1 = \pi_2$
$H_1\colon \pi_1 \neq \pi_2$

This is a two-tailed test because the only concern is whether or not there is a significant difference. The $Z$ distribution is used, and $Z = \pm 1.96$. The decision rule is:

Accept $H_0$ if CR falls between $\pm 1.96$.
*or*
Reject $H_0$ and accept $H_1$ if CR $< -1.96$ or CR $> +1.96$.

The critical ratio is computed as follows:

$$CR = \frac{p_1 - p_2}{\sqrt{\frac{p_1(100 - p_1)}{n_1 - 1} + \frac{p_2(100 - p_2)}{n_2 - 1}}}$$

$$= \frac{60 - 54.5}{\sqrt{\frac{(60)(40)}{199} + \frac{(54.5)(45.5)}{199}}}$$

$$= \frac{5.50}{4.95}$$

$$= 1.13$$

Conclusion: Since the CR value falls into the area of acceptance, we can conclude that there is no significant difference of opinion between the two cities.

2  The hypotheses are:

$H_0$: $\pi_1 = \pi_2$
$H_1$: $\pi_1 > \pi_2$

This is a right-tailed test to see if the urban percentage is significantly greater than the rural percentage. The rejection region is bounded by $Z = 1.64$. The decision rule is:

Accept $H_0$ if CR $\leqslant$ 1.64.
*or*
Reject $H_0$ and accept $H_1$ if CR > 1.64.

The critical ratio is computed as follows:

$$CR = \frac{p_1 - p_2}{\sqrt{\dfrac{p_1(100 - p_1)}{n_1 - 1} + \dfrac{p_2(100 - p_2)}{n_2 - 1}}}$$

$$= \frac{21.67 - 18.00}{\sqrt{\dfrac{(21.67)(78.33)}{299} + \dfrac{(18)(82)}{99}}}$$

$$= \frac{3.67}{4.54}$$

$$= .81$$

Conclusion: The CR falls into the acceptance region—i.e., we must reject the producer's claim and accept the $H_0$ that there is no significant difference in preference between urban and rural viewers. (Neither group is very enthusiastic about the producer's program!)

# CHAPTER 10

# COMPARISON OF THREE OR MORE SAMPLE MEANS: ANALYSIS OF VARIANCE

The previous chapter discussed a technique for determining whether there was a significant difference between the means of *two* independent samples. In this chapter, a technique which determines if there are significant differences between *three or more* sample means is discussed. This technique is called *analysis of variance* (ANOVA).

An analysis of variance permits a decision maker to conclude whether or not all means of the populations under study are equal based upon the degree of variability in the sample data. For example, if a plant supervisor wanted to find out if four stamping machines are producing parts within the same level of tolerance, the ANOVA technique could be used to analyze samples drawn from the output of each machine.

In the pages that follow, we will first consider the *hypotheses and assumptions* associated with the analysis of variance technique. Following this introductory material, an *overview of the reasoning behind analysis of variance* is presented. Finally, we will use an example problem to trace through the *steps in the ANOVA procedure.*

**THE HYPOTHESES AND ASSUMPTIONS IN ANALYSIS OF VARIANCE**

The *null hypothesis* in analysis of variance is that the independent samples are drawn from different populations with the same mean. In other words, the null hypothesis is always:

$$H_0:\ \mu_1 = \mu_2 = \mu_3 = \cdots = \mu_k$$

where $k$ = number of populations under study.

And the *alternative hypothesis* in any analysis of variance is:

$$H_1:\ \textit{Not all}\ \text{population means are equal.}$$

A careful reading of the alternative hypothesis will show you that if the alternative hypothesis is accepted, you may conclude that *at least one* population mean differs from the other population means. But the analysis of variance technique *cannot* tell you exactly *how many* population means differ; nor will it give you exact information about *which* means differ. For example, six populations could be under study, and if only one population mean differs from the other five means, which are equal, the null hypothesis may be rejected and the alternative hypothesis accepted.

All statistical techniques involve assumptions which must be met if the techniques are to be validly applied to a decision-making situation. In the case of analysis of variance, *the following assumptions must be made: First,* the samples are drawn randomly, and each sample is independent of the other samples. *Second,* the populations under study have distributions which approximate the normal curve. And *third,* the populations from which the sample values are obtained all have the same population variance ($\sigma^2$). That is, this third assumption is:

$$\sigma_1{}^2 = \sigma_2{}^2 = \sigma_3{}^2 = \cdots = \sigma_k{}^2$$

where $k$ = number of populations

Thus, *if* the null hypothesis is *true,* and *if* these three assumptions are valid, the net effect is conceptually equal to the case where all the samples are picked from the one population shown in Fig. 10-1a. But *if* the null hypothesis turns out to be *false,* and *if* the three assumptions still remain valid, the population means will not be equal. In this event, the samples in an application might be taken from populations

**FIGURE 10-1**

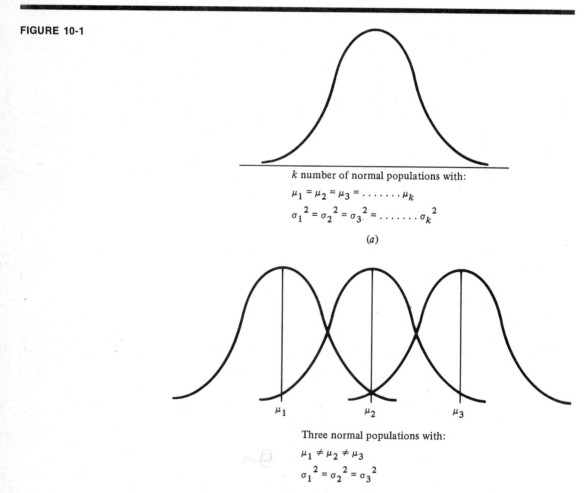

$k$ number of normal populations with:

$$\mu_1 = \mu_2 = \mu_3 = \ldots \ldots \mu_k$$
$$\sigma_1{}^2 = \sigma_2{}^2 = \sigma_3{}^2 = \ldots \ldots \sigma_k{}^2$$

(*a*)

Three normal populations with:

$$\mu_1 \neq \mu_2 \neq \mu_3$$
$$\sigma_1{}^2 = \sigma_2{}^2 = \sigma_3{}^2$$

(*b*)

such as those shown in Fig. 10-1*b*. Of course, these populations are still normally distributed, and they still have the same variance value. If there are any serious violations of the three assumptions, the appropriateness of the ANOVA concept becomes questionable.

**THE REASONING BEHIND ANALYSIS OF VARIANCE**

Before we describe the step-by-step procedure in conducting an analysis, you should understand the underlying conceptual framework of the technique. (Knowing the essence of the technique may help you to utilize it to its fullest potential.)

You may be surprised to learn that in order to determine if the means of several populations are equal, we examine estimates of $\sigma^2$, the *population variance*. That's right; estimates of $\sigma^2$ are analyzed so that conclusions about $\mu_k$ may be drawn. No, we aren't crazy; the logic appears in the following paragraphs.

In an analysis of variance, *two estimates* of the population variance, $\sigma^2$, are computed on the basis of two independent computational approaches. *A first*[1] *approach* is to compute an estimator of $\sigma^2$ that will remain appropriate *regardless* of any differences between population means. In other words, the means of the several populations may differ, but this estimator of $\sigma^2$ *will not* be affected by the possible fact that the $H_0$ is false. Because of this fact, this computed value cannot, by itself, be used to test the validity of the $H_0$. A second element is needed.

The *second approach* will result in the computation of an appropriate estimate of $\sigma^2$ *if, and only if,* the population means are equal. This approach produces an estimate which *will* contain the effects of any differences between the population means. If there are *no differences* between means, this computed value of $\sigma^2$ should not differ too much from the first value (which can now be used as a standard against which the second value can be evaluated).

In short, the preceding two paragraphs may be summarized with the following *decision rules:*

If the two computed estimates are approximately the same, we may conclude that there are probably no differences between the population means. Thus, the null hypothesis should be accepted.

*But*

If the estimate computed by the second approach is significantly different from the estimate of the first approach, we may conclude that the second estimate contains effects of population mean differences. Thus, the $H_0$ should be rejected.

Let's now briefly examine these two estimates of $\sigma^2$ in a little more detail.

---

[1] The word "first" is used here in a conceptual sense. As you will see later, it is not necessary for this value to be computed first in the ANOVA procedure.

## $\hat{\sigma}^2_{within}$: An Estimator of $\sigma^2$ Regardless of Population Mean Values

The first computed estimate of $\sigma^2$ is an average or mean of the variances found *within* each of the samples.[2] Although any one of the sample variances $(s^2)$ could, after minor modification,[3] be used as an unbiased estimator of $\sigma^2$, the average (arithmetic mean) of the variances of all the samples is generally used to estimate $\sigma^2$ because of the greater amount of data thus considered. Therefore, in the ANOVA procedure, a variance from each sample is computed, and the sample variances are then averaged to produce $\sigma^2_{within}$—an unbiased estimator of the population variance that remains appropriate regardless of the equality or inequality of the population mean.

## $\hat{\sigma}^2_{between}$: An Estimator of $\sigma^2$ if (and Only if) the $H_0$ Is True

This second approach to estimating $\sigma^2$ is based upon the variation *between* the sample means and is founded on the Central Limit Theorem. (Ha! And you had hoped that you had heard the last of that concept!) If the $H_0$ is true, then, as we saw in Fig. 10-1*a*, it is as though all the samples were selected from the same normal population distribution with the same $\mu$. And as we saw in Chapter 6, and as the Central Limit Theorem tells us, if the population is normally distributed, the sampling distribution of the sample means will also be normal. Furthermore, you will remember that the standard deviation of this sampling distribution—the standard error of the sample means—is found by this basic formula:

$$\sigma_{\bar{x}} = \frac{\sigma}{\sqrt{n}}$$

Now, if we square both sides of this basic equation, we get

$$\sigma_{\bar{x}}^2 = \frac{\sigma^2}{n}$$

which can be manipulated further to yield

$$n\sigma_{\bar{x}}^2 = \sigma^2$$

Thus, if we knew the square of the standard error $(\sigma_{\bar{x}}^2)$, we could compute the precise value of $\sigma^2$ merely by multiplying $\sigma_{\bar{x}}^2$ by the sample size.

"Well, so what?" you may be thinking, since you don't expect to have any idea of the size of $\sigma_{\bar{x}}^2$. Fortunately for us all, you will be able (by a formula to be presented in just a few pages) to effectively (1) compute an estimate of the square of the standard error $(\hat{\sigma}_{\bar{x}}^2)$ and (2) multiply this estimate by the sample size to effect an estimate of $\sigma^2$. This second approach to estimating $\sigma^2$ is designated $\hat{\sigma}^2_{between}$.

---

[2] Early in Chapter 4 (and in Fig. 4-2), a variance was computed. You might wish to review that material before going on.

[3] Based on the formula for $s$ given in Chapter 7, the formula for $s^2 = \dfrac{\Sigma(x - \bar{x})^2}{n}$. The following formula makes the necessary modification in this chapter to convert $s^2$ to an unbiased estimator of the population variance $(\hat{\sigma}^2)$:

$$\hat{\sigma}^2 = \frac{\Sigma(x - \bar{x})^2}{n - 1}$$

In summary, then, if the $H_0$ is true, the $\hat{\sigma}^2_{between}$ value should be an unbiased estimate of the population variance, and it should be approximately the same as the $\hat{\sigma}^2_{within}$ value. Should there be a significant difference between $\hat{\sigma}^2_{within}$ and $\hat{\sigma}^2_{between}$, however, it may be concluded that this difference is the result of differences between population means.

**The F Ratio and the F Distributions Tables**

When is the difference between the two $\sigma^2$ estimates statistically significant? When is the difference between $\hat{\sigma}^2_{within}$ and $\hat{\sigma}^2_{between}$ due to inequality between population means, and when is the difference simply due to random sampling error? For analytical purposes, a clever statistician has determined that the difference between $\hat{\sigma}^2_{within}$ and $\hat{\sigma}^2_{between}$ may be expressed as a ratio or computed $F$ value, where

$$F = \frac{\hat{\sigma}^2_{between}}{\hat{\sigma}^2_{within}} \qquad (10\text{-}1)$$

Ideally, from the standpoint of verifying the $H_0$, the computed $F$ ratio should have a value of 1. However, some disparity between the two $\sigma^2$ estimates may be expected because of sampling variation even when the $H_0$ *is* true. How much disparity, as reflected by the computed $F$ ratio, may be tolerated before the $H_0$ is rejected? The answer to this question is found in the $F$ distributions tables in Appendix 6.

The *maximum* value that the computed $F$ ratio may attain (at a chosen level of significance) before the $H_0$ must be rejected is specified in the $F$ distributions tables. Thus, conclusions concerning the $H_0$ in analysis of variance tests are based on comparisons of computed $F$ ratios with values from the $F$ distributions tables. If the computed $F$ ratio value $\le$ the table value, the $H_0$ is accepted; if the computed $F$ ratio $>$ the table value, the $H_0$ is rejected. Further comment on the use of Appendix 6 is given in a few pages.

**Self-testing Review 10-1**

1  What is the purpose of an analysis of variance?

2  There are three major assumptions in an analysis of variance. What are they?

3  What is the null hypothesis in an analysis of variance test?

4  When the $H_0$ is rejected, it may be concluded that all the population means are unequal. True or false?

5  The $\hat{\sigma}^2_{within}$ value is an unbiased estimator of $\sigma^2$ if, and only if, the population means are equal. True or false?

6  If $\hat{\sigma}^2_{within}$ is 6 and $\hat{\sigma}^2_{between}$ is 8, the $F$ ratio is .75. True or false?

**PROCEDURE FOR ANALYSIS OF VARIANCE**

The manager of the Nickel and Dime Savings Bank is reviewing the performance of the employees for possible salary increases and position promotions. In evaluating the tellers, the manager decided that the most important criterion of their performance was the number of customers served each day. The manager expected that each teller should handle approximately the same number of customers daily. Otherwise, each teller should be rewarded or penalized accordingly.

Six business days were sampled randomly, and the customer traffic for each of the bank's three tellers was recorded. The data are shown below.[4]

**Customer traffic data**

| Day | Teller 1 Ms. Munny | Teller 2 Mr. Coyne | Teller 3 Mr. Sentz |
|-----|-----|-----|-----|
| 1 | 45 | 55 | 54 |
| 2 | 56 | 50 | 61 |
| 3 | 47 | 53 | 54 |
| 4 | 51 | 59 | 58 |
| 5 | 50 | 58 | 52 |
| 6 | 45 | 49 | 51 |

## Stating the Null and Alternative Hypotheses

As you know by now, the first step in any hypothesis-testing situation is to state the null and alternative hypotheses. For the problem at hand, the *null hypothesis* is that the three tellers each serve the same average number of customers per day. That is, Ms. Munny, Mr. Coyne, and Mr. Sentz are assumed to have the same work load. Symbolically, we have

$$H_0\text{: } \mu_1 = \mu_2 = \mu_3$$

The *alternative hypothesis* is that not all the tellers are handling the same average number of customers per day. That is, at least one of the tellers is performing much better than the others, or, perhaps, at least one of the tellers is performing much worse than the others. Thus,

$H_1$: Not all the population means are equal.

## Specifying the Level of Significance

You will undoubtedly recall from preceding chapters that a criterion for rejection of the $H_0$ is necessary. In our case, if the bank manager is to make a judgment about the tellers' performance, and if the consequences of this judgment will be reflected directly in the tellers' paychecks, the manager should have some idea of the degree of error possible in the decision. Let's assume that the manager has stated that there should be at most only a 5 percent chance of erroneously rejecting a true $H_0$. Thus we will specify that $\alpha = .05$.

---

[4] For each teller there is a sample or group of data. Consequently, the terms "sample data" and "teller data" are used interchangeably in the subsequent paragraphs.

**Computing Estimates of $\sigma^2$: $\hat{\sigma}^2_{between}$ and $\hat{\sigma}^2_{within}$**

We are now into an essential phase of the ANOVA technique. Roll up your sleeves and concentrate.

**Computing $\hat{\sigma}^2_{between}$** The computed $\hat{\sigma}^2_{between}$ value is an estimate of $\sigma^2$ based upon the variation between sample means. Furthermore, the $\hat{\sigma}^2_{between}$ value is likely to yield a reliable estimate of the population variance if, and only if, the $H_0$ is true. *The following steps should be completed:*

1 For the sample data on each teller, a mean must be computed. The means are shown in Fig. 10-2 and have been designated $\bar{x}_1$, $\bar{x}_2$, and $\bar{x}_3$ for tellers 1, 2, and 3 respectively.

2 After computing the individual sample means, you should calculate the "total" or "grand" mean, $\bar{\bar{X}}$. This $\bar{\bar{X}}$ is simply the mean of all the sample values. It is computed using the following formula:

$$\bar{\bar{X}} = \frac{\text{Total of all sample items}}{\text{Number of sample items}} \tag{10-2}$$

In our example, and as shown in Fig. 10-2, the total mean is computed as follows:

$$\bar{\bar{X}} = \frac{\text{Total of all sample items}}{\text{Number of sample items}} = \frac{294 + 324 + 330}{18} = \frac{948}{18} = 52.67$$

**FIGURE 10-2 Data used in computing $\hat{\sigma}^2_{between}$**

| Day | Teller 1 Ms. Munny (1) | Teller 2 Mr. Coyne (2) | Teller 3 Mr. Sentz (3) |
|---|---|---|---|
| 1 | 45 | 55 | 54 |
| 2 | 56 | 50 | 61 |
| 3 | 47 | 53 | 54 |
| 4 | 51 | 59 | 58 |
| 5 | 50 | 58 | 52 |
| 6 | 45 | 49 | 51 |
| Totals | 294 | 324 | 330 |
| | $\bar{x}_1 = 49$ | $\bar{x}_2 = 54$ | $\bar{x}_3 = 55$ |

$$\bar{\bar{X}} = \frac{\text{Total of all sample items}}{\text{Number of sample items}} = \frac{294 + 324 + 330}{18} = 52.67$$

$$\hat{\sigma}^2_{between} = \frac{n_1(\bar{x}_1 - \bar{\bar{X}})^2 + n_2(\bar{x}_2 - \bar{\bar{X}})^2 + n_3(\bar{x}_3 - \bar{\bar{X}})^2}{k - 1}$$

$$= \frac{6(49 - 52.67)^2 + 6(54 - 52.67)^2 + 6(55 - 52.67)^2}{3 - 1}$$

$$= \frac{6(13.469) + 6(1.769) + 6(5.429)}{2}$$

$$= 62.0$$

**3** After these easy steps, we are now ready to determine the value of $\hat{\sigma}^2_{between}$. It is computed by the following formula:[5]

$$\hat{\sigma}^2_{between} = \frac{n_1(\bar{x}_1 - \bar{\bar{X}})^2 + n_2(\bar{x}_2 - \bar{\bar{X}})^2 + \cdots + n_k(\bar{x}_k - \bar{\bar{X}})^2}{k-1} \qquad (10\text{-}3)$$

where $n_1$ = number of items or observations in sample 1
$n_2$ = number of items or observations in sample 2
$n_k$ = number of items or observations in sample $k$
$k$ = number of samples under study
$\bar{x}_1$ = mean of sample 1
$\bar{x}_2$ = mean of sample 2
$\bar{x}_k$ = mean of sample $k$
$\bar{\bar{X}}$ = total mean or the average of all the items in the samples

From the use of this formula in Fig. 10-2, we can see that the computed value for $\hat{\sigma}^2_{between}$ is 62.0. This 62.0 is an appropriate estimate of $\sigma^2$ if, and only if, the null hypothesis is true.

The procedure for computing $\hat{\sigma}^2_{between}$ by focusing on the variance between sample means is summarized in Fig. 10-3.

---

[5] If you closely examine formula (10-3), you will notice that $\hat{\sigma}^2_{between}$ is basically $n\hat{\sigma}^2_{\bar{x}}$, which, as we have seen, is equal to $\hat{\sigma}^2$.

**FIGURE 10-3**

**Procedure for the computation of $\hat{\sigma}^2_{between}$**

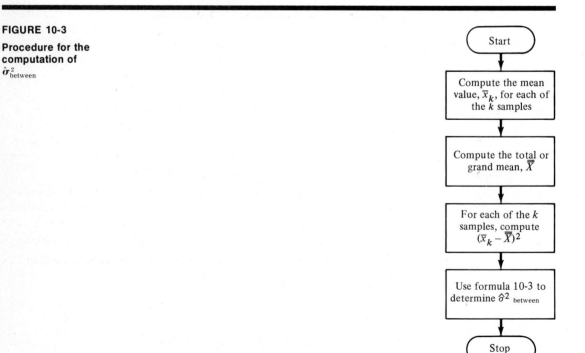

**Computing $\hat{\sigma}^2_{within}$** You will recall that $\hat{\sigma}^2_{within}$ is an estimate of $\sigma^2$ based upon an average of the variances *within* each sample. More specifically, $\hat{\sigma}^2_{within}$ is the mean of the unbiased estimates of $\sigma^2$ that have been obtained from each sample.

*If all samples are of equal size,* this approach to estimating the population variance may be carried out with the following formulas:

$$\hat{\sigma}^2_{within} = \frac{\hat{\sigma}_1^2 + \hat{\sigma}_2^2 + \hat{\sigma}_3^2 + \cdots + \hat{\sigma}_k^2}{k} \tag{10-4}$$

where $\hat{\sigma}^2 =$ the unbiased estimate of the population variance from each sample. That is, for each sample:

$$\hat{\sigma}^2 = \frac{\Sigma(x - \bar{x})^2}{n - 1} \tag{10-5}$$

$k =$ the number of samples

*If the samples are not of equal size,* we have

$$\hat{\sigma}^2_{within} = \frac{\Sigma d_1^2 + \Sigma d_2^2 + \Sigma d_3^2 + \cdots + \Sigma d_k^2}{T - k} \tag{10-6}$$

where $\Sigma d_1^2 =$ the sum of the squared differences—i.e., $\Sigma(x_1 - \bar{x}_1)^2$—for the first sample

$\Sigma d_2^2 =$ the sum of the squared differences—i.e., $\Sigma(x_2 - \bar{x}_2)^2$—for the second sample, etc.

$T =$ total number of all items in all samples (i.e., $n_1 + n_2 + n_3 + \cdots + n_k$)

$k =$ number of samples

Let's now work through the computational steps for $\hat{\sigma}^2_{within}$ for our bank teller example problem. (Since the samples are of equal size, we will use formulas (10-4) and (10-5). A self-testing review problem presented later in the chapter will have unequal sample sizes, and the application of formula (10-6) will be shown in the answer to this self-testing review problem.)

1   For *each sample,* we compute the *deviation* between each value within the sample and the mean of that sample—i.e., compute $x - \bar{x}$ for each sample. Figure 10-4 shows the deviation computations for each teller. For teller 1, Ms. Munny, the average customer traffic ($\bar{x}_1$) is 49. Thus, the deviation for day 1 is $45 - 49$, or $-4$; the deviation for day 2 is $56 - 49 = 7$; and so forth.

2   After the deviation of each observation from its sample mean is computed, each deviation is *squared*—i.e., $(x - \bar{x})^2$. These squared deviations are summed— $\Sigma(x - \bar{x})^2$—and the sums are labeled $\Sigma d^2$. That is, $\Sigma(x - \bar{x})^2 = \Sigma d^2$. For tellers 1, 2, and 3, the sums of the squared deviations are 90, 84, and 72 respectively.

3   The unbiased estimate of $\hat{\sigma}^2$ from *each sample under consideration* is then computed as follows, using formula (10-5):

$$\hat{\sigma}^2 = \frac{\Sigma(x - \bar{x})^2}{n - 1} = \frac{\Sigma d^2}{n - 1}$$

where $n =$ the size of the sample being considered

**FIGURE 10-4 Data used in computing $\hat{\sigma}^2_{within}$**

| Day | Teller 1<br>Ms. Munny<br>$x_1$ | $x_1 - \bar{x}_1$ | $(x_1 - \bar{x})^2$ | Teller 2<br>Mr. Coyne<br>$x_2$ | $x_2 - \bar{x}_2$ | $(x_2 - \bar{x}_2)^2$ | Teller 3<br>Mr. Sentz<br>$x_3$ | $x_3 - \bar{x}_3$ | $(x_3 - \bar{x}_3)^2$ |
|---|---|---|---|---|---|---|---|---|---|
| 1 | 45 | $-4$ | 16 | 55 | 1 | 1 | 54 | $-1$ | 1 |
| 2 | 56 | 7 | 49 | 50 | $-4$ | 16 | 61 | 6 | 36 |
| 3 | 47 | $-2$ | 4 | 53 | $-1$ | 1 | 54 | $-1$ | 1 |
| 4 | 51 | 2 | 4 | 59 | 5 | 25 | 58 | 3 | 9 |
| 5 | 50 | 1 | 1 | 58 | 4 | 16 | 52 | $-3$ | 9 |
| 6 | 45 | $-4$ | 16 | 49 | $-5$ | 25 | 51 | $-4$ | 16 |
|   | 294 |   | $\Sigma d_1^2 = 90$ | 324 |   | $\Sigma d_2^2 = 84$ | 330 |   | $\Sigma d_3^2 = 72$ |

$$\bar{x}_1 = 49 \qquad\qquad \bar{x}_2 = 54 \qquad\qquad \bar{x}_3 = 55$$

$$\hat{\sigma}_1^2 = \frac{90}{5} = 18 \qquad \hat{\sigma}_2^2 = \frac{84}{5} = 16.8 \qquad \hat{\sigma}_3^2 = \frac{72}{5} = 14.4$$

$$\hat{\sigma}^2_{within} = \frac{18 + 16.8 + 14.4}{3} = 16.4$$

In our case the unbiased estimates of the population variance are computed as follows:

$$\text{Teller 1} \qquad \hat{\sigma}_1^2 = \frac{\Sigma d_1^2}{n_1 - 1} = \frac{90}{5} = 18$$

$$\text{Teller 2} \qquad \hat{\sigma}_2^2 = \frac{\Sigma d_2^2}{n_2 - 1} = \frac{84}{5} = 16.8$$

$$\text{Teller 3} \qquad \hat{\sigma}_3^2 = \frac{\Sigma d_3^2}{n_3 - 1} = \frac{72}{5} = 14.4$$

Since the sample sizes are equal, $\hat{\sigma}^2_{within}$ is conveniently computed as follows, using formula (10-4):

$$\hat{\sigma}^2_{within} = \frac{\hat{\sigma}_1^2 + \hat{\sigma}_2^2 + \hat{\sigma}_3^2 + \cdots + \hat{\sigma}_k^2}{k}$$

$$= \frac{18 + 16.8 + 14.4}{3}$$

$$= \frac{49.2}{3}$$

$$= 16.4$$

Figure 10-5 summarizes the steps performed in this section. You will be delighted to know that most of the tedious computations are now behind us. All that remains now is for us to compute the $F$ ratio and then compare this value with an appropriate $F$ table figure. We will then be in a position to make a decision about the $H_0$.

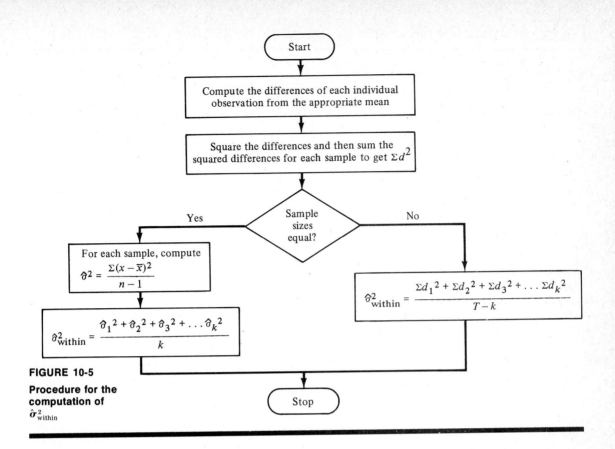

Start

Compute the differences of each individual observation from the appropriate mean

Square the differences and then sum the squared differences for each sample to get $\Sigma d^2$

Sample sizes equal?

Yes

No

For each sample, compute
$$\hat{\sigma}^2 = \frac{\Sigma(x - \bar{x})^2}{n - 1}$$

$$\hat{\sigma}^2_{within} = \frac{\hat{\sigma}_1{}^2 + \hat{\sigma}_2{}^2 + \hat{\sigma}_3{}^2 + \ldots \hat{\sigma}_k{}^2}{k}$$

$$\hat{\sigma}^2_{within} = \frac{\Sigma d_1{}^2 + \Sigma d_2{}^2 + \Sigma d_3{}^2 + \ldots \Sigma d_k{}^2}{T - k}$$

Stop

**FIGURE 10-5**
**Procedure for the computation of** $\hat{\sigma}^2_{within}$

**Computing the Critical F Ratio: $CR_F$**

Now that we have computed $\hat{\sigma}^2_{between}$ and $\hat{\sigma}^2_{within}$, we are ready to compare these two $\sigma^2$ estimates and calculate the magnitude of the difference between the two. In short, we are now ready to compute a value called the *computed F ratio*, or $CR_F$. As we saw earlier in formula (10-1),

$$CR_F = \frac{\hat{\sigma}^2_{between}}{\hat{\sigma}^2_{within}}$$

On the basis of our problem data and our recent calculations, we may compute the $F$ ratio as follows:

$$CR_F = \frac{62.0}{16.4} = 3.78$$

Thus, it is obvious that the estimate of $\hat{\sigma}^2_{between}$ is almost four times as large as the estimate of $\hat{\sigma}^2_{within}$. But is this calculated $CR_F$ of such magnitude that we may conclude that the $H_0$ is not true? In order to answer this question, we must know the *critical value* that separates the area of acceptance from the area of rejection. That is, we must know the critical value shown in Fig. 10-6. If the computed $F$ ratio $\lessgtr$ the critical value shown in Fig. 10-6, it will fall into the area of acceptance and the $H_0$

**231**

CHAPTER 10/COMPARISON OF THREE OR MORE SAMPLE MEANS: ANALYSIS OF VARIANCE

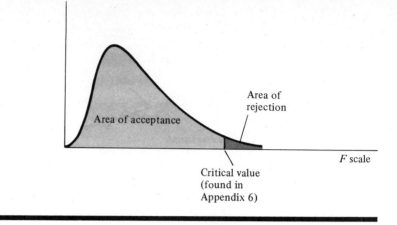

**FIGURE 10-6**

**An F distribution for a given number of samples and a given number of observations in the samples**

will be accepted. And if the computed $F$ value > this critical value, it will fall into the area of rejection and the $H_0$ will be rejected. The critical value against which the computed $F$ ratio is compared is found in the $F$ distributions tables in Appendix 6. If a hypothesis test is being made at the .05 level, the critical value is simply the point on the $F$ scale beyond which an F ratio value would be expected to fall only 5 times in 100 if the $H_0$ were true.

As you can see in Fig. 10-6, an $F$ distribution is skewed to the right. And as you can also see in the Fig. 10-6 caption, each $F$ distribution is determined by the number of samples and the number of observations in the samples. This means that there is a *different* $F$ distribution for every possible combination of sample number and sample size. And since there would be different table critical values for each possible combination and for each selected level of significance, the size of an $F$ distributions table could quickly become unmanageable. Therefore, in Appendix 6, only a selected number of possible combinations are presented, and only the .01 and .05 levels of significance are available. Let's now see how to use Appendix 6.

## Using the F Distributions Tables

In using the $F$ distributions tables in Appendix 6, you must first determine the degrees of freedom *(df)* for both the numerator and the denominator of the computed $F$ ratio. The *df for the numerator is computed as follows:*

$$df_{num} = k - 1 \qquad (10\text{-}7)$$

where $k$ = the number of samples

And the *df for the denominator is computed as follows:*

$$df_{den} = T - k \qquad (10\text{-}8)$$

where $T$ = the total number of items in all samples, or $n_1 + n_2 + n_3 + \cdots + n_k$
       $k$ = the number of samples

The degrees of freedom for our bank teller example, then, are determined as follows:

$$df_{num} = 3 - 1 = 2$$

$$df_{den} = 18 - 3 = 15$$

(As a check, it might be worth noting here that the total of the $df$ in the numerator and denominator should equal $T - 1$ — i.e., $df_{num} + df_{den} = T - 1$. Since $15 + 2 = 18 - 1$, we can conclude that our math is correct.)

We are now ready to find the critical $F$ value from the appropriate table for the specific combination of $df_{num}$, $df_{den}$, and $\alpha$ that is present in our teller problem. The critical $F$ value in our case is based upon (1) 2 degrees of freedom in the numerator, (2) 15 degrees of freedom in the denominator, and (3) a level of significance of .05. In using the $F$ tables in Appendix 6, we must first locate the table with the relevant $\alpha$. (In this problem, it is the first table in that appendix.) Next, we must locate the critical value of $F$ where the degrees of freedom for the numerator (found at the top of the *columns*) and the degrees of freedom for the denominator (shown to the left of the *rows*) intersect. For our example, the critical $F$ value is 3.68:

$$F_{(2, \ 15, \ \alpha \ = \ .05)} = 3.68$$

The final step in the ANOVA procedure is to draw a statistical conclusion concerning the validity of the null hypothesis.

**Making the Statistical Decision**

As indicated earlier, the following *decision rule* applies in the analysis of variance test:

> If $CR_F > F$ value found in the table, reject $H_0$ and accept the alternative hypothesis.

If the computed $F$ ratio is greater than the table $F$ value, it may be concluded that the difference between the two estimates of the population variance, $\hat{\sigma}^2_{between}$ and $\hat{\sigma}^2_{within}$, is so large that such a magnitude of difference has an extremely unlikely chance of occurring. Therefore, such a magnitude of difference implies that it is likely that the $H_0$ is not valid.

Since in our example problem we have $CR_F = 3.78$, and since the critical table value $= 3.68$, we must conclude that the $H_0$ is unlikely and that the alternative hypothesis should be accepted. At least one of the tellers among Munny, Coyne, and Sentz is likely to be handling more or less work than the others. Additional research is needed to precisely define the nature of the difference in work performance.

Congratulations! You have made it through a rather detailed chapter. Figure

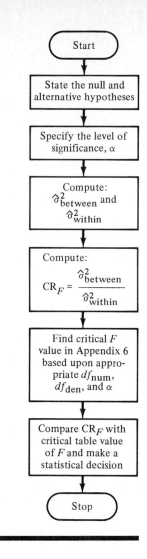

**FIGURE 10-7**

**Procedure for analysis of variance**

The flowchart contains the following steps:

Start

State the null and alternative hypotheses

Specify the level of significance, $\alpha$

Compute: $\hat{\sigma}^2_{between}$ and $\hat{\sigma}^2_{within}$

Compute: $CR_F = \dfrac{\hat{\sigma}^2_{between}}{\hat{\sigma}^2_{within}}$

Find critical $F$ value in Appendix 6 based upon appropriate $df_{num}$, $df_{den}$, and $\alpha$

Compare $CR_F$ with critical table value of $F$ and make a statistical decision

Stop

10-7 provides an outline of the analysis of variance procedure which you have now successfully digested.

**Self-testing Review 10-2**

1 The Jingle and Cliché Ad Agency has been commissioned to design a package for a new powdered soft-drink mix. According to Mr. Jones of Jingle and Cliché, a good package should lead to good sales. The creative staff, after 3 months' work, has produced three attractive packages. One is gold, another is red, and the third is orange. Prior to making a final package decision, however, the ad agency managers decided that each package should be placed on the shelves of supermarkets in a city for 14 business days. This was done, and the package sales for each day were recorded. The sales data are shown below.

**Package sales**

| Day | Gold package | Red package | Orange package |
|-----|-----|-----|-----|
| 1 | 10 | 12 | 9 |
| 2 | 19 | 13 | 18 |
| 3 | 17 | 25 | 36 |
| 4 | 22 | 39 | 18 |
| 5 | 25 | 44 | 25 |
| 6 | 29 | 37 | 21 |
| 7 | 32 | 36 | 38 |
| 8 | 31 | 38 | 31 |
| 9 | 29 | 35 | 33 |
| 10 | 33 | 27 | 28 |
| 11 | 31 | 42 | 29 |
| 12 | 32 | 22 | 32 |
| 13 | 28 | 36 | 30 |
| 14 | 27 | 25 | 31 |

Using these data, conduct an analysis of variance at the .05 level of significance.

2   Professor Krusher overheard one of his students saying, "Krusher's exams are so tough, it makes no difference whether you take them after a good night's rest, at the end of a tiring day, or after 36 hours without sleep! You get the same grade." Professor Krusher decided to conduct an experiment to see whether the student's statement had any validity. He initially recruited 36 student volunteers and divided the 36 students into three groups of 12. Each group was administered a typical 1-hour Krusher exam. One group was fresh and alert when they took the exam; another group took the exam at the end of a long day; and the third group took the test in a groggy state. During the testing, one student in the tired second group could not complete the exam, and two students in the groggy group fell asleep. The scores on the exam were:

| | Group | | |
|---|---|---|---|
| | Alert | Tired | Groggy |
| Exam scores | 43 | 56 | 84 |
| | 67 | 67 | 78 |
| | 76 | 69 | 69 |
| | 55 | 62 | 64 |
| | 71 | 61 | 67 |
| | 62 | 57 | 65 |
| | 39 | 59 | 66 |
| | 65 | 68 | 63 |
| | 61 | 72 | 58 |
| | 53 | 66 | 70 |
| | 58 | 64 | |
| | 62 | | |

Using these data, conduct an analysis of variance at the .01 level of significance.

The purpose of the analysis of variance technique discussed in this chapter is to enable a decision maker to compare three or more independent sample means to determine if there are statistically significant differences between the means of the populations from which the samples were selected. The following three assumptions must be met before the ANOVA technique can be applied to a decision-making situation: (1) the samples are random and independent of each other, (2) the population distributions approximate the normal distribution, and (3) the variances of all populations are equal.

As is the case with other hypothesis-testing procedures, the ANOVA technique begins with a statement of the null and alternative hypotheses; it then requires that the test be made at a suitable level of confidence. Following these steps, two estimates of the population variance are computed by two independent computational approaches. In one approach, the $\hat{\sigma}^2_{within}$ approach, $\sigma^2$ is estimated by computing the variances found in each sample. In the other approach, $\hat{\sigma}^2_{between}$, an estimate of $\sigma^2$ is computed by measuring the variation between the sample means. The two estimates of $\sigma^2$ are then used to compute an $F$ ratio. Ideally, if the $H_0$ is true, this ratio will have a value of 1.00, since the two estimates of $\sigma^2$ would yield the same results. Realistically, however, sampling variation will normally cause the $F$ ratio to exceed 1.00 even when the $H_0$ *is* true. The amount by which the computed $CR_F$ value may be permitted to exceed 1.00 before the $H_0$ is rejected is determined in the $F$ distributions tables in Appendix 6. If a computed $CR_F$ value is found to *exceed* the appropriate $F$ table value at a specified level of confidence, the ANOVA test procedure concludes with the rejection of the $H_0$; if the $CR_F$ value $\leq$ the table value, the $H_0$ is accepted.

**Important Terms and Concepts**

1  Analysis of variance (ANOVA)

2  $\hat{\sigma}^2_{within}$ computational approach to estimating $\sigma^2$

3  $\hat{\sigma}^2_{between}$ computational approach to estimating $\sigma^2$

4  $F$ ratio

5  Total or grand mean $(\bar{\bar{X}})$

6  $\Sigma d^2$

7  $CR_F$

8  $F$ distributions

9  $df_{num}$

10  $df_{den}$

11  $\bar{\bar{X}} = \dfrac{\text{Total of all sample items}}{\text{Number of sample items}}$

12  $\hat{\sigma}^2_{between} = \dfrac{n_1(\bar{x}_1 - \bar{\bar{X}})^2 + n_2(\bar{x}_2 - \bar{\bar{X}})^2 + \cdots + n_k(\bar{x}_k - \bar{\bar{X}})^2}{k - 1}$

13  $\hat{\sigma}^2 = \dfrac{\Sigma(x - \bar{x})^2}{n - 1} = \dfrac{\Sigma d^2}{n - 1}$

14  $\hat{\sigma}^2_{within} = \dfrac{\hat{\sigma}_1^2 + \hat{\sigma}_2^2 + \hat{\sigma}_3^2 + \cdots + \hat{\sigma}_k^2}{k}$   (For samples of equal size)

$$15 \quad \hat{\sigma}^2_{\text{within}} = \frac{\Sigma d_1^2 + \Sigma d_2^2 + \Sigma d_3^2 + \cdots + \Sigma d_k^2}{T - k} \qquad \text{(For samples of unequal size)}$$

$$16 \quad CR_F = \frac{\hat{\sigma}^2_{\text{between}}}{\hat{\sigma}^2_{\text{within}}}$$

$$17 \quad df_{\text{num}} = k - 1$$

$$18 \quad df_{\text{den}} = T - k$$

**Problems**

1  A high school track coach has learned of two new training techniques which are designed to reduce the time needed to run a mile. Three samples of novice runners have been selected for an experiment. Group A has trained under the old approach, group B has trained under one of the new techniques, and group C has trained under the other new technique. After a month of training, each runner was timed in a mile run, and the following times were recorded (in minutes).

| Group A | Group B | Group C |
| --- | --- | --- |
| 4.81 | 4.43 | 4.38 |
| 4.62 | 4.50 | 4.29 |
| 5.02 | 4.32 | 4.33 |
| 4.65 | 4.37 | 4.36 |
| 4.58 | 4.41 | 4.74 |
| 4.52 | 4.39 | 4.42 |
| 4.73 | 4.64 | 4.40 |

Conduct an analysis of variance at the .05 level to determine if $\mu_A = \mu_B = \mu_C$.

2  A consumer advocate group wants to determine if the three leading brands of aspirin are really different from each other in terms of speed of relief. Consumers have been randomly selected and assigned to use either brand A, brand B, or brand C. Each subject was instructed to take the recommended dosage when a headache began, and each subject was told to record the number of minutes that elapsed before relief occurred. The following data (in minutes) have been gathered:

| Brand A | Brand B | Brand C |
| --- | --- | --- |
| 7.3 | 6.7 | 7.4 |
| 8.5 | 7.1 | 7.8 |
| 6.4 | 9.0 | 6.9 |
| 7.9 | 8.4 | 8.5 |
| 6.7 | 7.8 | 7.4 |
| 9.1 | 6.9 | 8.2 |
| 7.4 | 8.7 | 7.4 |

Conduct an analysis of variance at $\alpha = .05$. Is there a significant difference in speed of relief?

3  A manufacturer of canned beans wonders if there is truly any difference in sales based upon the height of the shelves on which the product is displayed.

With the cooperation of three retail stores, an experiment has been conducted. One store placed the cans at eye level on the shelves; another store placed the cans at waist level; and the third store placed the cans at knee level. Sales data over 8 days were recorded:

**Sales of cans of beans**

| Eye level | Waist level | Knee level |
| --- | --- | --- |
| 98 | 106 | 103 |
| 106 | 105 | 95 |
| 111 | 98 | 87 |
| 85 | 93 | 94 |
| 108 | 96 | 92 |
| 86 | 98 | 82 |
| 83 | 97 | 87 |
| 109 | 104 | 83 |

Conduct an analysis of variance at $\alpha = .05$. Is there a significant difference in sales based on the shelf location of the product?

4  A large accounting firm wants to determine if the accuracy of its employees is related to the school from which the employees graduated. Accountants representing four schools were randomly selected, and the number of errors committed by each accountant over a 2-week period was recorded as shown below:

| School A | School B | School C | School D |
| --- | --- | --- | --- |
| 14 | 17 | 19 | 23 |
| 16 | 16 | 20 | 12 |
| 17 | 18 | 22 | 21 |
| 13 | 15 | 21 | 10 |
| 22 | 16 | 18 | 9 |
| 9 | 12 | 19 | 15 |
| 10 | 14 | 15 | 16 |

Conduct an analysis of variance at $\alpha = .01$. Is there a significant difference in accuracy?

5  A manufacturer wants to know if the four machines in operation are performing with equal efficiency. Samples have been drawn from the machines, and the deviations of the samples from specifications have been recorded in millimeters as shown below:

| Machine A | Machine B | Machine C | Machine D |
| --- | --- | --- | --- |
| 50 | 66 | 50 | 70 |
| 50 | 61 | 75 | 75 |
| 55 | 57 | 65 | 73 |
| 45 | 72 | 60 | 80 |
| 61 | 68 | 55 | 72 |
| 56 | 55 | 52 | 78 |

What statistical conclusion can be reached at the .01 level?

**6** A large retailer must make a choice between three sales locations within a shopping mall. The retailer is wondering if the daily traffic count is the same for all locations. The following data are traffic counts for a 10-day period:

| Location X | Location Y | Location Z |
|------------|------------|------------|
| 643 | 249 | 458 |
| 542 | 404 | 513 |
| 569 | 378 | 475 |
| 552 | 337 | 482 |
| 607 | 426 | 539 |
| 514 | 298 | 491 |
| 576 | 345 | 468 |
| 585 | 362 | 487 |
| 581 | 425 | 464 |
| 600 | 376 | 476 |

At the .05 level, is there a significant difference in traffic count at the three locations?

**7** What can be concluded about a three-sample experiment from the following information?
$n_1 = 13$, $n_2 = 12$, $n_3 = 8$, and $\alpha = .05$
$\hat{\sigma}^2_{between} = 164$    and    $\hat{\sigma}^2_{within} = 43$

**8** Draw a statistical conclusion based upon the following data:
$n_1 = 20$, $n_2 = 16$, $n_3 = 19$, $n_4 = 21$, and $\alpha = .05$
$\hat{\sigma}^2_{between} = 158$    and    $\hat{\sigma}^2_{within} = 54$

**9** At $\alpha = .01$, what decision can be reached if the following facts are known?
$n_1 = 22$, $n_2 = 22$, $n_3 = 22$, $n_4 = 23$, and $n_5 = 26$
$\hat{\sigma}^2_{between} = 374$    and    $\hat{\sigma}^2_{within} = 93$

**10** The sales manager of Itty Bitty Machines wants to determine if a dress code will have any effect on sales. In an experiment, salespersons were selected to wear either attire A, attire B, or attire C in visiting clients and prospects. The following sales for a 4-week period have been recorded:

| Attire A | Attire B | Attire C |
|----------|----------|----------|
| 26 | 19 | 22 |
| 37 | 24 | 33 |
| 41 | 31 | 34 |
| 35 | 28 | 19 |
| 29 | 23 | 25 |
| 33 | 25 | 29 |
| 40 | 24 | 31 |
|    | 29 |    |
|    | 32 |    |

What statistical decision can be made at the .05 level?

**11** As a part of his research on smoking addiction, Nick O. Teen is interested in the abilities of cigarette, cigar, and pipe smokers to refrain from lighting up. Subjects were randomly selected in each category, and each subject was asked to wait as long as possible between smokes. The time interval was recorded in minutes for each smoker. The following data have been obtained:

| Cigarette smokers | Cigar smokers | Pipe smokers |
|---|---|---|
| 6 | 13 | 8 |
| 13 | 22 | 17 |
| 7 | 12 | 14 |
| 19 | 14 | 23 |
| 8 | 17 | 18 |
| 9 | 19 | 12 |
| 12 | 20 | 11 |
| 23 | 11 | 15 |
| 16 | | 12 |
| 10 | | 27 |
| 25 | | 31 |
| 8 | | |

What may be concluded about the ability of different smokers to refrain from smoking? Use $\alpha = .05$.

**12** A major distributor of cameras suspects that consumers would be insensitive to price changes for the highest-quality camera. To test this suspicion, four retail outlets have been selected, and each outlet has sold the highest-quality camera at one of four predetermined prices. After some time, the following weekly unit sales at each store were reported:

| Price I | Price II | Price III | Price IV |
|---|---|---|---|
| 3 | 5 | 10 | 8 |
| 5 | 4 | 9 | 4 |
| 7 | 6 | 4 | 5 |
| 9 | 5 | 7 | 7 |
| 4 | 8 | 2 | 6 |
| 2 | 7 | 6 | 9 |
| 10 | 6 | 8 | 6 |
| 8 | 5 | 8 | |
| | | 11 | |

What conclusion may be drawn at $\alpha = .05$?

**13** A college professor believes that the season of the year affects the amount of time students spend studying. In a year-long experiment, students were selected randomly during the different seasons and were asked to estimate the average number of hours they spent per week studying. The hourly estimates of the students were:

| Summer | Fall | Winter | Spring |
|--------|------|--------|--------|
| 4 | 6 | 7 | 7 |
| 3 | 8 | 11 | 5 |
| 6 | 7 | 12 | 6 |
| 7 | 9 | 8 | 4 |
| 5 | 6 | 13 | 4 |
| 3 | 8 | 6 | 3 |
| 4 | 5 | 5 | 4 |
|   | 4 | 4 | 7 |
|   |   | 7 |   |

What may the professor conclude at the .05 level?

14 Due to a clerical error on the part of a chicken farmer, four truckloads of chickens have arrived simultaneously at a processing plant. The receiving supervisor is thus confronted with the problem of selecting one of the truckloads of chickens. If the weights of the chickens in all four trucks can be assumed to be equal, any truck may be selected. Samples have been drawn randomly from each truck. (Equal samples were not possible because the crates on each truck were loaded differently.) The weights, in pounds, are as follows:

| Truck 1 | Truck 2 | Truck 3 | Truck 4 |
|---------|---------|---------|---------|
| 4.3 | 3.7 | 4.1 | 3.4 |
| 3.7 | 3.6 | 3.9 | 4.1 |
| 3.8 | 4.0 | 3.4 | 4.2 |
| 4.2 | 3.8 | 4.2 | 3.9 |
| 3.9 | 3.7 | 3.8 | 4.0 |
| 3.5 |     |     | 3.7 |

At the .05 level, what decision should the supervisor make?

**Topics for Discussion**

1 What is an analysis of variance?

2 What are the three major assumptions in an analysis of variance?

3 What is the most important assumption in an analysis of variance? Why?

4 What does acceptance of the alternative hypothesis in an analysis of variance mean?

5 You have just overheard a classmate make the following comment: "The procedure for computing $\hat{\sigma}^2_{within}$ is the same in all situations in analysis of variance." Discuss this comment.

6 "In conducting an analysis of variance, as soon as we see that $CR_F$ is less than 1, we may accept the null hypothesis and assume that there is no need for further computations." Discuss this statement.

7 Find $F_{(3, 14, .05)}$, $F_{(3, 16, .05)}$, $F_{(3, 18, .05)}$, and $F_{(3, 20, .05)}$.

8 In comparing the four $F$ values found in the preceding question, what may be

concluded about the relationship between the total sample size and the $F$ value for the hypothesis test?

**10-1**

1   The purpose of an analysis of variance is to determine whether or not all means of the populations under study are equal.

2   **a**   The samples are random and independent of each other.
   **b**   The population distributions approximate the normal distribution.
   **c**   The variances of all populations are equal.

3   The null hypothesis states that all population means are equal.

4   False. Acceptance of the alternative hypothesis simply means that at least one mean differs from the other means.

5   False. The $\hat{\sigma}^2_{within}$ value is an unbiased estimator of $\sigma^2$ regardless of the truth of the null hypothesis.

6   False. $F = 1.33$.

**10-2**

1   **a**   The hypotheses are: $H_0$: The packages are equally attractive and will result in equal average sales—i.e., $\mu_1 = \mu_2 = \mu_3$. $H_1$: Not all population means are equal.
   **b**   $\alpha$ is specified at .05.
   **c**   Computation of $\hat{\sigma}^2_{between}$:
   $\bar{x}_1$ = mean of gold package sales = 26.07
   $\bar{x}_2$ = mean of red package sales = 30.79
   $\bar{x}_3$ = mean of orange package sales = 27.07
   $\bar{\bar{X}} = 27.98$

   $$\hat{\sigma}^2_{between} = \frac{14(-1.91)^2 + 14(2.81)^2 + 14(-.91)^2}{2}$$

   $$= \frac{173.20}{2}$$

   $$= 86.60$$

   **d**   Computation of $\hat{\sigma}^2_{within}$:
   Gold package: $\Sigma d_1^2 = 596.93$
   Red package: $\Sigma d_2^2 = 1,358.36$
   Orange package: $\Sigma d_3^2 = 834.93$
   Gold package: $\hat{\sigma}_1^2 = 45.92$
   Red package: $\hat{\sigma}_2^2 = 104.49$
   Orange package: $\hat{\sigma}_3^2 = 64.23$

   $$\hat{\sigma}^2_{within} = \frac{45.92 + 104.49 + 64.23}{3}$$

   $$= 71.55$$

**e** Computation of $CR_F$:

$$CR_F = \frac{86.60}{71.55} = 1.21$$

**f** Using the $F$ table:
$$F_{(2, 39, \alpha = 05)} \approx 3.24$$

**g** Since the computed $F$ ratio is less than 3.24, we cannot reject the null hypothesis. We may conclude that all the packages have the same effect on sales.

**2 a** The hypotheses are: $H_0$: The students will perform equally on the exam— i.e., $\mu_1 = \mu_2 = \mu_3$. $H_1$: Not all population means are equal.

**b** $\alpha$ is specified at .01.

**c** Computing $\hat{\sigma}^2_{\text{between}}$, we have:
$\bar{x}_1$ = mean score of alert group = 59.33
$\bar{x}_2$ = mean score of tired group = 63.73
$\bar{x}_3$ = mean score of groggy group = 68.40

And $\bar{\bar{X}} = \dfrac{2,097}{33} = 63.55$

Finally,

$$\hat{\sigma}^2_{\text{between}} = \frac{12(-4.22)^2 + 11(.18)^2 + 10(4.85)^2}{2}$$

$$= \frac{213.70 + .36 + 235.22}{2}$$

$$= 224.64$$

**d** Computing $\hat{\sigma}^2_{\text{within}}$, we have:
$\Sigma d_1{}^2 = 1,262.67$ = sum of squared deviations for alert group
$\Sigma d_2{}^2 = \phantom{0}268.18$ = sum of squared deviations for tired group
$\Sigma d_3{}^2 = \phantom{0}514.40$ = sum of squared deviations for groggy group
Thus,

$$\hat{\sigma}^2_{\text{within}} = \frac{1,262.67 + 268.18 + 514.40}{33 - 3}$$

$$= \frac{2,045.25}{30}$$

$$= 68.18$$

**e** Computing $CR_F$:

$$CR_F = \frac{224.64}{68.18} = 3.29$$

**f** Using the $F$ table:

$F_{(2, 30, \alpha = .01)} = 5.39$

**g** Since the $CR_F$ of $3.29 < 5.39$, we conclude that there is no real difference between the group means; you are likely to get the same grade on a Krusher exam whether you are alert or groggy!

# CHAPTER 11

# COMPARISON OF SEVERAL SAMPLE PERCENTAGES: CHI-SQUARE ANALYSIS

**LEARNING OBJECTIVES**

After studying this chapter, working the problems, and answering the discussion questions, you should be able to:

☞ Explain the purpose of (*a*) the *k*-sample hypothesis test of percentages and (*b*) the goodness of fit test.

☞ Describe the steps in the *k*-sample testing procedure, and use them to arrive at statistical decisions concerning the percentages of three or more populations.

☞ Describe the steps in the goodness of fit test, and use them to arrive at statistical decisions about whether or not a population under study follows a uniform distribution.

**CHAPTER OUTLINE**

In Chapter 9, we first considered two-sample hypothesis tests of *means,* and then in Chapter 10 we used analysis of variance as a technique to test for the existence of significant differences between *three or more* sample means. In Chapter 9, we also studied two-sample hypothesis tests of *percentages.* Now in this chapter (and this probably comes as no surprise to you), one of the uses we will make of the concepts of *chi-square analysis* will be to test the hypothesis that *three or more* independent samples have all come from populations that have the same *percentage* of a given characteristic.

When three or more sample percentages must be compared, the techniques described in Chapter 9 are not adequate for the job. However, the chi-square testing procedure that will be presented in this chapter *could be used* to test the significance of the difference between *two* sample percentages, and the results would be exactly the same as those obtained in Chapter 9. In this chapter, however, we will be concerned with problems involving the comparison of three or more samples.

In the pages that follow, we will first present an *overview of chi-square distributions and chi-square testing.* This orientation is then followed by a discussion of the procedures involved in conducting a *k-sample hypothesis test of percentages.* Finally, the steps in another procedure — the *chi-square test for goodness of fit* — are presented.

## CHI-SQUARE DISTRIBUTIONS AND CHI-SQUARE TESTING: AN OVERVIEW

The symbol for the Greek letter chi (which is pronounced as the first two letters in the word "kind") is $\chi$. Therefore, the symbol that we will use in this discussion of chi-square distributions and chi-square testing is $\chi^2$.

In any hypothesis-testing situation, we are always concerned with the computation of the value of some critical ratio or statistic for which we know the appropriate distribution. For example, in the tests that have been considered in the previous three chapters, the appropriate distributions at one time or another were the normal distribution, the *t* distribution, and the *F* distribution. But none of these distributions is appropriate if we want to test the significance of the differences that may exist between three or more sample percentages. When such a test is made, a $\chi^2$ *distribution* is needed.

## The $\chi^2$ Distributions

Since, as you will soon see, a $\chi^2$ value is found by adding squared numbers, it will always have a positive sign. As a matter of fact, the *scale* of possible $\chi^2$ values extends from zero indefinitely to the right in a positive direction. Like the binomial, *t*, and *F* probability distributions that we have already studied, the $\chi^2$ probability distribution is *not* a single probability curve. Rather, there is an entire family of $\chi^2$ curves, and the *shapes* of these curves vary according to the number of *degrees of freedom* (*df*) that exists in a given problem. The *df* value, in fact, is the only parameter in a $\chi^2$ distribution, and the mean of any $\chi^2$ distribution is equal to the *df* value.[1]

---

[1] For $\chi^2$ distributions with over 2 *df*, the mode or peak of the curve will be at $df - 2$. Thus, the mode for a $\chi^2$ curve with 7 *df* will be a $\chi^2$ value of 5.

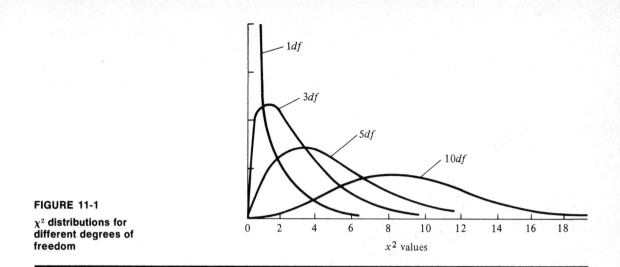

**FIGURE 11-1**

$\chi^2$ **distributions for different degrees of freedom**

Figure 11-1 illustrates the shapes of several $\chi^2$ probability distributions with different degrees of freedom. As you will notice, the $\chi^2$ curves with small degrees of freedom—e.g., from 3 to 10 *df*—are skewed to the right. As the *df* number continues to increase, however, the $\chi^2$ distributions begin to take on the appearance of a normal curve.

**An Introduction to Chi-Square Testing**

Chi-square distributions are used in a procedure that involves the comparison of the differences between the *sample frequencies* of occurrence or percentages that are *actually observed* and the hypothetical or theoretical *population frequencies* of occurrence or percentages that are *expected* if the hypothesis is true.[2] The *steps in the general $\chi^2$ testing procedure are as follows:*

1  *Formulate the null and alternative hypotheses.* The null hypothesis in a particular application, for example, might be that there is *no* significant difference between the frequencies of occurrence or percentages that are of interest in the several populations under study. The alternative hypothesis, on the other hand,[3] might then be that *not all* the population percentages are equal.

2  *Select the level of significance* to be used in the particular testing situation.

3  Take random samples from the populations, and *record the observed frequencies* that are actually obtained.

---

[2] In Chapter 8, the one-sample hypothesis tests of percentages that we made basically involved the comparison of an observed frequency—i.e., the sample percentage—with an expected frequency or hypothetical population percentage.

[3] The following is the supplication of an executive who, after listening to his staff statistician, had once again published an embarrassingly inaccurate market forecast: "Lord, please find me a one-armed statistician . . . so I won't always hear 'on the other hand. . . .' "

**4** *Compute the frequencies or percentages that would be expected if the $H_0$ is true.*

**5** *Use the observed (sample) and expected (hypothetical population) frequencies to compute a $\chi^2$ value with the following formula:*

$$\chi^2 = \sum \frac{(f_o - f_e)^2}{f_e} \qquad (11\text{-}1)$$

where $f_o$ = an observed (sample) frequency

$f_e$ = an expected (hypothetical) frequency if the $H_0$ is true

If the observed frequencies and the expected frequencies are *identical*, formula (11-1) shows us that the computed $\chi^2$ value will be *zero*. From the standpoint of verifying the null hypothesis, a computed $\chi^2$ value of zero would be ideal, since our sample data would be exactly what we had expected. But you are sophisticated enough by now in matters statistical to know that it is very unlikely that the $f_o$ and $f_e$ values in an application would be identical even when the $H_0$ is actually true. Sampling variation would, after all, usually cause some discrepancy between the $f_o$ and $f_e$ values.

**6** *Compare the value of $\chi^2$ computed in step 5 with a $\chi^2$ table value* (found for the specified level of significance from the appropriate $\chi^2$ distribution) to determine if the computed $\chi^2$ value is *significantly* above zero. The maximum value that the computed $\chi^2$ statistic may attain at a chosen level of significance before the $H_0$ must be rejected is specified in Appendix 7. (Further comment on the use of Appendix 7 is given in a few pages.) This table value is the critical $\chi^2$ value that separates the area of acceptance from the area of rejection. That is, this table value is the critical value shown in Fig. 11-2. If the computed $\chi^2$ value $\leq$ the critical value shown in Fig. 11-2, it will fall into the area of acceptance and the $H_0$ will be accepted. But if the computed $\chi^2$ value > the table value, this will be cause for rejecting the $H_0$. If a hypothesis test is being made at the .05 level, the critical table value is simply the point on the $\chi^2$ scale beyond

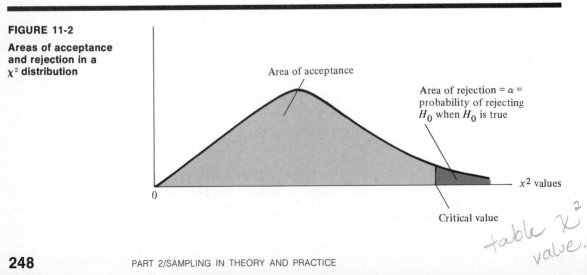

**FIGURE 11-2**

**Areas of acceptance and rejection in a $\chi^2$ distribution**

Area of acceptance

Area of rejection = $\alpha$ = probability of rejecting $H_0$ when $H_0$ is true

$x^2$ values

0

Critical value

*table $\chi^2$ value.*

which a computed $\chi^2$ value would be expected to fall only 5 times in 100 if the $H_0$ were true. (Since there is a separate $\chi^2$ distribution for each $df$ value, the size of a detailed $\chi^2$ distributions table would very quickly become too large for our purposes. Therefore, in Appendix 7, only a selected number of $df$ and $\alpha$ values are available.)

Now that you have some idea of the general $\chi^2$ hypothesis-testing procedure, let's see how this testing concept may be applied to situations in which three or more sample percentages are to be evaluated.

## k-SAMPLE HYPOTHESIS TEST OF PERCENTAGES

Our purpose in this section is to use the $\chi^2$ testing procedure to analyze the significance of the percentage differences that may exist among three or more—i.e., $k$—independent samples. This type of hypothesis test can perhaps be best explained by the use of an example problem.

Three candidates are running for sheriff of Lawless County. These aspirants to public office are Larson E. Bound, Graff D. Lux, and Emma Nocruk. Bound's campaign manager has conducted candidate preference polls in the county's three towns. The results of these samples of county voters are shown in the *contingency table* in Fig. 11-3.[4] As you can see, the samples of voters are classified by town of residence and by candidate preference. In planning future campaign strategies, Bound's manager would like to know if there is a significant difference in voter preference between the three towns.

## Steps in the Hypothesis-testing Procedure

**Formulate the null and alternative hypotheses** The *null hypothesis* basically is that the population percentages favoring each of the three candidates will be unchanged from town to town. This does not mean that our $H_0$ is that each candidate has an equal population percentage of 33.33. Rather, our $H_0$ is to the effect that the population percentage of voters favoring Bound is the same in all three

---

[4] The table is called a contingency table because it shows all the cross-classifications of the variables being studied—i.e., it accounts for all contingencies.

**FIGURE 11-3 Survey of voters classified by town of residence and candidate preference**

| | Towns | | | |
|---|---|---|---|---|
| | White Lightning | Casino City | Smugglersville | |
| Bound | 50 | 40 | 35 | 125 ⎤ row |
| Lux | 30 | 45 | 25 | 100 ⎬ totals |
| Nocruk | 20 | 45 | 20 | 85 ⎦ |
| | 100 | 130 | 80 | 310 |
| | column totals | | | grand total |

towns, the $\pi$ favoring Lux is the same regardless of location of residence, and the $\pi$ favoring Nocruk is equal in the three locations. Of course, the population percentage favoring Lux can be different from the $\pi$ favoring Bound in the context of our null hypothesis. Thus, our $H_0$ may be stated as follows:

$H_0$: The population percentage favoring each candidate is the same from town to town.

The *alternative hypothesis* is:

$H_1$: The population percentage favoring each candidate is *not* the same from town to town.

The null and alternative hypotheses could also be expressed in the following terms:

$H_0$: The candidate preference of voters is *independent*[5] of their town of residence.

$H_1$: The candidate preference of voters is dependent on (or related to) their place of residence.

**Selecting the level of significance** This step does not differ from similar steps in other hypothesis-testing procedures. As noted earlier, however, a detailed table of $\chi^2$ distributions could quickly become very large. Thus, there are only a limited number of $\alpha$ values in Appendix 7. We will assume in our example problem that Bound's campaign manager has specified that the test be made at the .05 level of significance.

**Determine the observed (sample) frequencies ($f_o$)** The actual data from the random samples of voters taken in the three towns are shown in Fig. 11-3.

**Compute the expected frequencies ($f_e$)** If our null hypothesis is true, and if the population percentage of voters favoring Bound is the same in each of the three towns, we should be able to compute the number of sample responses favoring Bound that would be expected in each of the locations. A total of 310 voters were polled, and 125 of these voters expressed a preference for Bound. Since the 125 "votes" received by Bound is 40.32 percent of the total cast in the three towns — $(125/310) \times 100 = 40.32$ percent — Bound should be favored by 40.32 percent of those interviewed in each of the three towns if the $H_0$ is true. Thus, of the 100 people interviewed in White Lightning, we would expect 40.32 percent (or 40.3) of them to favor Bound. Similarly, of the 130 persons polled in Casino City, we would anticipate that 40.32 percent (or 52.3) of them would prefer Bound. And we would expect 32.4

---

[5] This type of $\chi^2$ hypothesis test is frequently referred to as a *test of independence* because the $H_0$ tested is essentially that sample data are being classified in independent ways.

people in Smugglersville (40.32 percent of the 80 persons interviewed) to be in Bound's corner.

The same analysis can also be used to compute the $f_e$ values for the other candidates in each of the towns. (Since 100/310 or 32.26 percent of all those sampled showed a preference for Lux, for example, we would expect that Lux would be favored by about 32.26 of the 100 persons interviewed in the White Lightning sample.) But *the computations of the expected frequencies are somewhat easier to follow* if we refer to the contingency table in Fig. 11-3 and compute the hypothetical or expected frequencies for each *cell* in the table. A cell is formed by the intersection of a column and a row. Since there are 3 rows and 3 columns in Fig. 11-3, there are 3 × 3 or 9 cells. (Tables with *r* rows and *c* columns are often referred to as *r* × *c* tables; in our example, we have a 3 × 3 table.)

The *expected frequencies may be computed for each cell of a contingency table by the following formula:*

$$f_e = \frac{(\text{row total})(\text{column total})}{\text{grand total}} \tag{11-2}$$

The use of this formula may be illustrated by computing the number of persons who would be *expected* to favor Bound in the town of White Lightning if the $H_0$ is true. From Fig. 11-3, you can see that the total for the row in which this cell is located is 125, the total for the column for this cell is 100, and the grand total is 310. Thus, the $f_e$ value for the cell is computed as follows:

$$f_e = \frac{(125)(100)}{310} = 40.3$$

The same procedure can be used to get the $f_e$ values for all the other cells in the table. Figure 11-4 duplicates all the observed frequencies from Fig. 11-3 and shows the expected frequencies for each cell in parentheses.

**FIGURE 11-4 Observed and expected frequencies**

| | White Lightning | Casino City | Smugglersville | | |
|---|---|---|---|---|---|
| | | Towns | | | |
| Bound | 50 (40.3) | 40 (52.3) | 35 (32.4) | 125 | row totals |
| Lux | 30 (32.3) | 45 (41.9) | 25 (25.8) | 100 | |
| Nocruk | 20 (27.4) | 45 (35.8) | 20 (21.8) | 85 | |
| | 100 | 130 | 80 | 310 | |
| | | column totals | | grand total | |

**FIGURE 11-5 Computation of $\chi^2$**

| Row/column (cell) | $f_o$ (1) | $f_e$ (2) | $f_o - f_e$ (3) | $(f_o - f_e)^2$ (4) | $\dfrac{(f_o - f_e)^2}{f_e}$ (5) |
|---|---|---|---|---|---|
| 1-1 | 50 | 40.3 | 9.7 | 94.09 | 2.335 |
| 1-2 | 40 | 52.3 | −12.3 | 151.29 | 2.893 |
| 1-3 | 35 | 32.4 | 2.6 | 6.76 | .208 |
| 2-1 | 30 | 32.3 | −2.3 | 5.29 | .164 |
| 2-2 | 45 | 41.9 | 3.1 | 9.61 | .229 |
| 2-3 | 25 | 25.8 | −.8 | .64 | .025 |
| 3-1 | 20 | 27.4 | −7.4 | 54.76 | 1.998 |
| 3-2 | 45 | 35.8 | 9.2 | 84.64 | 2.364 |
| 3-3 | 20 | 21.8 | −1.8 | 3.24 | .149 |
|  | 310 | 310.0 | 0 |  | 10.365 |

**Compute the $\chi^2$ value using $f_o$ and $f_e$** The computed $\chi^2$ value, you will recall, is found by formula (11-1):

$$\chi^2 = \sum \frac{(f_o - f_e)^2}{f_e}$$

where $f_o$ is the observed frequency of a cell and $f_e$ is the hypothetical expected frequency of the cell

Figure 11-5 illustrates the use of this formula. The columns numbered 1 and 2 in Fig. 11-5 reproduce the data from Fig. 11-4 in a more convenient format.[6] The *steps to compute $\chi^2$* (shown in Fig. 11-5) *are:*

1   Subtract the $f_e$ value from the $f_o$ value—i.e., $f_o - f_e$—for each cell in the contingency table and record the difference[7] (as shown in column 3 of Fig. 11-5).

2   Square the $f_o - f_e$ differences to get $(f_o - f_e)^2$ (column 4, Fig. 11-5).

3   Divide each of these squared differences—$(f_o - f_e)^2$—by the $f_e$ value for each cell to get $\dfrac{(f_o - f_e)^2}{f_e}$ (column 5, Fig. 11-5).

4   Add the $\dfrac{(f_o - f_e)^2}{f_e}$ values to get the computed $\chi^2$ value (column 5, Fig. 11-5). As you can see, the computed $\chi^2$ value for our example problem is 10.365.

**Compare the computed $\chi^2$ value with the table $\chi^2$ value** It was pointed out earlier that the computed $\chi^2$ value should not differ significantly from zero if the $H_0$ is to be accepted. Our computed $\chi^2$ value of 10.365 is obviously not zero, but

---

[6] As a check on your arithmetic, note that $\Sigma f_o = \Sigma f_e$.

[7] As another check on your math, note also that $\Sigma(f_o - f_e)$ must equal zero.

is it close enough to zero to fall into the area of acceptance shown in Fig. 11-2, page 248? Or does our value of 10.365 exceed the critical value point on the $\chi^2$ scale in Fig. 11-2 and fall into the area of rejection? To answer these questions, we must know the $\chi^2$ table value from Appendix 7 that separates the area of acceptance from the area of rejection. We will then be in a position to compare the computed $\chi^2$ value with the table $\chi^2$ value and make a decision about the $H_0$.

*In order to locate the table value* in Appendix 7 and thus make the comparison, *you must know* (1) the desired level of significance and (2) the number of degrees of freedom for the particular problem. We already know that $\alpha$ is .05 for our example problem, but what about the number of degrees of freedom? The answer to this question is that *for an $r \times c$ contingency table the df value is found by the following formula:*

$$df = (r - 1)(c - 1) \tag{11-3}$$

where $r$ = number of rows in the table

$c$ = number of columns in the table

Thus, in a $6 \times 4$ table, the *df* will equal $(6 - 1)(4 - 1)$, or 15. Figure 11-6 may give you an intuitive understanding of the meaning of this *df* figure of 15. For a $6 \times 4$ table, it is actually necessary to compute only 15 $f_e$ values (for the cells in Fig. 11-6 indicated by the check marks). The other nine $f_e$ values for the nine remaining cells in a $6 \times 4$ table (the ones indicated by the X's in Fig. 11-6) are then automatically determined by the row and column totals. In other words, the first three cells in *row 1* are "free" to accept many different values, but the last cell in row 1 is constrained by the row total of 100. Likewise, the first five cells in *column 1* are "free," but the last cell in the column is constrained by the column total. Thus, there are 15 *df* in a $6 \times 4$ table.

For our example table with 3 rows and 3 columns, we have

$$\begin{aligned} df &= (r - 1)(c - 1) \\ &= (3 - 1)(3 - 1) \\ &= 4 \end{aligned}$$

**FIGURE 11-6 The 15 *df* in a 6 × 4 contingency table (A *df* is indicated by a check mark.)**

|  | (1) | (2) | (3) | (4) | Row totals |
|---|---|---|---|---|---|
| (1) | ✔ | ✔ | ✔ | X | 100 |
| (2) | ✔ | ✔ | ✔ | X | 120 |
| (3) | ✔ | ✔ | ✔ | X | 90 |
| (4) | ✔ | ✔ | ✔ | X | 230 |
| (5) | ✔ | ✔ | ✔ | X | 180 |
| (6) | X | X | X | X | 145 |
| Column totals | 200 | 150 | 180 | 335 | 865 |

From Appendix 7, then, we can find the critical $\chi^2$ table value at the intersection of the .05 column (the $\alpha$ specified in our example) and the 4 $df$ row. This table value is 9.488, and, of course, it is less than our computed $\chi^2$ value of 10.365. The table value of 9.488 means that if the $H_0$ is true, the probability of obtaining a computed $\chi^2$ value as large as 9.488 is only .05. Therefore, the probability of getting a computed $\chi^2$ value as large as 10.365 *is less than .05.*

**Make the statistical decision** The *decision rule* in a $\chi^2$ hypothesis test is as follows:

Accept the $H_0$ if the computed $\chi^2$ value $\leqq$ the appropriate table value.

*or*

Reject the $H_0$ if the computed $\chi^2$ value $>$ the appropriate table value.

Since our computed $\chi^2$ value of 10.365 $>$ our table value of 9.488, we will reject the $H_0$ and conclude that the population percentage favoring each candidate for sheriff of Lawless County is not the same from town to town. Bound's campaign manager may decide, as a result of this conclusion, to conduct campaign activities in a more selective fashion.

A summary of the steps in a $k$-sample hypothesis test of percentages is outlined in Fig. 11-7.

**Self-testing Review 11-1**

1 **a** If a $\chi^2$ distribution has 8 degrees of freedom, the mean of this distribution will be 8, and the mode will be 6. True or false?
  **b** If a test is being conducted at the .01 level, the critical $\chi^2$ table value for the distribution described in **a** is 18.475. True or false?

2 The $H_0$ in a $k$-sample hypothesis test of percentages is basically that there is no significant difference between the percentages that are of interest in the several populations under study. True or false?

3 The computed $\chi^2$ value will be zero if the $H_0$ is true. True or false?

4 If the computed $\chi^2$ value $<$ the table $\chi^2$ value, it will fall into the area of rejection. True or false?

5 The student government at Radical University conducted a sample survey to determine student opinion on a proposed new constitution. The results of the survey are summarized below. Is there a significant difference, at the .05 level, between the percentages of each class approving the proposal?

|  | Freshmen | Sophomores | Juniors | Seniors |
|---|---|---|---|---|
| Approve | 40 | 30 | 30 | 20 |
| Disapprove | 30 | 30 | 20 | 30 |

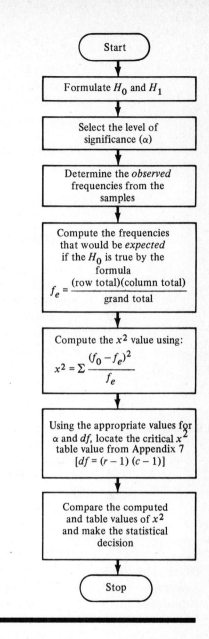

**FIGURE 11-7**

**Procedure for a**
**$k$-sample hypothesis**
**test of percentages**

The flowchart contents:

Start

Formulate $H_0$ and $H_1$

Select the level of significance ($\alpha$)

Determine the *observed* frequencies from the samples

Compute the frequencies that would be *expected* if the $H_0$ is true by the formula
$$f_e = \frac{\text{(row total)(column total)}}{\text{grand total}}$$

Compute the $x^2$ value using:
$$x^2 = \Sigma \frac{(f_0 - f_e)^2}{f_e}$$

Using the appropriate values for $\alpha$ and $df$, locate the critical $x^2$ table value from Appendix 7
$$[df = (r - 1)(c - 1)]$$

Compare the computed and table values of $x^2$ and make the statistical decision

Stop

## ANOTHER CHI-SQUARE TEST: GOODNESS OF FIT

A second important $\chi^2$ analysis procedure is the test for "goodness of fit." Such a test can be used, for example, to determine if a population under study "fits" or follows one with a known distribution of values. That is, the test can be used to evaluate the *goodness of fit* between the population under study and normal, binomial, or uniform distribution. (A *uniform* distribution is one in which the frequencies are equal or uniform.)

Since the purpose of a goodness of fit test is to decide if the sample results are

FIGURE 11-8
Results of
Bitter Bottling's
taste test

| Brand | Number preferring brand |
|-------|-------------------------|
| A | 50 |
| B | 65 |
| C | 45 |
| D | 70 |
| E | 70 |
| | 300 |

consistent with the results that would have been obtained if a random sample had been selected from a population with a known distribution, the *null hypothesis* is essentially that the population being analyzed fits some distribution pattern—e.g., the $H_0$ might be that the population distribution is uniform. The *alternative hypothesis* will then simply be that the population does not fit the specified distribution— e.g., $H_1$: The population distribution is *not* uniform.

Let's now use the following problem situation to show how $\chi^2$ concepts can be used to test the goodness of fit between sample data and a *uniform* distribution.[8] As you will see, the steps in the testing procedure follow a familiar pattern.

The Bitter Bottling Company has developed "Featherweight," the cola with fewer calories and less taste. To evaluate this new product, the marketing manager has given a taste test to 300 people. Each person in the sample was asked to taste Featherweight and four other diet cola brands. To avoid prejudice, the actual brand labels were replaced by the letters A, B, C, D, and E. The results of the sample are shown in Fig. 11-8.

**Steps in the Goodness of Fit Testing Procedure**

**Formulate the null and alternative hypotheses** The *null hypothesis* is:

$H_0$: The population distribution is uniform—i.e., the cola brands are preferred by an equal percentage of the population.

And the *alternative hypothesis* is:

$H_1$: The population distribution is *not* uniform—i.e., the taste preference frequencies are not equal.

---

[8] The test for goodness of fit between sample data and a uniform distribution is the simplest and perhaps the most useful. The procedure in testing other known distributions would differ from our example primarily in the ways the $f_e$ and $df$ values are determined. We will not consider tests of other distributions.

**Select the level of significance** Let's assume that Bitter's marketing manager wants to conduct the test at the .05 level of significance.

**Determine the observed (sample) frequencies ($f_o$)** The $f_o$ values are given in Fig. 11-8.

**Compute the expected frequencies ($f_e$)** If the null hypothesis is true, we would expect an equal or uniform number of people to prefer each of the five brands. That is, there is no significant difference in taste preference, and one-fifth or 20 percent of the tasters should prefer brand A, 20 percent should prefer brand B, etc. Thus, the $f_e$ value for each brand should be 20 percent of 300, or 60.

**Compute the $\chi^2$ value using $f_o$ and $f_e$** Once again, the appropriate formula is:

$$\chi^2 = \sum \frac{(f_o - f_e)^2}{f_e}$$

Figure 11-9 shows the computation of $\chi^2$ for our goodness of fit example problem. The computed value is 9.168.

**Compare the computed $\chi^2$ value with the table $\chi^2$ value** The number of degrees of freedom for our example is 4. This is because the sum of the sample observations is 300, and this total must also be the sum of the expected frequencies. Therefore, one of the five brand preference categories is constrained by the total. In fitting a uniform distribution, then, the *df* value is one less than the number of classes or categories.

Once the *df* value is known, the $\chi^2$ table value can be determined. At the .05 level, and with 4 degrees of freedom, the table value found in Appendix 7 is 9.488 — a figure that is slightly larger than our computed $\chi^2$ statistic of 9.168.

**Make the statistical decision** The decision rule for the goodness of fit test is the same as the rule given earlier for the *k*-sample test of percentages. Thus, we will *accept* our $H_0$ because the computed $\chi^2$ value < the table value. At the .05 level, we cannot reject the hypothesis that all the cola brands were preferred by an equal

**FIGURE 11-9 Computation of $\chi^2$ for goodness of fit test**

| Brand | Number preferring ($f_o$) | $f_e$ | $f_o - f_e$ | $(f_o - f_e)^2$ | $\dfrac{(f_o - f_e)^2}{f_e}$ |
|-------|---------------------------|-------|-------------|-----------------|------------------------------|
| A | 50 | 60 | −10 | 100 | 1.667 |
| B | 65 | 60 | 5 | 25 | .417 |
| C | 45 | 60 | −15 | 225 | 3.750 |
| D | 70 | 60 | 10 | 100 | 1.667 |
| E | 70 | 60 | 10 | 100 | 1.667 |
|   | 300 | 300 | 0 | | 9.168 |

percentage of the population. We must conclude that at the .05 level Featherweight is not significantly better tasting than the other brands.

1 The chi-square goodness of fit test follows the same steps as the $k$-sample test, although there are some differences within the steps. True or false?

2 In a test to determine if a population distribution is uniform, the $df$ value is one less than the number of classes being considered. True or false?

3 If the null hypothesis is that the sales produced in six sales districts are equal or uniform, and if the test is to be made at the .10 level, the $\chi^2$ table value in this test is 11.070. True or false?

4 In the preceding question, should the $H_0$ be accepted if the computed $\chi^2$ value had been 10.625?

5 An examination of an automobile agency's sales records shows that on 70 business days only one new truck was sold. However, two trucks per day were sold on each of 60 business days; three trucks per day were sold on 40 days; and there were daily sales of four trucks on 30 days. That is,

| Truck sales per day | Number of sales days |
|---|---|
| 1 | 70 |
| 2 | 60 |
| 3 | 40 |
| 4 | 30 |

At the .01 level, test the hypothesis that the demand for trucks is uniformly distributed.

## SUMMARY

Chi-square analysis may be used to test the hypothesis that three or more independent samples have all come from populations that have the same percentage of a given characteristic. This application of $\chi^2$ analysis—the $k$-sample hypothesis test of percentages—has been considered in this chapter. And $\chi^2$ analysis can also be used to determine if a population under study fits or follows one with a known distribution of values. This application of $\chi^2$ concepts has been used in the chapter to test the goodness of fit between sample data and a uniform distribution.

Regardless of the application, however, the steps followed in one $\chi^2$ testing situation are very similar to the steps followed in another. The null and alternative hypotheses will, of course, be phrased differently depending on the type of application. And the expected frequencies and number of degrees of freedom will be computed in different ways depending on the logic of the situation. But the $\chi^2$ test statistic will generally be computed in the same way. This computed value must always then be compared with a critical table value, and the statistical decision will be to accept the $H_0$ if the computed value $\leq$ the table value.

## Important Terms and Concepts

1 Chi-square distributions

2 Degrees of freedom ($df$)

3 Observed frequencies ($f_o$)

4 Expected frequencies ($f_e$)

5 $\chi^2 = \sum \dfrac{(f_o - f_e)^2}{f_e}$

6 $k$-Sample hypothesis test of percentages

7 Contingency table

8 Cell

9 $f_e = \dfrac{\text{(row total)(column total)}}{\text{grand total}}$

10 $df = (r-1)(c-1)$

11 Goodness of fit test

12 Uniform distribution

## Problems

1 The following table shows the number of good and defective parts produced on each work shift at a manufacturing plant. Using the .05 level of significance, test the hypothesis that there is no significant difference between the percentages of defective parts produced on the three shifts.

|  | First shift | Second shift | Third shift |
|---|---|---|---|
| Good | 90 | 70 | 60 |
| Defective | 10 | 20 | 20 |

2 A survey of rural and urban television viewing populations is made to determine television programming preferences, and the following results are obtained:

| | Types of programs preferred | | | |
|---|---|---|---|---|
| | Western | Comedy | Mystery | Variety |
| Urban | 80 | 100 | 100 | 60 |
| Rural | 70 | 70 | 50 | 40 |

At the .05 level, test the hypothesis that there is no difference in program preference between urban and rural residents.

3 A sample of 300 consumers was asked to choose among six brands of coffee. The results of the survey are as follows:

| Brand | Number preferring |
|---|---|
| A | 40 |
| B | 55 |
| C | 60 |
| D | 40 |
| E | 60 |
| F | 45 |
|  | 300 |

Using a .01 level of significance, test the hypothesis that a uniform distribution describes the coffee-drinking population.

**4** Charlie "Crank" Schaff, the automotive writer for the local newspaper, believes that front fenders of cars are more likely to be damaged in accidents than rear fenders. In a sampling of body shops, Charlie observed the following damage:

| Fender damaged | Number of damaged fenders |
|---|---|
| Right front | 150 |
| Left front | 120 |
| Right rear | 125 |
| Left rear | 105 |
| | 500 |

At the .10 level, would you agree with Charlie?

**5** George Gullible has a bet with a fraternity brother that if the 1, 2, or 3 face of a die turns up on a single roll, he will pay the brother 10 cents. If, on the other hand, the 4, 5, or 6 face is rolled, George will win 10 cents. The brother is to supply the die. A third fraternity member has borrowed and experimented with the die, and the results were as follows:

| Die face | Frequency of appearance |
|---|---|
| 1 | 115 |
| 2 | 100 |
| 3 | 125 |
| 4 | 95 |
| 5 | 85 |
| 6 | 80 |

At the .05 level, is George going to be rolling with a fair die?

**6** Integrated circuits are supplied to a computer manufacturer by two vendors. Each circuit is tested by the manufacturer for four common defects before it is accepted. The following data, representing test results over a 2-week period, have been supplied to the purchasing department by the quality control manager:

| Vendor | Defective circuits by type of defect | | | |
|---|---|---|---|---|
| | 1 | 2 | 3 | 4 |
| A | 60 | 80 | 40 | 30 |
| B | 30 | 32 | 25 | 20 |

At the .01 level, would you accept the hypothesis that the percentages of common defects are the same for both vendors?

**7** Still seeking public office, Polly Tician is planning her campaign strategy on a particular issue and wants to know if there is a significant difference in the percentage of voters who favor the issue between urban, suburban, and rural voters. The following data have been gathered:

|  | Urban | Suburban | Rural |
|---|---|---|---|
| In favor | 78 | 65 | 60 |
| Opposed | 42 | 45 | 50 |

Should the null hypothesis that the percentage favoring the issue is the same for the urban, suburban, and rural voters be accepted at the .10 level?

**Topics for Discussion**

**1** In what ways are the $t$, $F$, and $\chi^2$ probability distributions similar?

**2** "In any hypothesis-testing situation, we are always concerned with the computation of the value of some critical ratio or statistic for which we know the appropriate probability distribution." Discuss this statement in the context of $\chi^2$ procedures.

**3 a** Discuss the general steps to be followed in conducting a $k$-sample hypothesis test of percentages.
**b** In conducting a goodness of fit test.

**4** "From the standpoint of verifying the null hypothesis, a $\chi^2$ computed value of zero would be ideal." Explain why this statement is true.

**5** Given the statement in question **4**, explain why it is very unlikely that you would get a computed value of zero even if the $H_0$ is true.

**6** A critical $\chi^2$ table value is found to be 9.236 in a hypothesis test conducted at the .10 level.
**a** Explain the meaning of 9.236.
**b** Assume that the computed $\chi^2$ value in the test is 12.312. Make the appropriate statistical decision, and explain the reason for your decision.

**7** Discuss the procedure for computing the expected frequencies in the 6 × 4 contingency table shown in Fig. 11-6.

**8 a** Discuss the purpose of a $k$-sample hypothesis test of percentages.
**b** Discuss the purpose of a goodness of fit test.

**Answers to Self-testing Review Questions**

**11-1**

**1 a** This statement is true.
**b** This is incorrect. The table value is 20.090 when $\alpha = .01$ and $df = 8$.

**2** This is a true statement.

**3** This is false. Sampling variation is likely to lead to discrepancies between $f_o$ and $f_e$ even when the $H_0$ is true.

**4** Another false statement. The computed value would be in the area of acceptance, and the $H_0$ would be accepted.

**5** **a** The hypotheses are:

   $H_0$: The population percentage of students approving the proposed constitution is the same from class to class—i.e., student approval is independent of class standing.

   $H_1$: The population percentage of students approving the proposed constitution is not the same from class to class—i.e., student approval is dependent on (or related to) class standing.

**b** A .05 value of $\alpha$ is specified.

**c** The $f_e$ values for each cell are shown below in parentheses:

|  | Freshmen | Sophomores | Juniors | Seniors |
|---|---|---|---|---|
| Approve | 40 | 30 | 30 | 20 |
|  | (36.5) | (31.3) | (26.1) | (26.1) |
| Disapprove | 30 | 30 | 20 | 30 |
|  | (33.5) | (28.7) | (23.9) | (23.9) |

**d** The computed $\chi^2$ value is 5.017.

**e** The $df$ value is computed as follows:
$$df = (r-1)(c-1)$$
$$= (2-1)(4-1)$$
$$= 3$$
Therefore, the $\chi^2$ table value at $df = 3$ and $\alpha = .05$ is 7.815.

**f** The statistical decision is:

   Accept $H_0$, since the computed $\chi^2$ value of 5.017 < the table value of 7.815.

## 11-2

**1** This is true.

**2** This is also true.

**3** Finally, a false statement. The table value of 11.070 would be correct for the .05 level and 5 $df$. At the .10 level, however, the figure should be 9.236.

**4** No. With a computed value of 10.625, and with a correct table value of 9.236, the $H_0$ should be rejected.

**5** **a** The hypotheses are:

   $H_0$: The population distribution is uniform.

   $H_1$: The population distribution is not uniform.

**b** The level of significance is specified at .01.

**c**  If the $H_0$ is true, there would be 50 days in which one truck was sold, 50 days in which two trucks were sold, etc.

**d**  The computed $\chi^2$ value is 20.00.

**e**  The *df* value = 3. Therefore, the $\chi^2$ table value at $df = 3$ *and* $\alpha = .01$ is 11.345.

**f**  The statistical decision is:

Reject $H_0$ since the $\chi^2$ computed value > the table value.

# COPING WITH CHANGE

It was pointed out in Chapter 1 that statistical theories and methods could be useful in (1) describing relationships between variables, (2) performing the decision-making process, and (3) providing a conceptual foundation to analyze and cope with change. There are sections in one or more of the chapters in this part that deal with *each* of these uses of statistical techniques. As indicated by the part title, however, we are particularly interested here in the *measurement* and *prediction of change*.

To plan is to decide in advance on a future course of action. Therefore, virtually all plans and decisions are based on expectations about future events and/or relationships. And thus virtually all planners and decision makers are required to employ some forecasting process or technique to arrive at a future expectation. *One* forecasting approach, of course, is to assume that nothing will change and that no new plans are needed. In the short run, this may frequently be the approach followed, but the assumption about a lack of change seldom remains valid for long, although it is possible for gradual change to go unnoticed for some time. In Chapter 12, therefore, we look at procedures which have been developed to *measure relative changes* in economic conditions over time.

A *second* forecasting approach consists of (1) analyzing past empirical data to detect reasonably dependable patterns and then (2) projecting these patterns into the future to arrive at future expectations. In Chapter 13 we examine some of the techniques employed in this approach.

Finally, a *third* forecasting approach is to identify and analyze reasonably predictable independent factors which are having a significant influence on the dependent variable to be predicted. A prediction of values of the dependent variable may then be made on the basis of (1) the closeness of the relationship between the variables and (2) the values of the independent variable(s). A brief introduction to this forecasting approach involving the use of only two variables is the subject of Chapter 14.

The chapters included in Part 3 are:

# CHAPTER 12

# MEASURING CHANGE: INDEX NUMBERS

**LEARNING OBJECTIVES**

After studying this chapter, working the problems, and answering the discussion questions, you should be able to:

☞ Explain the uses and advantages of index numbers.

☞ Describe the types of index numbers in current use.

☞ Construct index numbers by the use of the aggregative and average of relatives methods.

☞ Discuss some of the possible problems and pitfalls associated with the use of index numbers.

**CHAPTER OUTLINE**

ADVANTAGES OF INDEX NUMBERS

TYPES OF INDEX NUMBERS
Price Indexes
Self-testing Review 12-1
Quantity Indexes
Value Indexes

AGGREGATIVE METHOD OF INDEX NUMBER CONSTRUCTION
Implicitly Weighted Aggregative Indexes
Self-testing Review 12-2

The Need for Explicit Weights
Weighted Aggregative Indexes
Self-testing Review 12-3

AVERAGE OF RELATIVES METHOD OF INDEX NUMBER CONSTRUCTION
Implicitly Weighted Average of Relatives Indexes
Self-testing Review 12-4
Weighted Average of Relatives Indexes
Self-testing Review 12-5

Index numbers are generally used to measure changes occurring over some period of time. Let us assume, for example, that the famous overworked pedagogue, Professor Staff, has an unhealthy interest in the price of the chalk he uses in teaching his statistics course. Figure 12-1 shows the prices for a box of chalk for the years 1976–1980.

**FIGURE 12-1**

| Year | Price of chalk per box |
|------|------------------------|
| 1976 | $0.60 |
| 1977 | 0.60 |
| 1978 | 0.65 |
| 1979 | 0.70 |
| 1980 | 0.75 |

In its most basic form, an index number is simply a *percentage relative* that shows the relationship between two values. Thus, we can compare the prices of chalk from Fig. 12-1 by (1) dividing each year's price by the price in 1976 and (2) converting each relative figure obtained into a percentage as shown below:

| Year | Price in year ÷ price in 1971 | Price relative | Percent relative or index number |
|------|-------------------------------|----------------|----------------------------------|
| 1976 | $0.60 ÷ $0.60 = | 1.000 × 100 = | 100.0 |
| 1977 | 0.60 ÷ 0.60 = | 1.000 × 100 = | 100.0 |
| 1978 | 0.65 ÷ 0.60 = | 1.083 × 100 = | 108.3 |
| 1979 | 0.70 ÷ 0.60 = | 1.167 × 100 = | 116.7 |
| 1980 | 0.75 ÷ 0.60 = | 1.250 × 100 = | 125.0 |

The rightmost column shows the relationship between yearly chalk prices in a simple index number form.

Several points can be illustrated by this example:

**1** A *base period* (the year 1976) was *arbitrarily* selected, and this *base period has an index number value of 100.*

**2** The computed index numbers are percentages, but the percent sign is seldom, if ever, used.

**3** The index number value for 1978 (108.3) does not show the price of chalk in either 1976 or 1978; nor does it show the fact that there was a $0.05 difference in price. Rather, the index number only shows the *relative change.*

**4** The *interpretation* of the 1978 index number (108.3) is that the price of chalk in 1978 was 108.3 percent of the 1976 price; alternatively, we could say that there was an 8.3 percent *increase* in chalk price between 1976 and 1978. (Between 1976 and 1980, of course, there was a 25 percent increase in price.)

Although a series of index numbers *can* be constructed from data pertaining to a *single* variable (after all, we have just used data pertaining to a single variable to produce a series of chalk price indexes), *most* of the index numbers commonly used today are prepared from data about *several* different items and form a *composite* index number series.[1] In general usage, therefore, an *index number may be defined as a measure of how much a composite group changes on a relative basis with respect to a base period, usually defined to have a value of 100.*

Composite index numbers[2] may be prepared in different ways, as we shall see later, but the points illustrated by the chalk price example will usually remain valid. Let us now, in the remainder of this chapter, consider the (1) *advantages of index numbers,* (2) *types of index numbers,* (3) *aggregative method of index number construction,* (4) *average of relatives method of index number construction,* (5) *important index numbers in current use,* and (6) *problems and pitfalls with index numbers.*

## ADVANTAGES OF INDEX NUMBERS

Index numbers provide a means of *measuring, summarizing,* and *communicating* the nature of changes that occur from time to time or from place to place. When compared with raw data, *index numbers may have the following advantages:*

**1** *They may simplify data and aid in communication.* A *single* composite index number may be used, for example, to give an indication of some overall change in an economic variable. The Consumer Price Index (CPI) is a popular index number series that periodically measures the cost of about 400 dissimilar items that urban workers buy. Thus, if the CPI for a particular month was 195.3, and if the base period was 1967, the single index value would indicate the average relationship between prices in the two time periods. (In this example, it would take $19.53 to buy the same quantities of items that could have been purchased

---

[1] Index numbers prepared from changes in a single data item are called *simple* index numbers; those constructed from changes in a number of different items are called *composite* index numbers.

[2] For the remainder of this chapter, the term "index number" will refer to a composite index number.

for $10.00 in 1967.) Obviously, the single value of 195.3 is simpler to understand and deal with than the prices of 400 items which, in the case of the CPI, come from over 50 different urban areas.

2 *They may facilitate comparisons.* Changes in series of items expressed in different absolute measurements (e.g., in dollars, tons, cubic feet, bales) may be compared by converting the different measurements to relative values.

3 *They may be used to show typical seasonal variations.* A series of index numbers may be prepared by a company to show the variations in retail store sales during each month for several years. When compared with an average monthly sales index of 100, the variations in seasonal sales patterns become obvious. An index of 145 for December would show, for example, that retail sales during this peak month were 45 percent greater than they were during an average month of the year. We shall consider this topic again in the next chapter.

## TYPES OF INDEX NUMBERS

Most of the frequently used index numbers may be classified as (1) *price indexes,* (2) *quantity indexes,* or (3) *value indexes.*[3]

### Price Indexes

The first recorded price index was prepared in Italy in 1764 by G. R. Carli to compare Italian prices in 1750 for oil, grain, and wine with the prices in 1500 for the same basic items. Computed for relative price changes only, price indexes are useful to a wide spectrum of decision makers, including (1) business managers who are keenly aware of the important role of price in consumer purchasing decisions, (2) government economists and policy makers who must attempt to plan budgets and avoid the perils of both price inflation[4] and economic recession, and (3) union leaders who are very sensitive to the effect of price changes on the economic well-being of their union members.[5]

A union leader might wish to determine, for example, what has been happening to the *purchasing power* of the dollar prior to entering into wage negotiations. A proposed raise in the money earnings of workers may be misleading, unless price changes affecting what the earnings will buy are known. What is needed by the union leader, in short, are data on real earnings—dollar earnings adjusted for changes in price levels. Fortunately for the union negotiator, a *purchasing power index* may be easily computed by using an appropriate price index. "How?" you

---

[3] A few of the most important index numbers in current use are discussed later in the chapter.

[4] Will Rogers had little encouragement for those struggling with this problem. "If I had to populate an asylum with people certified insane," he said, "I'd just pick 'em from all those who claim to understand inflation."

[5] In describing methods of index number construction later, we will be concerned *primarily with price indexes* because (1) the principal techniques of index number preparation can be well illustrated with price indexes, and (2) you are likely to come into more frequent contact with price indexes than with the other types. However, we will also briefly consider the computation of quantity indexes. We need not calculate value indexes here.

ask (with commendable curiosity). In this way: If we take, say, the June 1978 CPI of 195.3 (where 1967 is the base of 100) and (1) convert it back to the price relative of 1.953 and then (2) compute the reciprocal of 1.953 (i.e., $1 \div 1.953$), we will get a value of 0.512 which, if it is multiplied by 100 to produce a purchasing power index number in conventional form, becomes 51.2. What does this number mean? It means that there is no difference in purchasing power between $100 in June 1978 and $51.20 in 1967. A union negotiator would, of course, want to convert old wage rates and proposed new wage rates into *comparable* real earnings terms before reaching any agreements.

**Self-testing Review 12-1**

The monthly money wages of a worker for 1970, for 1974, and for June 1978, along with the CPI values for these periods, are:

| Time period | Money wages (monthly) | Consumer Price Index (1967 = 100) |
| --- | --- | --- |
| 1970 | $600 | 116.3 |
| 1974 | 750 | 147.7 |
| June 1978 | 925 | 195.3 |

1   Prepare a purchasing power index for each time period.

2   What has happened to the worker's purchasing power between 1970 and 1974? Between 1970 and June 1978?

**Quantity Indexes**

Price indexes, as we have just seen, have important uses. There are numerous variables subject to change, however, that cannot and/or should not be expressed in terms of price changes. Rather, these variables should perhaps be measured in terms of relative change in the physical volume of such activities as production and construction. *Quantity indexes, therefore, have been developed to show relative change in physical units.* For example, production of some good — say electric motors — is a measure of industrial output which may be best expressed as a physical count of motors produced if that is the information needed by a decision maker. The number of books sold by a publisher and the number of housing starts within some geographic area are additional examples for which physical activity values may be more useful than price data. The methods used to prepare price indexes are easily adapted to quantity index construction.

**Value Indexes**

Since a measure of *value* is obtained by multiplying price by quantity, a *value index deals with a variable which has elements of change in both price and quantity.* Dollar sales volume, for example, is the product of price and quantity sold. When dollar volume changes, the change could be the result of a change in price, a change in quantity sold, or a change in both of these components. Similarly, changes in income received could result from changes in the rate of pay, from changes in the level of output, or from changes in both factors. An example of a value index is

**FIGURE 12-2**

Professor Staff's essential commodities for teaching statistics

| Commodity | Unit of purchase | Unit price | | | | |
|---|---|---|---|---|---|---|
| | | 1976 | 1977 | 1978 | 1979 | 1980 |
| Chalk | 12-piece box | $ 0.60 | $ 0.60 | $ 0.65 | $ 0.70 | $ 0.75 |
| Red pencils | ½ dozen box | 0.72 | 0.75 | 0.80 | 0.85 | 0.92 |
| Statistics book | 1 workbook | 9.50 | 10.00 | 10.00 | 10.50 | 11.00 |
| Aspirin | 1 bottle | 0.59 | 0.59 | 0.65 | 0.69 | 0.75 |
| | | $11.41 | $11.94 | $12.10 | $12.74 | $13.42 |

the one showing the value of construction contracts awarded, which is published monthly by the F. W. Dodge Corporation, a McGraw-Hill division.

Regardless of whether they deal primarily with price, quantity, or value, most index numbers are constructed by either an *aggregative* or an *average of relatives* method. Let us now examine each of these techniques used to prepare index number series.

## AGGREGATIVE METHOD OF INDEX NUMBER CONSTRUCTION

The term "aggregate" means the sum of a series of values. An aggregative index number,[6] then, is one for which the basis for computation in each time period is the sum of the usually quoted units of measure of the components being considered. In the discussion that follows, we will look at (1) *implicitly weighted aggregative indexes*, (2) *the need for explicit weights, and* (3) *weighted aggregative indexes.*

## Implicitly Weighted Aggregative Indexes

**Price index** Let us assume that the mild-mannered statistics instructor we met earlier, Professor Ogive Staff, is interested in measuring how the price of certain commodities has changed during his past 5 years of teaching statistics. Figure 12-2 gives the items regarded as essential by Professor Staff and the prices of these items from 1976 to 1980.

An implicitly weighted aggregative price index may be found by using the following formula:

$$PI_n = \frac{\Sigma p_n}{\Sigma p_0} \cdot 100 \qquad (12\text{-}1)$$

where $PI_n$ = price index for a given time period $n$

$p_n$ = prices in period $n$ of the components in the series

$p_0$ = prices in the base period of the components in the series

Thus, if we select the prices of Professor Staff's commodities in 1976 to represent

---

[6] Students sometimes irreverently term this an aggravative index number.

**FIGURE 12-3**

Implicitly weighted aggregative price index for Professor Staff's commodities

| Year | Price index (1976 = 100) |
|------|--------------------------|
| 1976 | 100.0 |
| 1977 | 104.6 |
| 1978 | 106.0 |
| 1979 | 111.7 |
| 1980 | 117.6 |

the base period  we can compute the price index for selected time periods covered by the series as shown below:

$$PI_{1976} = \frac{\Sigma p_{1976}}{\Sigma p_0} \cdot 100 = \frac{\$11.41}{\$11.41} \cdot 100 = 100.0$$

$$PI_{1980} = \frac{\Sigma p_{1980}}{\Sigma p_0} \cdot 100 = \frac{\$13.42}{\$11.41} \cdot 100 = 117.6$$

The complete price index series for our example is shown in Fig. 12-3.

**Quantity index** An implicitly weighted aggregative quantity index *could* be computed using the formula $QI = \Sigma q_n / \Sigma q_0 \cdot 100$, where $q_n$ and $q_0$ obviously refer to *quantities* in different periods rather than prices. However, this formula is seldom used to measure changes in quantities because (to take one possible case) it would be meaningless to add together the quantities in a time period that are expressed in different units such as ounces, pounds, tons, and bales.

**Self-testing Review 12-2**

1  Compute an implicitly weighted aggregative price index series for the data given below, using the prices in 1979 as the base.

| Commodity | 1979 Unit price | 1980 Unit price | 1981 Unit price |
|-----------|-----------------|-----------------|-----------------|
| A | $21 | $23 | $24 |
| B | 40 | 44 | 48 |
| C | 10 | 9 | 10 |
| D | 25 | 25 | 28 |

2  Interpret the meaning of your price index series.

## The Need for Explicit Weights

Perhaps you have wondered what was meant by the words "implicitly weighted" in the preceding paragraphs. Let us see now if we can clarify this matter. In our example of the prices of Professor Staff's commodities, we did not attach any *relative importance* to the various commodities. But, of course, Professor Staff probably does *not* consider each item to be equally important. Rather, he may attach a greater priority or *weight* to aspirin than to red pencils. But as you can see from Fig. 12-2, the price of a unit of pencils is greater than the price of a unit of aspirin. Thus, by using formula (12-1) we have automatically or *implicitly* given *greater* weight to red pencils than to aspirin.

When weights are *not* assigned to the various items in a series, the index produced is often called an unweighted index. It is more appropriate to call it an implicitly weighted index, however, because (in the case of a price index) greater weight will implicitly be given to higher-priced items than to lower-priced items. To summarize, if explicit weights were *not* assigned to the individual components in a composite consumer price series in a logical way — on the basis of the relative importance of the components — an item such as lawn mower blades would be given more influence than bread or milk. Obviously, however, people buy much greater *quantities* of bread than lawn mower blades, and thus they consider an increase in the price of bread more significant than an increase in the price of blades. To give bread its proper place in a composite price index series, a *system of explicit weights* is needed that will take into account the *typical quantities* of each item consumed. In the next section we will examine methods of constructing weighted aggregative indexes.

## Weighted Aggregative Indexes

**Price index** In order to assign proper priority to each of the items included in an index number series, a logical weighting system must be used. The weights employed depend on the nature and purpose of the computed index. In the case of price indexes, as we saw above, the usual weighting scheme is to take into account the *typical quantities used*[7] of each of the items in the series. By multiplying the price of

---

[7] In some situations, the base period quantities ($q_0$) are used to represent typical quantities; in other situations, the quantities in the given time period ($q_n$) are used as weights. The technical considerations involved in selecting the appropriate weights are generally beyond the scope of this book, but we will briefly consider weighting problems in a later section.

### FIGURE 12-4

Professor Staff's essential commodities: data for weighted aggregative price index series

| Commodity | Unit of purchase | Typical annual consumption ($q_t$) | Price | | | | |
|---|---|---|---|---|---|---|---|
| | | | 1976 | 1977 | 1978 | 1979 | 1980 |
| Chalk | 12-piece box | 4 | $0.60 | $ 0.60 | $ 0.65 | $ 0.70 | $ 0.75 |
| Red pencils | 1/2 dozen | 1/2 | 0.72 | 0.75 | 0.80 | 0.85 | 0.92 |
| Statistics book | 1 workbook | 1 | 9.50 | 10.00 | 10.00 | 10.50 | 11.00 |
| Aspirin | 1 bottle | 6 | 0.59 | 0.59 | 0.65 | 0.69 | 0.75 |

an item by the quantities of the item consumed in a time period, we get a dollar value that is indicative of the overall importance placed on the item.

Let us now illustrate how a weighted aggregative price index may be prepared by referring once again to Professor Staff's commodities. The data in Fig. 12-4 are the same as those in Fig. 12-2, with one important addition: We have now included a column indicating the typical annual consumption of each of the items. (You will note the much greater consumption of aspirin than of red pencils.)

The formula for computing a weighted aggregative price index may be expressed as follows:

$$PI_n = \frac{\Sigma(p_n q_t)}{\Sigma(p_0 q_t)} \cdot 100 \qquad (12\text{-}2)$$

where $PI_n$ = price index in a given time period $n$

$p_n$ = price in a given period $n$

$p_0$ = price in the base period

$q_t$ = typical number of units produced or consumed during the time periods considered

The application of this formula is not difficult. To get the *numerator*, we merely:

1  Multiply each commodity price in a *given period* by the corresponding typical quantities consumed.

2  Total the products of prices and quantities for the period.

And to get the *denominator*, we:

1  Multiply each commodity price in the *base period* by the corresponding typical quantities consumed.

2  Total the products of base period prices and typical quantities.

Figure 12-5 shows the results of these computations. Thus, if we designate *1976 as the base year*, we can put the totals for selected time periods from Fig. 12-5 into formula (12-2) as shown below:

**FIGURE 12-5**

Professor Staff's essential commodities: Value sums — 1976–1980

| Commodity | 1976 $(p_{1976}q_t)$ | 1977 $(p_{1977}q_t)$ | 1978 $(p_{1978}q_t)$ | 1979 $(p_{1979}q_t)$ | 1980 $(p_{1980}q_t)$ |
|---|---|---|---|---|---|
| Chalk | $ 2.40 | $ 2.40 | $ 2.60 | $ 2.80 | $ 3.00 |
| Red pencils | 0.36 | 0.38 | 0.40 | 0.42 | 0.46 |
| Statistics book | 9.50 | 10.00 | 10.00 | 10.50 | 11.00 |
| Aspirin | 3.54 | 3.54 | 3.90 | 4.14 | 4.50 |
| | $15.80 | $16.32 | $16.90 | $17.86 | $18.96 |

$\Sigma(p_n q_t)$

$\Sigma(p_0 q_t)$ ------------------------→ $15.80

Divided by

**FIGURE 12-6**

Weighted aggregative price index
series (1976 = 100)

| Year | Price index series |
| --- | --- |
| 1976 | 100.0 |
| 1977 | 103.3 |
| 1978 | 107.0 |
| 1979 | 113.0 |
| 1980 | 120.0 |

$$PI_{1976} = \frac{\Sigma(p_{1976}q_t)}{\Sigma(p_0q_t)} \cdot 100 = \frac{\$15.80}{\$15.80} \cdot 100 = 100.0$$

$$PI_{1980} = \frac{\Sigma(p_{1980}q_t)}{\Sigma(p_0q_t)} \cdot 100 = \frac{\$18.96}{\$15.80} \cdot 100 = 120.0$$

The complete price index series for this example is shown in Fig. 12-6.

**Quantity index** What is actually being measured by a weighted quantity index is the change in the value (quantity times price) of a group of items between a base period and a given time period. Measured changes in value are attributed to *quantity changes only,* however, because the prices used are constant or unchanging. Of course, this is also what we did in computing a weighted aggregative price index: We attributed changes in value to price changes only because we kept the typical quantity figures unchanged. Thus, if we exchanged the $p$'s and the $q$'s in formula (12-2), we would get the following formula for the weighted aggregative quantity index:

$$QI_n = \frac{\Sigma(q_n p_t)}{\Sigma(q_0 p_t)} \cdot 100 \qquad (12\text{-}3)$$

where $QI_n$ = quantity index for time period $n$
$q_n$ = quantity in period $n$
$q_0$ = quantity in the base period
$p_t$ = typical price during the time periods considered

Since the procedure for computing a weighted aggregative quantity index is exactly the same as the procedure outlined above for a weighted price index, it is not necessary here to work through an example. You should test your understanding of the procedure, however, by preparing the quantity index in the following review exercise.

1   Compute a weighted aggregative *price index* number series for the data given (base = 1979 = 100.0):

| Commodity | Unit of price quotation | Typical annual consumption | Price 1979 | Price 1980 | Price 1981 |
|-----------|------------------------|---------------------------|------|------|------|
| A | 1 dozen | 4 | $ 1.00 | $ 1.20 | $ 1.30 |
| B | 100 pounds | 3 | 10.00 | 10.00 | 12.00 |
| C | 1 each | 10 | 8.00 | 9.50 | 9.75 |

2   Interpret your answers.

3   Compute a weighted aggregative *quantity index* series for the data given (base = 1981 = 100.0):

| Commodity | Unit of production | Typical price | Quantity produced 1981 | Quantity produced 1982 | Quantity produced 1983 |
|-----------|-------------------|---------------|------|------|------|
| A | Dozen | $1.00 | 40 | 48 | 52 |
| B | Pound | 2.00 | 800 | 860 | 900 |
| C | Carton | 1.50 | 500 | 550 | 500 |

4   Interpret your answers.

**AVERAGE OF RELATIVES METHOD OF INDEX NUMBER CONSTRUCTION**

The use of the average of relatives method is the second common approach followed in the preparation of index numbers. (In the early paragraphs of this chapter we constructed price relatives for the chalk used by Professor Staff for the years 1976–1980.) Essentially, the average of relatives method involves (1) dividing the price or quantity in a particular period by the price or quantity in the base period and (2) calculating the arithmetic mean of the *resulting* price or quantity *relatives*.

The average of relatives method, like the aggregative method, may be used to prepare *implicitly* or *explictly* weighted indexes.

**Implicitly Weighted Average of Relatives Indexes**

**Price index**   Are you ready to take another look at the by now overworked data associated with Professor Staff? Well read on anyway! (However, we will reduce the arithmetic tedium by considering only the years 1976 and 1980.)

Figure 12-7 shows the calculation of the price index by the average of relatives method using the following formula:

**FIGURE 12-7**

Construction of an implicitly weighted average of relatives price index for Professor Staff's essential commodities (1976 = 100)

| Commodity | 1976 | | 1980 | |
|---|---|---|---|---|
| | Price $(p_0)$ | Relative $\left(\dfrac{p_{1976}}{p_0} \cdot 100\right)$ | Price | Relative $\left(\dfrac{p_{1980}}{p_0} \cdot 100\right)$ |
| Chalk | $0.60 | 100.0 | $ 0.75 | 125.0 |
| Red pencils | 0.72 | 100.0 | 0.92 | 127.8 |
| Statistics book | 9.50 | 100.0 | 11.00 | 115.8 |
| Aspirin | 0.59 | 100.0 | 0.75 | 127.1 |

$$\Sigma\left(\frac{p_n}{p_0} \cdot 100\right) = \qquad 400.0 \qquad\qquad\qquad 495.7$$

$$PI = \frac{\Sigma\left(\frac{p_n}{p_0} \cdot 100\right)}{n} = \quad \frac{400}{4} \text{ or } \underline{100} \qquad\qquad \frac{495.7}{4} \text{ or } \underline{123.9}$$

$$PI_n = \frac{\Sigma\left(\frac{p_n}{p_0} \cdot 100\right)}{n} \qquad\qquad (12\text{-}4)$$

where $p_n/p_0 \cdot 100$ = price relative in period $n$
$\quad\quad n$ = total number of commodities

Thus, by the average of relatives procedure, the prices in 1980 are 123.9 percent of the prices in 1976—i.e., there has been an average relative increase of 23.9 percent.

**Quantity index** The formula for computing an implicitly weighted average of relatives quantity index is:

$$QI_n = \frac{\Sigma\left(\frac{q_n}{q_0} \cdot 100\right)}{n} \qquad\qquad (12\text{-}5)$$

where $q_n/q_0 \cdot 100$ = quantity relative in period $n$
$\quad\quad n$ = total number of commodities

Obviously, with the exception that we have substituted *quantities* for period $n$ and for the base period *in place of prices,* the procedure for solving a quantity index is identical with the procedure outlined in Fig. 12-7. Therefore, it is not necessary here to work through an example. You may test your understanding of the procedure, however, by preparing the quantity index in the following review exercise.

**Self-testing Review 12-4**

1  Using the price data in Fig. 12-4, compute the implicitly weighted average of relatives price index for Professor Staff's commodities for the years 1978 and 1979.

**2** Compute an implicitly weighted average of relatives price index for the data shown (base = 1980 = 100):

| | Relatives | | |
|---|---|---|---|
| Commodity | 1980 $\left(\frac{p_n}{p_0} \cdot 100\right)$ | 1981 $\left(\frac{p_n}{p_0} \cdot 100\right)$ | 1982 $\left(\frac{p_n}{p_0} \cdot 100\right)$ |
| A | 100.0 | 105.0 | 110.0 |
| B | 100.0 | 110.0 | 120.0 |
| C | 100.0 | 120.0 | 110.0 |
| | 300.0 | 335.0 | 340.0 |

**3** Compute an implicitly weighted average of relatives quantity index for the data shown (base = 1980 = 100):

| Product | 1980 Quantity produced | 1981 Quantity produced |
|---|---|---|
| A | 3,000 | 2,900 |
| B | 12,000 | 14,000 |

**Weighted Average of Relatives Indexes**

A major problem in using the implicitly weighted average of relatives indexes described above is, of course, their failure to provide a measure of relative importance for the various commodities, products, or services being considered. Thus, we now need to examine methods of constructing *weighted* average of relatives indexes.

**Price index** The importance assigned to the relative price change for each item to be included in the average of relatives index is generally measured by the dollar amount typically spent on the item during the time periods considered. And since a dollar amount spent is the product of price and quantity, the weight used in an average of relatives index is customarily expressed as the *value* of an item that is consumed, bought, or sold.[8]

The general formula for a weighted average of relatives price index is:

---

[8] We can't simply use quantities alone as weights as we did earlier in the chapter because if we *multiply relatives* (which are likely to be in percentage form and which have no units) *by the quantities* in a series expressed in such units as pounds, tons, or bales, the resulting products of these multiplications would also be in the different units and could thus not be added logically.

**FIGURE 12-8**

Computation of weighted average of relatives price index for Professor Staff's essential commodities (base = 1976 = 100)

| Commodities (1) | Prices | | Price relatives | | Typical annual consumption | Value weights | Weighted price relatives | |
|---|---|---|---|---|---|---|---|---|
| | 1976 ($p_0$ and $p_t$) (2) | 1980 ($p_n$) (3) | $\left(\dfrac{p_{1976}}{p_0}\cdot 100\right)$ (4) | $\left(\dfrac{p_{1980}}{p_0}\cdot 100\right)$ (5) | ($q_t$) (6) | ($p_tq_t$) (col. 2 × col. 6) (7) | 1976 (col. 4 × col. 7) (8) | 1980 (col. 5 × col. 7) (9) |
| Chalk | $0.60 | $ 0.75 | 100.0 | 125.0 | 4 | $ 2.40 | 240 | 300 |
| Red pencils | 0.72 | 0.92 | 100.0 | 127.8 | 0.5 | 0.36 | 36 | 46 |
| Statistics book | 9.50 | 11.00 | 100.0 | 115.8 | 1 | 9.50 | 950 | 1,100.1 |
| Aspirin | 0.59 | 0.75 | 100.0 | 127.1 | 6 | 3.54 | 354 | 449.9 |
| | | | | | | →$15.80 | →1,580 | →1,896 |

$$\frac{\sum\left[\left(\dfrac{p_n}{p_0}\cdot 100\right)(p_tq_t)\right]}{\sum(p_tq_t)}$$

Therefore, $PI_{1976} = \dfrac{1580}{\$15.80} = 100.0$ and $PI_{1980} = \dfrac{1896}{\$15.80} = 120.0$

$$PI_n = \frac{\sum\left[\left(\dfrac{p_n}{p_0} \cdot 100\right)(p_t q_t)\right]}{\sum(p_t q_t)} \qquad\qquad (12\text{-}6)$$

where $p_n/p_0 \cdot 100 =$ price relative for period $n$

$\qquad\quad p_t =$ typical price during the time periods considered

$\qquad\quad q_t =$ typical number of units produced or consumed during the time periods studied

$\qquad\quad p_t q_t =$ typical dollar expenditures (value weights) on the items during the time periods studied

Although you were hoping it wouldn't happen, we are once again going to use Professor Staff's essential commodities to illustrate the use of formula (12-6). Only the price data for the years 1976 — the base period — and 1980 will be used. Figure 12-8 shows the computational procedures. You will notice in column 2 that we are assuming that a typical price is the price in the base period — i.e., $p_t = p_0$. The *numerator* for the price index for the year 1976 is the total in column 8; the *numerator* for the index for 1980 is the total in column 9; and the *denominator* for the index for both years is the total in column 7. By now you should be able to interpret the meaning of the index of 120.0 for 1980.

If you compare the 1980 index number just computed with the price index for 1980 computed by the weighted aggregative method (see Fig. 12-6), you will notice that they are equal. This isn't surprising because formulas (12-2) and (12-6) are algebraically identical *if* the typical prices $(p_t)$ are assumed to be the prices in the base year $(p_0)$. This can be easily shown below if we change the $p_t$ symbols in formula (12-6) to $p_0$'s and cancel the $p_0$'s in the numerator:

$$PI_n = \frac{\sum\left[\left(\dfrac{p_n}{\cancel{p_0}} \cdot 100\right)(\cancel{p_0}q_t)\right]}{\sum(p_0 q_t)} = \frac{\sum(p_n q_t)}{\sum(p_0 q_t)} \cdot 100$$

Formula (12-6) = formula (12-2)

Why, then, if these two index construction methods yield the same results, do we bother with the weighted average of relatives approach? (The formula is, after all, a real bear!) The answer, of course, is that the two methods are *not* identical if $p_t \neq p_0$, and it is often desirable to use prices other than those in the base period to represent what is typical during the time periods studied.

**Quantity index** You will recall that exchanging the $p$'s and $q$'s in the weighted aggregative price index formula produced the weighted aggregative quantity index formula. We will get a similar result if we exchange the $p$'s and $q$'s in formula (12-6). In other words, the general formula for a weighted average of relatives quantity index is:

$$QI_n = \frac{\sum\left[\left(\dfrac{q_n}{q_0} \cdot 100\right)(q_t p_t)\right]}{\sum(q_t p_t)} \qquad\qquad (12\text{-}7)$$

With the exception that quantity relatives are computed in place of price relatives in the numerator of formula (12-7), the arithmetic *procedures* are identical to those used in preparing a price index. Therefore, it is not necessary here to work through an example quantity index problem. However, you may test your understanding of the procedure by preparing the quantity index values in the following review section.

1 Using the price and quantity data in Fig. 12-4, compute the weighted average of relatives price index for Professor Staff's commodities for the years 1978 and 1979, using 1978 as the base period and the prices in 1978 as typical prices.

2 Compute a weighted average of relatives quantity index for the data given below. (Assume that the base year is 1980 and the typical quantities produced are those in 1980.)

| Commodity | Unit of production | Typical price | Quantity produced | |
|---|---|---|---|---|
| | | | 1980 | 1981 |
| X | Dozen | $1.00 | 48 | 52 |
| Y | Pound | 2.00 | 860 | 900 |
| Z | Carton | 1.50 | 550 | 500 |

## IMPORTANT INDEX NUMBERS IN CURRENT USE

We frequently learn about the movements of certain index number series by reading newspapers or watching television. Responsible persons in the news media see that index number information is communicated because they correctly believe that the public has a right to know the nature of general economic changes and the ways in which such changes may be affecting individuals. (Voters frequently use such information about economic changes as an important decision-making factor on Election Day.) And, of course, economists and administrators with a need for more detailed and specialized index number information for planning, decision making, and control purposes may choose from a wealth of published index number series. Sources of such series are the *Survey of Current Business, Federal Reserve Bulletin, Monthly Labor Review,* and *Business Conditions Digest,* to name just a few. Let us now very briefly describe a few of the more commonly used index number series.

## Price Indexes

The *Consumer Price Index* (CPI) published by the United States Bureau of Labor Statistics measures the average change in prices of many types of consumer goods and services. Actually, two CPIs are available: one index for *all* urban consumers represents about 80 percent of the population, and the other index includes only urban wage earners and clerical workers. (Professional, managerial, and technical workers; retirees; and self-employed and unemployed persons are included in the first index but not in the second.) A sample of the prices of about 400 items (a fixed market basket of goods and services) is taken each period (usually monthly) from 56 urban areas. (These areas are given weights according to the size of the working population in each area.) Price data from the 56 urban areas are then combined to form a composite national index. The items in the current series were selected because of their frequency of purchase and their importance relative to total expenditures. However, the composition of the market basket *does* gradually change; old items are dropped when they are no longer sold in volume, and new items are added when they become sufficiently important.

The weighted average of relatives method is used to construct the CPI, and constant weights based on studies made in 1972–1973 are employed. The current base period is 1967. The weighting system and base years have been revised several times in the past to improve the usability of the index.

The CPI is the most widely known and used of all index numbers; it may even possibly be the most important statistic published regularly by the federal government. One important use of the CPI is to serve as a basis for adjusting union wage rates to take into account changes in consumer prices. A clause (often called an *escalator clause*) may be written into a union contract, for example, specifying that an automatic wage rate increase of a given amount will be added to a union member's pay when the CPI increases by a specified amount. The CPI for selected periods is shown in Fig. 12-9.

The *Producer Price Index* (PPI) published by the United States Bureau of Labor Statistics measures changes in the general price level of goods at their *first commercial transaction level*—i.e., in prices received at the *producer* level. (For years the PPI was referred to as the *Wholesale Price Index*, but this name often led to the erroneous assumption that it measured prices received by wholesalers and other middlemen in the channels of distribution.)

**FIGURE 12-9**

Consumer and producer price indexes for selected periods (1967 = 100)

| Year | Producer price index | Consumer price index |
|------|------|------|
| 1925 | 53.3 | 52.5 |
| 1930 | 44.6 | 50.0 |
| 1935 | 41.3 | 41.1 |
| 1940 | 40.5 | 42.0 |
| 1945 | 54.6 | 53.9 |
| 1950 | 81.8 | 72.1 |
| 1955 | 87.8 | 80.2 |
| 1960 | 94.9 | 88.7 |
| 1965 | 96.6 | 94.5 |
| 1967 | 100.0 | 100.0 |
| 1970 | 110.4 | 116.3 |
| 1971 | 113.9 | 121.3 |
| 1972 | 119.1 | 125.3 |
| 1973 | 134.7 | 133.1 |
| 1974 | 160.1 | 147.7 |
| 1975 | 174.9 | 161.2 |
| 1976 | 182.9 | 170.5 |
| 1977 | 194.2 | 181.5 |
| 1978 (July) | 210.6 | 196.7 |

The PPI is calculated as a weighted average of relatives price index, has a current base period of 1967, and currently includes about 2,800 commodities. Commodity data are broken down into groups and subgroups by type of finished product, by durability of product, by stage of processing, and by other classifications. The value weights for the composite indexes are derived from the shipments (sales) of commodities in a particular year. These weights are revised periodically, the last time being in 1976. The PPI is used for a number of purposes, including market analysis and the preparation of escalator clauses in long-term commodity purchase and sales contracts. Values of the PPI for selected periods are shown in Fig. 12-9.

**Quantity Index**

The *Index of Industrial Production* (IIP) is published monthly by the Board of Governors of the Federal Reserve System and measures changes in the output quantities of plants, mines, and electrical and gas utilities. It does not cover farm production, construction, or transportation activities or the various trade and service industries. Component series are combined using the weighted average of relatives quantity method to prepare the IIP. The value weights used are based on the concept of *value added* — the difference between the value of production and the cost of materials or supplies consumed. The base period is currently 1967. As an indicator of changes taking place in the economy's output, the IIP is widely used to support administrative planning decisions. An industrial producer, for example, can compare the company's performance with that of competitors and make those decisions that seem appropriate.

**PROBLEMS AND PITFALLS WITH INDEX NUMBERS**

To paraphrase somebody, making the statement "Index numbers are highly sophisticated and extremely accurate measures of change that are based on exhaustive samples and rigorous mathematical theories" is no problem—but it is wrong! The truth is that although most index number series are *not deliberately misleading*, they are far from perfect. It is beyond the scope of this book to go into the limitations of index numbers in great detail; however, there are (1) a number of major *problems associated with the construction of indexes* and (2) *several pitfalls associated with their use* that should be mentioned. As potential consumers of statistical data, you should be aware of these problems and pitfalls.

**Problems of Index Number Construction**

Some of the major problems encountered by statisticians in constructing index numbers are:

1  *The problem of selecting a sample.* Because the most popular price and quantity indexes are used in such a large number of ways, it is difficult (if not impossible) to select the items to be included in an index in such a way that the results are equally meaningful to those who use (or unknowingly misuse) the index. The CPI, we have seen, focuses attention on prices paid by *urban* persons for about 400 items (out of an identified total of about 2,000 items). Thus, the CPI is not equally applicable to small-community families and farm families.

And the decision on which 400 items will be in the sample selected is based on the *judgment* of government statisticians; thus the sample is not a random sample.

2  *The problem of assigning appropriate weights.* In the weighted index formulas presented in this chapter, we have used typical prices and quantities for weighting purposes. But what is typical or appropriate for one period and for one purpose may quickly become inappropriate for another time and for another use. There is often a lag between the time that a weighting system should be changed and the time when it is actually revised. Isn't it possible, for example, that the typical quantities purchased (and used to weight a price index) could change significantly during periods of rapid price increases when consumption patterns are being altered? How typical would the quantity weights then be? To illustrate, at this writing the CPI is based on spending habits for 1972–1973 —a period before the energy crisis and rapid escalation in housing costs forced many people to revise their buying habits.

3  *The problem of choosing an appropriate base period.* The criteria for determining an appropriate base period include (1) the recency and normalcy of the period—i.e., how well it is remembered and how well it avoids both the peaks and troughs that periodically occur in economic activity—and (2) the need for comparability with other popular indexes. The difficulty of defining a fairly recent normal period for a composite index made up of hundreds of items is obvious. Yet some period must be used, even if it is not particularly suited for some items. The year 1967 is currently used in many indexes—in some cases because it is fairly recent and meets the normalcy criterion, and in other cases, perhaps, merely because of the desire to have an index that can be more easily compared with those using the 1967 period.

## Pitfalls in the Use of Index Numbers

Some of the pitfalls encountered by users of index numbers may be attributed to their failure to:

1  *Obtain adequate knowledge of indexes.* Few serious users of index numbers have as good an understanding of the kinds of indexes available and of the methods employed in their construction as they should have. (Careful study of the publications of the organizations that print the indexes is necessary to gain a specialized understanding.) If, by summarizing a few of the problems of index construction, we have motivated you to make a more thorough study of an index before you use it, we have performed a useful service.

2  *Consider possible bias introduced through the passage of time.* This pitfall is related to the problem of assigning appropriate weights discussed in the preceding section. As time passes, an index number may understate or overstate the extent of the change being measured because of shifts in the relative importance of (but not the weights of) the items being studied. And, of course, the use of a biased index could lead to unfortunate decisions.

3  *Consider qualitative changes.* Technological change over time may lead to quality changes in products being compared. When this happens, it is very dif-

ficult, if not impossible, to make satisfactory adjustments in a price index to reflect adequately the qualitative differences. Automobiles are not the same today as they were 10 years ago; nor, for that matter, are automobile tires. If a $40 tire delivers 50 percent more mileage today than a comparably priced tire of several years ago, how is this qualitative change reflected in a price index? Well, in the CPI, if there is no change in selling price, the tires are considered the same. Thus, some analysts are of the opinion that over a number of years the increases in the CPI may overstate actual price increases in terms of a truly fixed basket of items.

## SUMMARY

Index numbers are generally used to measure change over some period of time, and they provide a means of summarizing and communicating the nature of changes that occur. They are also useful in simplifying data, facilitating comparisons, and showing typical seasonal variations.

Most of the frequently used index numbers may be classified as (1) price indexes, (2) quantity indexes, or (3) value indexes. One use of a price index is to serve as a basis for preparing a purchasing power index.

Regardless of whether they deal primarily with price, quantity, or value, most index numbers are constructed by either an aggregative or an average of relatives method. Examples of implicitly weighted aggregative and average of relatives price indexes were given in the chapter. Another term sometimes used in place of "implicitly weighted" is "unweighted," but this is not strictly correct. A system of explicit weights is generally needed to take into account the relative importance of the various items in an index. Examples of weighted aggregative indexes were presented in the chapter, as were examples of weighted average of relatives indexes.

Among the more important price indexes in current use are the Consumer Price Index and the Producer Price Index. These indexes, along with the Index of Industrial Production (a quantity index), are briefly described in the later pages of the chapter. In addition, a number of the major problems associated with the construction of indexes and several of the pitfalls associated with their use have been summarized.

## Important Terms and Concepts

1 Index number
2 Percentage relative
3 Base period
4 Simple index number
5 Composite index number
6 Price index
7 Quantity index
8 Value index
9 Purchasing power index

10 Aggregative index number
11 Implicitly weighted indexes
12 Explicit weights
13 Average of relatives index number
14 Weighted index number series
15 Value weights
16 Consumer Price Index (CPI)
17 Market basket

18 Escalator clause

19 Producer Price Index (PPI)

20 Index of Industrial Production (IIP)

21 Value added

22 $PI_n = \dfrac{\Sigma(p_n)}{\Sigma(p_0)} \cdot 100$

23 $PI_n = \dfrac{\Sigma(p_n q_t)}{\Sigma(p_0 q_t)} \cdot 100$

24 $QI_n = \dfrac{\Sigma(q_n p_t)}{\Sigma(q_0 p_t)} \cdot 100$

25 $PI_n = \dfrac{\Sigma\left(\dfrac{p_n}{p_0} \cdot 100\right)}{n}$

26 $QI_n = \dfrac{\Sigma\left(\dfrac{q_n}{q_0} \cdot 100\right)}{n}$

27 $PI_n = \dfrac{\Sigma\left[\left(\dfrac{p_n}{p_0} \cdot 100\right)(p_t q_t)\right]}{\Sigma(p_t q_t)}$

28 $QI_n = \dfrac{\Sigma\left[\left(\dfrac{q_n}{q_0} \cdot 100\right)(q_t p_t)\right]}{\Sigma(q_t p_t)}$

**Problems**

1 Using 1975 as a base, compute a simple price index series for 1975–1980 with the data given below.

| Year | Price per pound of corn |
|------|-------------------------|
| 1975 | $0.10 |
| 1976 | 0.09 |
| 1977 | 0.11 |
| 1978 | 0.13 |
| 1979 | 0.13 |
| 1980 | 0.15 |

2 An old survey has recently been found which shows the consumption habits of a member of the Signa Phi Nuthen fraternity at a nondescript college. The following data had been gathered, and in a fit of nostalgia you decide to apply your substantial expertise in the area of index numbers.

| Commodity | Unit of price quotation | 1970 | | 1971 | | 1972 | |
|-----------|------------------------|-------|----------|-------|----------|-------|----------|
| | | Price | Quantity | Price | Quantity | Price | Quantity |
| Shirts | 1 each | $6.00 | 10 | $6.67 | 11 | $7.55 | 10 |
| Books | 1 each | 8.75 | 11 | 9.35 | 12 | 10.15 | 13 |
| Beer | 1 mug | 0.30 | 105 | 0.30 | 127 | 0.35 | 153 |
| Gasoline | 1 gallon | 0.28 | 380 | 0.32 | 130 | 0.39 | 364 |

Assuming that the base year is 1970 and the typical quantities are the quantities in 1970:

a Construct a price index series using the implicitly weighted aggregative method.

b Construct a price index series using the weighted aggregative method.

3  The Bureau of Business Research of a university must prepare a report for a statewide publication. The purpose of this report is to show how prices and quantities of three products have changed over time. The bureau obtained the following information from its files:

| Product | 1978 | | 1979 | | 1980 | | 1981 | |
|---------|------------|----------------|------------|----------------|------------|----------------|------------|----------------|
|         | Unit price | Quantity sold | Unit price | Quantity sold | Unit price | Quantity sold | Unit price | Quantity sold |
| A | $0.30 | 960 | $0.35 | 975 | $0.45 | 900 | $0.47 | 965 |
| B | 0.75 | 135 | 0.70 | 100 | 0.80 | 115 | 0.78 | 140 |
| C | 1.00 | 290 | 1.05 | 280 | 1.07 | 285 | 1.05 | 295 |

Using 1978 as a base, compute the following for all years:
a  A simple price index series for product A
b  An implicitly weighted aggregative price index
c  A weighted aggregative price index (weighted with base period quantities)
d  A weighted aggregative quantity index (weighted with base period prices)
e  An implicitly weighted average of relatives price index
f  A weighted average of relatives price index (weighted with prices and quantities in 1978)
g  A weighted average of relatives quantity index (weighted with prices and quantities in 1978)

**Topics for Discussion**

1  What advantages are there in expressing data in the form of index numbers?
2  What is the difference between a composite index number series and a simple index number series?
3  What are some of the possible uses of (a) price indexes, (b) quantity indexes, and (c) value indexes?
4  a  What is the distinction between money wages and real wages?
   b  How may price indexes be used to compare purchasing power between two periods?
5  a  What is meant by the term "weighting"?
   b  Why are explicit weights needed with index numbers?
6  Why can't we just use typical quantities to weight an average of relatives price index?
7  Discuss the possible uses of (a) the Consumer Price Index, (b) the Producer Price Index, and (c) the Index of Industrial Production.
8  What are some possible problems in index number construction?
9  What are some possible pitfalls in the use of index numbers?

**10** Discuss the role of index numbers in the figure presented below.

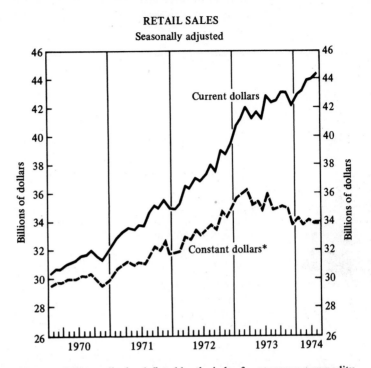

RETAIL SALES
Seasonally adjusted

*Current-dollar retail sales deflated by the index for consumer commodity prices: 1969 = 100.
Source: United States Department of Commerce, Bureau of the Census.

**Answers to Self-testing Review Questions**

**12-1**

**1** Purchasing power index for 1970:
$1 \div 1.163 = .85984$
Purchasing power index for 1974:
$1 \div 1.477 = .67705$
Purchasing power index for June 1978:
$1 \div 1.953 = .51203$

**2** Real earnings (in 1967 dollars) are:
$1970 = \$600 \times .85984 = \$515.90$
$1974 = \$750 \times .67705 = \$507.79$
In spite of an increase of $150 in money wages between 1970 and 1974, the worker has suffered about an $8 loss in purchasing power. And in terms of 1967 dollars, earnings for June 1978 are $473.63 ($925 × .51203). Thus, in real terms the worker's income in June 1978 is about $42 less than it was in 1970 even though money wages have increased by $325 over the years!

**12-2**

**1** Implicitly weighted aggregative price index number series:

$$PI_{1979} = \frac{\$96}{\$96} \cdot 100 = 100.0$$

$$PI_{1980} = \frac{\$101}{\$96} \cdot 100 = 105.2$$

$$PI_{1981} = \frac{\$110}{\$96} \cdot 100 = 114.6$$

**2** There was an increase of 5.2 percent and 14.6 percent, respectively, in prices between 1979 and 1980, and between 1979 and 1981.

**12-3**

**1** Weighted aggregative price index series:

$$PI_{1979} = \frac{\$114.00}{\$114.00} \cdot 100 = 100.0$$

$$PI_{1980} = \frac{\$129.80}{\$114.00} \cdot 100 = 113.9$$

$$PI_{1981} = \frac{\$138.70}{\$114.00} \cdot 100 = 121.7$$

**2** There was an increase in price of 13.9 percent between 1979 and 1980; the price increase between 1979 and 1981 was 21.7 percent.

**3** Weighted aggregative quantity index series:

$$QI_{1981} = \frac{\$2,390}{\$2,390} \cdot 100 = 100.0$$

$$QI_{1982} = \frac{\$2,593}{\$2,390} \cdot 100 = 108.5$$

$$QI_{1983} = \frac{\$2,602}{\$2,390} \cdot 100 = 108.9$$

**4** Quantities produced increased 8.5 percent and 8.9 percent, respectively, between the base period and the years 1982 and 1983.

**12-4**

**1** $PI_{1978} = \dfrac{434.9}{4} = 108.7$

$PI_{1979} = \dfrac{462.2}{4} = 115.6$

**2** $PI_{1980} = \dfrac{300}{3} = 100.0$

$$PI_{1981} = \frac{335.0}{3} = 111.7$$

$$PI_{1982} = \frac{340.0}{3} = 113.3$$

**3** $\quad QI_{1980} = \frac{200}{2} = 100.0$

$$QI_{1981} = \frac{213.4}{2} = 106.7$$

**12-5**

**1** Weighted average of relatives price indexes:

$$PI_{1978} = \frac{\$1,690}{\$16.90} = 100.0$$

$$PI_{1979} = \frac{\$1,786.70}{\$16.90} = 105.7$$

**2** Weighted average of relatives quantity indexes:

$$QI_{1980} = \frac{\$259,300}{\$2,593} = 100.0$$

$$QI_{1981} = \frac{\$260,274.90}{\$2,593} = 100.4$$

# CHAPTER 13

# FORECASTING TOOLS: TIME-SERIES ANALYSIS

**LEARNING OBJECTIVES**

After studying this chapter, working the problems, and answering the discussion questions, you should be able to:

☞ Explain (*a*) the reasons for studying time series and (*b*) the possible problems that may be encountered in the use of time-series analysis.

☞ Identify the components of a time series.

☞ Compute and project a linear secular trend.

☞ Compute and use a seasonal index.

☞ Summarize how time-series analysis may be used in forecasting.

**CHAPTER OUTLINE**

Administrators in all types of organizations engage in the activity of planning in order to cope with future changes. The planning function looks to the future; to plan is to make decisions in advance about a future course of action. Obviously, then, planning and decision making involve expectations of what the future will bring. Thus, whether they employ complex quantitative methods or merely rely on intuitive hunches, administrators are required to look down the road and make *forecasts* before they (1) select both short- and long-run strategies and goals, (2) develop policies and procedures that will help accomplish future objectives or counter anticipated threats, and (3) revise earlier plans and decisions in the light of changing conditions.[1] Among the statistical forecasting tools used by managers in planning and in coping with change are *time-series analysis* (the subject of this chapter) and *regression analysis* (the subject of the next chapter).

A *time series* is nothing more than a set (or series) of numerical values of a particular variable listed in chronological order. In Fig. 13-1, for example, the variable is private multifamily housing starts in the United States, and the time represented is the period from 1975 through early 1978. *Time-series analysis,* then, involves classifying and studying the patterns of movement of the values of the variable over regular intervals of time (and the time intervals may be expressed in such terms as months or quarters as well as years).

In the following pages of this chapter, we shall first consider the *reasons for studying time series.* We will then look at (1) an *overview of time-series components,* (2) the subject of *trend analysis,* (3) the topic of *seasonal variation,* (4) the *identification of cyclical and irregular components,* (5) the use of *time-series analysis in forecasting,* and (6) the possible *problems in the use of time-series analysis.*

---

[1] Once plans are implemented, an organization may be more or less committed to them for a variable period of time. The term "planning horizon" is sometimes used to refer to the period in which planners are committed to their decisions. In some cases, the planning horizon may be as short as 1 day—e.g., your plans as to what you will wear to class tomorrow based on a weather forecast—but in business decisions involving, say, capital expenditures, the planning horizon may be measured in many months or years, and the accuracy of the forecast becomes a matter of much greater concern.

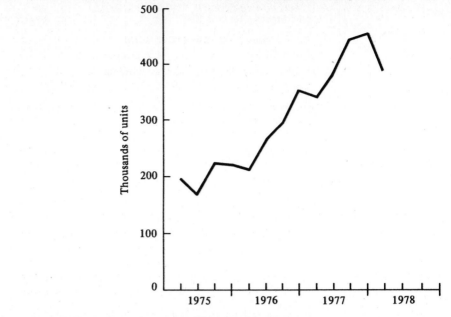

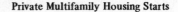

**FIGURE 13-1**

**Private multifamily housing starts in the United States**

## WHY STUDY TIME SERIES?

In the chapters in Part 2, we dealt with the subject of sampling and statistical inference—a subject employing a mature theoretical base to support a number of popular and proven application techniques. In time-series analysis, however, we are dealing with a subject which (1) *will not* give us future estimate figures to which we can confidently assign a high probability and (2) uses a popular model that is *rather crude* and may be expected to yield only approximate results at best. In light of these disturbing facts, you may indeed ask yourself the question, "Why study time series?"

*Time series are worthy of our attention for the following reasons:*

1  *They may enhance understanding of past and current patterns of change.* A careful study of past events and experiences may give us greater insight into the dynamic forces affecting the patterns of change. Historians are fond of saying that "those ignorant of the past are condemned to repeat it." The decision maker who understands why past changes have occurred may be less likely to repeat past mistakes.

2  *They may provide clues about future patterns to aid in forecasting.* Historians are also fond of saying that "all that is past is prologue." *If*, after studying a time series, a decision maker has reason to believe that an identifiable past pattern *will persist*, he or she can then build a forecast by projecting the past pattern into the future. Of course, subjective judgment is bound to be involved in any future estimate, but since decision makers do not have the choice of whether to engage in forecasting or to abstain from it, it is considered a better

approach to use thoughtfully projected time-series data than to rely solely on speculative hunches.

To summarize, it is only natural that time series are seriously studied. There is a need to understand the dynamic forces at work in the economy, and time-series analysis can enhance this understanding; and there is no choice but to make forecasts, and time-series analysis may reveal persistent patterns. Yet in our study of time series we should keep a proper perspective. Perhaps the following comment best sums up this perspective:

> *In our desire to find a logical integration of different stories told by the time series of economics, we must not be unmindful of a conversation between Alice and the White King. Alice had been asked by the King to look down the road and tell him if she could see his two messengers. "I see nobody on the road," said Alice. "I only wish I had such eyes," the King remarked in a fretful tone. "To be able to see Nobody! and at that distance too."*[2]

## TIME-SERIES COMPONENTS: AN OVERVIEW

It is possible, of course, to concentrate on the broad picture presented by the total values of a particular time series. More frequently in time-series analysis, however, the analyst will single out one or more of the *components* which, superimposed and acting in unison, tend to represent the effects on the series of a number of diverse factors—e.g., change in technology, population, consumer buying habits, weather conditions, capital investment, productivity—of an economic, natural, and cultural nature that may account for changes in the series over time. The four components which may be identified in a time series are (1) *secular trend,* (2) *seasonal variation,* (3) *cyclical fluctuations,* and (4) *irregular movements.* It is convenient in time-series analysis to use the symbols $T$, $S$, $C$, and $I$ to represent these trend, seasonal, cyclical, and irregular elements. Although in later pages we will consider each of these components in more detail, it is probably appropriate here to pause briefly to examine the nature of these elements.

## Secular Trend

The *secular trend* is the smooth or regular long-term (i.e., secular) growth or decline of a series.[3] In Fig. 13-2a, we see a time-series variable[4] (identified by the letter $Y$) plotted over a 15-year period. A trend line has been superimposed on the

---

[2] Harold T. Davis, *The Analysis of Economic Time Series,* The Cowles Commission for Research in Economics, Monograph 6, The Principia Press, Bloomington, Ind., 1951, p. 580.

[3] What does "long-term" mean? You certainly can ask embarrassing questions. The exact length is probably impossible to determine but rather varies depending on the series. In the case of an economic time series, however, it is often argued that the period should be long enough to include two or more business cycles so that a persistent pattern may have time to emerge.

[4] The variable represented in Fig. 13-2 happens to be the production of chemicals and related products.

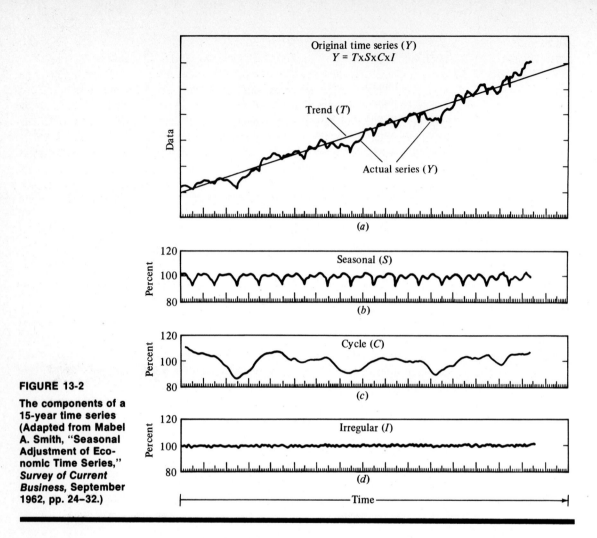

**FIGURE 13-2**

**The components of a 15-year time series (Adapted from Mabel A. Smith, "Seasonal Adjustment of Economic Time Series,"** *Survey of Current Business,* **September 1962, pp. 24–32.)**

original data. In this case, there is a *growth trend* which can be satisfactorily represented by a *straight line;* in other situations, however, the data may show a *declining* pattern (e.g., farm employment in the United States has declined from over 7 million in the early 1950s to less than 4 million in the 1970s), and the appropriate trend line may be *curvilinear* rather than straight.[5] Included among the underlying factors responsible for an average long-term growth or decline are (1) *population changes* (which have an obvious impact on the sales of producers of food, clothing, and shelter) and (2) *technological innovations* (buggy manufacturers have never recovered from the development of the automobile).

---

[5] We shall limit our study in this book to those situations in which the trend can be represented by a straight line.

## Seasonal Variations

Variations in a time series that are *periodic* in nature and that *recur regularly* within a period of 1 year (or less) are called *seasonal variations* (see Fig. 13-2*b*). As the name implies, *climatic conditions* are one important factor responsible for seasonal variations. Construction activities, sales of heating fuel and suntan lotion, the production of agricultural products—all these variables are obviously related to the weather. In addition, however, seasonal variations may be attributed to *social customs* and *religious holidays*. Any department store manager knows that there will be recurring patterns of sales activity in the days (and weeks) prior to Christmas, Easter, Labor Day, the first day of a new school year, etc. Figure 13-3 shows the number of passengers traveling across the North Atlantic on certain airlines between 1973–1974 and 1977–1978. The recurring pattern (including the February valley) is obvious.

**FIGURE 13-3**

**Seasonal patterns in North Atlantic scheduled air traffic (Source:** *Aviation Week & Space Technology,* **Oct. 23, 1978, p. 143.)**

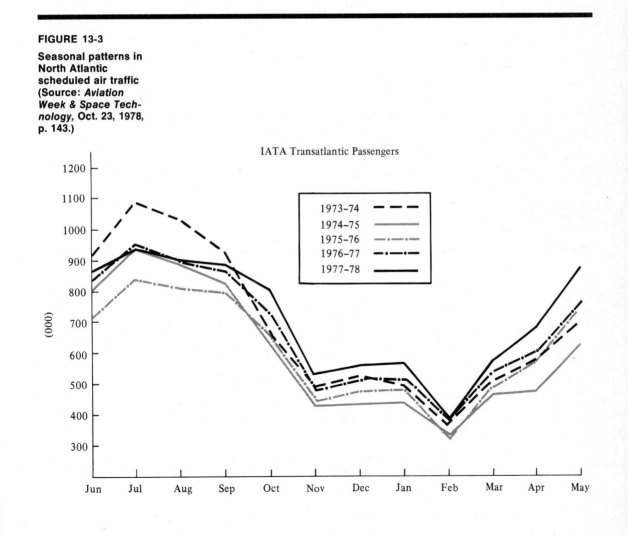

IATA Transatlantic Passengers

## Cyclical Fluctuations

Cyclical fluctuations, like seasonal variations, are *periodic* in nature and generally involve recurring up-and-down movement (see Fig. 13-2c). Unlike seasonal variation, however, the business cycle extends over a period of *several years* and *cannot* be counted on to repeat itself with relatively predictable regularity. Different factors tend to account for each new cycle. A great deal of effort has gone into the study of the periods of prosperity or stability, recession or depression, and recovery that characterize a complete cycle, and there have probably been at least as many theories to explain the cycle as there have been cycles. Thus far, however, there has been no generally satisfactory model developed to explain or forecast cyclical fluctuations adequately.

## Irregular Movements

There are often movements in a series that occur over varying but usually brief periods of time, that follow no regular pattern, and that are unpredictable in nature. These irregular movements (see Fig. 13-2d) may be caused by quite random factors or special events such as labor strikes, floods and droughts, hurricanes and other acts of God, armed conflicts, changes in governments, and a thousand other possibilities. As a practical matter, if a time-series variation cannot be attributed to a trend, seasonal, or cyclical element, it is lumped into the irregular or residual category.

## An Approach to Time-Series Analysis

From the preceding sections it is possible to conclude that if the original data ($Y$) representing a time series are made up of the four components ($T$, $S$, $C$, and $I$) mentioned above, we can describe the relationship between the components and the original data as follows (see Fig. 13-2a):

$$Y = T \times S \times C \times I \tag{13-1}$$

Using this time-series model, an analyst may, for example, (1) compute measures of, say, the trend and seasonal elements in order to analyze these elements and then *eliminate* them from the original data so that (2) the cyclical and irregular components may be separately identified and studied. That is, the following decomposition of the time series may be made:

$$\frac{\cancel{T} \times \cancel{S} \times C \times I}{\cancel{T} \times \cancel{S}} = C \times I$$

In the following pages we will follow this time-series analysis approach; i.e., we will consider (1) the *computation and projection of secular trend*, (2) the *measurement and use of seasonal patterns*, and (3) the *identification of cyclical and irregular fluctuations*.

## TREND ANALYSIS

As noted above, an important part of time-series analysis is the study of secular trend. In this section we will consider (1) the *reasons for measuring trend*, (2) a *method of computing the trend component*, and (3) the *use of trend in prediction or forecasting*.

| **Reasons for Measuring Trend** | We measure trend for several of the same general reasons that we study time series. That is, the *trend component is studied for the following purposes:* |
|---|---|

1  *To describe historical patterns.* The sales trend of one firm, for example, may be measured and compared with the trend patterns exhibited by competitors.

2  *To project persistent patterns into the future.* If (and this is a very important "if") an analyst believes that there is reason to expect a past trend to continue into the future, the measurement of that trend may be used as the basis for a future estimate or forecast.

3  *To eliminate the trend component.* An analyst may be primarily concerned with the cyclical element in a time series. In order to isolate the cyclical component, however, the effects of trend must be measured and eliminated from the original data.

| **Linear Trend Computation** | Although there are other techniques of measuring a linear trend (e.g., the *freehand* or eyeball *method* of using a ruler to draw a straight line through a graph of the data), we shall use the more objective and most popular *method of least-squares* approach, which *mathematically* fits the line to the data. Before looking at the trend computations, however, it is necessary here to discuss briefly (1) the *equation for a straight line* and (2) the *properties of the linear trend line*. |
|---|---|

**The straight-line equation** Figure 13-4a shows a straight line (identified by the symbol $Y_t$) of the type that we will soon be computing. In order to define this line rigidly, we must know *two things about it. First,* we must know the value of the *Y intercept* — i.e., the value (read on the Y axis) of *a* in Fig. 13-4a when X is at the origin or equal to zero. And *second,* we need to know the *slope of the line* (*b*), which may be found as shown in Fig. 13-4a by (1) taking a segment of the line, (2) measuring the change in one unit of time on the X axis, (3) measuring the corresponding change in $Y_t$ on the Y axis, and (4) dividing the change in $Y_t$ by the change in time. In Fig. 13-4a, the slope of the line would have a *positive* value; in Fig. 13-4b, the slope would be a *negative* value. In both cases, however, the formula for $Y_t$ (and the formula that we will be using to compute the straight-line trend) is:

$$Y_t = a + bx \tag{13-2}$$

where $Y_t$ = trend value for a given time period

     *a* = value of $Y_t$ when X is at the origin

     *b* = slope of the line, or the increase or decrease in $Y_t$ for each change of one unit of time

     *x* = any time period selected

Thus, if we selected a time period $X_1$ as shown in Fig. 13-4c, drew a vertical line up to the trend line ($Y_t$), and then drew a horizontal line to the Y axis, the value of $Y_1$ would be the value of the trend component for the $X_1$ time period.

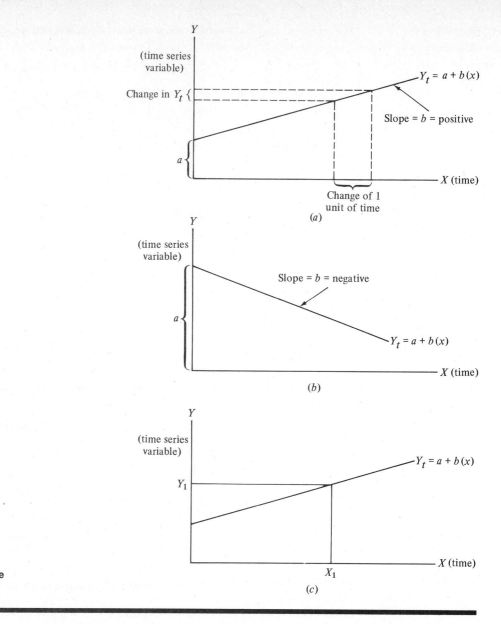

**FIGURE 13-4**

**The straight-line equation**

**Properties of the linear trend line** There are *two properties* of a trend line computed by the method of least squares. The *first* of these properties can be demonstrated in Fig. 13-5, which shows the trend component ($Y_t$) fitted to the time-series data ($Y$) that are plotted on a chart. At time $X_1$, the value of the actual data ($Y$) is greater than the value of $Y_t$. (The values of both $Y$ and $Y_t$ are read on the $Y$ axis.) Thus, if we subtract $Y_t$ from $Y$, we will get a *positive* value. At time $X_2$, however, the reverse is true—i.e., the $Y_t$ value is larger than the $Y$ value, and so the difference $Y - Y_t$ is *negative*. The fact is that the trend line will be fitted to the data in such a way

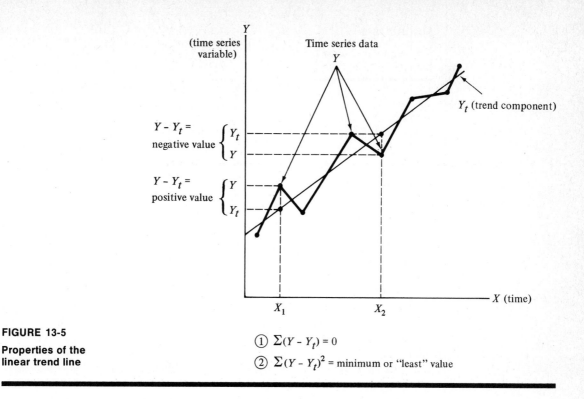

**FIGURE 13-5**

**Properties of the linear trend line**

① $\Sigma(Y - Y_t) = 0$

② $\Sigma(Y - Y_t)^2$ = minimum or "least" value

that the sum of the deviations of the $Y$ values about the trend line will be zero. Therefore, the *first property* of the trend line is:

$$\Sigma(Y - Y_t) = 0$$

And the sum of the *squares* of the deviations is less than would be the case if any other straight line were substituted for the $Y_t$ line and the process of computing and squaring deviations were carried out. In other words, the *second property* is:

$$\Sigma(Y - Y_t)^2 = \text{a minimum or least value}[6]$$

**Trend computation—odd number of years** We will use the hypothetical data in Fig. 13-6 for purposes of illustrating a linear trend computation. As you can see, annual sales of the Buckaroo Lingerie Company have generally been increasing, although there were slumps in 1976 and 1980. (In 1980 a new line of girdles and bras was developed, and the *former* sales manager, Hoagy "Tex" Chauvinist, approved an extensive advertising campaign to market these products built around the tactless slogan "We round up the strays." This campaign, combined with general economic

---

[6] And so the name "method of least squares." The two properties of the arithmetic mean, you will recall, are $\Sigma(X - \mu) = 0$ and $\Sigma(X - \mu)^2 = $ minimum value. Therefore, the least-squares linear trend line is to a time series what a mean is to a frequency distribution.

**FIGURE 13-6**

Annual sales of Buckaroo
Lingerie Company

| | |
|------|----|
| 1975 | 10 |
| 1976 | 8 |
| 1977 | 10 |
| 1978 | 12 |
| 1979 | 16 |
| 1980 | 12 |
| 1981 | 16 |

conditions, led to predictable results: sales declined as shown in Fig. 13-6; Tex was removed as sales manager and given the sales territory for a radius of 20 miles around Cut and Shoot, Texas[7]; and a new sales manager — Forrestal D. Zaster — was appointed.)

Figure 13-7 shows the work sheet used to compute the trend component for the Buckaroo Company for the years 1975–1981. (You will notice that the first two columns are unchanged from Fig. 13-6.) To simplify the computations, we have *coded* the 7-year time period by restating the calendar year numbers ($X$) in terms of the number of years from the *middle of the time period*. In other words, in our example we have shifted the *origin* from the beginning of the time period to the *middle of 1978*. Then, *in column 3 of Fig. 13-7, we have merely determined the deviations from the origin to the middle of each year* in yearly units. Thus, the middle of 1975 is $-3$ years from the middle of 1978 (the origin), and the middle of 1980 is $+2$ years from the origin. Having performed this little trick to code time, we can now compute the linear trend equation. The formulas for $a$ and $b$ in the trend equation are:

$$a = \frac{\Sigma Y}{n} \qquad (13\text{-}3)$$

where $Y$ = time-series variable, and $n$ is the number of years,

$$b = \frac{\Sigma(xY)}{\Sigma(x^2)} \qquad (13\text{-}4)$$

where $x$ = *coded time values* rather than the actual years ($X$)

The value for $a$ in our example is 84/7 or 12.00 (million dollars). The value for $b$ is found in Fig. 13-7 by dividing the sum of column 4 by the sum of column 5. Thus, $b = 1.143$ (million dollars). What are the meanings of these computed values of $a$ and $b$? The $12 million of $a$ means that the computed value of $Y_t$ will be $12 million in sales in the middle of 1978. Why? Because, as we saw earlier, $a$ is the value of the $Y_t$ line when $x$ is zero or at the origin, and $x$ is zero in the middle of 1978.

---

[7] Yes, there really is a Cut and Shoot, Texas! According to M. J. Quimby in *Scratch Ankle, USA,* "Cut and Shoot, Texas, got its name from a community dispute said to have arisen over the pattern for a new church steeple."

**FIGURE 13-7**

Annual sales of Buckaroo Lingerie Company

| Years (X) (1) | Company sales (millions of $) (Y) (2) | Coded time (x) (3) | (xY) (cols. 2 × 3) (4) | (x²) (5) | $Y_t$ (for years indicated, in millions of $) (6) |
|---|---|---|---|---|---|
| 1975 | 10 | −3 | −30 | 9 | 8.571 |
| 1976 | 8 | −2 | −16 | 4 | 9.714 |
| 1977 | 10 | −1 | −10 | 1 | 10.857 |
| 1978 | 12 | 0 | 0 | 0 | 12.000 |
| 1979 | 16 | 1 | 16 | 1 | 13.143 |
| 1980 | 12 | 2 | 24 | 4 | 14.286 |
| 1981 | 16 | 3 | 48 | 9 | 15.429 |
| | 84 | 0 | 32 | 28 | 84.000 |

$$a = \frac{\Sigma Y}{n} = \frac{84}{7} = 12.00 \text{ (millions of \$)}$$

$$b = \frac{\Sigma(xY)}{\Sigma(x^2)} = \frac{32}{28} = 1.143 \text{ (millions of \$)}$$

$Y_t = 12.00 + 1.143(x)$

Origin (where $x = 0$): the middle of 1978
Coded time $(x)$ unit : 1 year
　　　　　$Y$ data : Company sales in millions of dollars

The $1.143 million value for $b$ means that for *each change of 1 year* in $x$, the sales trend component will change by $1.143 million.

Column 6 in Fig. 13-7 indicates the trend values in millions of dollars for each of the 7 years. To compute $Y_t$ for 1975, you would use the following approach:

$$Y_{t_{1975}} = a + bx$$
$$= 12.00 + 1.143 \, (-3)$$
$$= 12.00 + (-3.429)$$
$$= 8.571, \text{ the trend for Buckaroo sales in 1975 (millions of dollars)}$$

Other $Y_t$ values may be found in the same way by substituting the appropriate value of $x$ in the trend equation. Alternatively, once you have the first year's trend, you can simply add to it the value of $b$ to get the next year's trend since, by definition here, $b$ is the change in $Y_t$ for each change of one unit of time.[8]

Figure 13-8 shows the original Buckaroo sales data and the trend component

---

[8] You will note that the sums of columns 2 and 6 are equal. This is always the case because of the first property of the trend line; that is, if $\Sigma(Y - Y_t) = 0$, then $\Sigma Y - \Sigma Y_t$ must obviously also be zero.

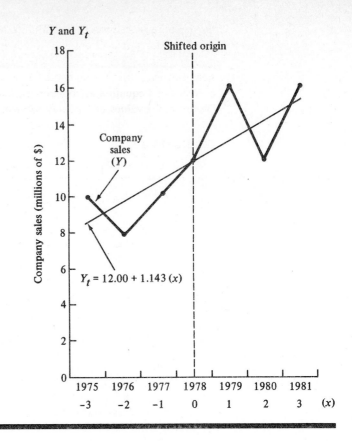

Y and $Y_t$

Shifted origin

Company sales (millions of $)

Company sales (Y)

$Y_t = 12.00 + 1.143\,(x)$

| 1975 | 1976 | 1977 | 1978 | 1979 | 1980 | 1981 |
|------|------|------|------|------|------|------|
| -3 | -2 | -1 | 0 | 1 | 2 | 3 | (x) |

**FIGURE 13-8**

found in column 6 of Fig. 13-7. You may study this figure to verify comments made in previous paragraphs.

**Trend computation—even number of years**  In the preceding paragraphs we computed the Buckaroo sales trend for an odd number (7) of years. But, of course, the available data could just as easily cover an even number of years. Let us assume, for example, that we now add the 1982 sales data for Buckaroo to the data for the previous 7 years (see Fig. 13-9). Figure 13-9 shows *two alternative approaches* to computing trend for an even number of years.

The approach in Fig. 13-9a is no different from the one we have just considered, and so we need not dwell at length on it here. You will notice that since the middle of the time period is now *between* the years 1978 and 1979, the origin will be shifted to January 1, 1979. Then, in the coded time column of Fig. 13-9a, the deviations from the origin to the *middle of each year* in coded *yearly* units will be as indicated. For example, the middle of 1980 is 18 months from January 1, 1979, or 1 1/2 yearly units. The remaining computations in Fig. 13-9a are performed exactly as they were in Fig. 13-7. You will note that the values of *a* and *b* in the trend equation have changed as a result of the additional sales data.

An alternative to the approach shown in Fig. 13-9a is presented in Fig. 13-9b. Up to now we have been coding time in *yearly* units, but this is an arbitrary decision. We could just as easily use 2-year intervals or 1/2-year units. As a matter of fact, we *are* coding time in *1/2-year intervals* in Fig. 13-9b, and this is the *only difference* between the procedures in Fig. 13-9a and b. By redefining the x unit in Fig. 13-9b to represent 1/2 year, we are able to eliminate the use of fractions in the coded time column, and for many this tends to simplify the remaining computations. You will note, however, that the computed $Y_t$ values are exactly the same in Fig. 13-9a and b. The trend equation in both approaches will obviously have the same value of a; however, the values of b appear to be different. In Fig. 13-9a, the change in $Y_t$ for

**FIGURE 13-9**

Annual sales of Buckaroo Lingerie Company

| Years | Company sales (millions of $) (Y) | Coded time (x) | xY | $x^2$ | $Y_t$ (for years indicated, millions of $) |
|---|---|---|---|---|---|
| 1975 | 10.00 | −3-1/2 | −35 | 12.25 | 8.86 |
| 1976 | 8.00 | −2-1/2 | −20 | 6.25 | 9.86 |
| 1977 | 10.00 | −1-1/2 | −15 | 2.25 | 10.86 |
| 1978 | 12.00 | − 1/2 | − 6 | .25 | 11.86 |
| 1979 | 16.00 | 1/2 | 8 | .25 | 12.86 |
| 1980 | 12.00 | 1-1/2 | 18 | 2.25 | 13.86 |
| 1981 | 16.00 | 2-1/2 | 40 | 6.25 | 14.86 |
| 1982 | 14.86 | 3-1/2 | 52 | 12.25 | 15.86 |
| | 98.86 | 0 | 42 | 42.00 | 98.88 |

$$a = 98.86/8 = 12.36$$
$$b = 42/42 = 1.00$$
$$Y_t = 12.36 + 1.00(x)$$
$$Y_{t_{1975}} = 12.36 + 1.00(-3\text{-}1/2)$$
$$= 8.86$$

(a)  Origin: January 1, 1979
   x unit: 1 year
   Y unit: millions of dollars of sales

| Years | Company sales (millions of $) (Y) | Coded time (x) | xY | $x^2$ | $Y_t$ (for years indicated, millions of $) |
|---|---|---|---|---|---|
| 1975 | 10.00 | −7 | −70 | 49 | 8.86 |
| 1976 | 8.00 | −5 | −40 | 25 | 9.86 |
| 1977 | 10.00 | −3 | −30 | 9 | 10.86 |
| 1978 | 12.00 | −1 | −12 | 1 | 11.86 |
| 1979 | 16.00 | 1 | 16 | 1 | 12.86 |
| 1980 | 12.00 | 3 | 36 | 9 | 13.86 |
| 1981 | 16.00 | 5 | 80 | 25 | 14.86 |
| 1982 | 14.86 | 7 | 104 | 49 | 15.86 |
| | 98.86 | 0 | 84 | 168 | 98.88 |

$$a = 98.86/8 = 12.36$$
$$b = 84/168 = .50$$
$$Y_t = 12.36 + .50(x)$$
$$Y_{t_{1975}} = 12.36 + .50(-7) = 8.86$$

(b)  Origin: January 1, 1979
   x unit: 1/2 year
   Y unit: millions of dollars of sales

each change in time (measured in 1-year units) is 1.00; and in Fig. 13-9*b*, the change in $Y_t$ for each change in time (now measured in 1/2-year units) is 0.50. Thus, there is really no difference in *b*. This fact is demonstrated in both parts of Fig. 13-9 by the computation of $Y_t$ for 1975, where there is obviously no difference between multiplying 1.00 by $-3.5$ and multiplying 0.50 by $-7$. The choice of which approach to use is a matter of personal preference.

**Trend Projection**

Let us now assume that Forrestal D. Zaster, the Buckaroo sales manager, wishes to use the past sales pattern as an aid in developing future sales forecasts. If Mr. Zaster can assume that the past pattern is likely to persist into the future, he can use the linear trend equation to project the trend line. Suppose that Mr. Zaster wants a trend projection for 1985, given the data in Fig. 13-9*b*. The coded time value for 1982 in Fig. 13-9*b* is 7. The value of *x* for 1983 would be 9; for 1984 it would be 11; and for 1985, *x* would be 13—i.e., the middle of 1985 would be 13 1/2-year periods from the origin of January 1, 1979. Therefore,

$$Y_{t_{1985}} = 12.36 + 0.50 \ (13)$$
$$= 12.36 + 6.5$$
$$= 18.86 \text{ million dollars of projected sales}$$

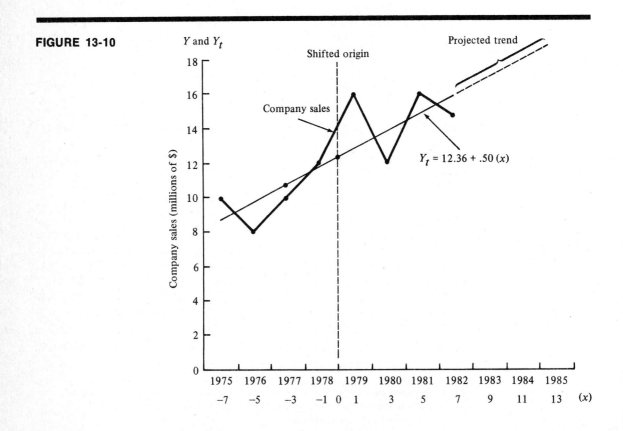

**FIGURE 13-10**

$Y_t = 12.36 + .50 \ (x)$

The same procedure could be used to forecast $Y_t$ for any other future period. Of course, Mr. Zaster should realize that the very precise-looking forecast figure computed above is merely a point of departure or a beginning value that quite possibly should be modified by many subjective considerations before a final forecast figure is obtained. Figure 13-10 shows the Buckaroo sales data through 1982 and the computed and projected trend component. A study of this figure should give you a better grasp of the comments made above.

**Self-testing Review 13-1**

1   The total annual production of saddles by the Callis X. Tremity Saddle Company for the years 1971–1981 is shown below.

| Year | Annual saddle production (hundreds) |
|------|-------------------------------------|
| 1971 | 8 |
| 1972 | 9 |
| 1973 | 12 |
| 1974 | 11 |
| 1975 | 14 |
| 1976 | 17 |
| 1977 | 18 |
| 1978 | 22 |
| 1979 | 24 |
| 1980 | 25 |
| 1981 | 28 |

  **a**   What are the trend equation values for the above data?
  **b**   What would be the projected trend value for 1983?
  **c**   Explain the meaning of the value computed in problem **1b**.

2   Given the trend equation and the other information below, forecast the value of $Y_t$ for 1983.

$$Y_t = 10 + 3x \qquad \text{Origin: the middle of 1979}$$
$$x \text{ unit: 1 year}$$

**SEASONAL VARIATION**

In addition to the computation and projection of secular trend, another important part of time-series analysis is the study of seasonal variation. Of course, since by definition seasonal variations occur within a period of 1 year or less, we must have data reported periodically (weekly, monthly, quarterly, etc.) throughout the year.[9]

The measurement or description of seasonal patterns is accomplished by constructing an *index of seasonal variation*. (Having read Chapter 12 on index numbers with enormous enthusiasm and astounding comprehension, you will be pleased at this bit of information.) A very crude seasonal index could be prepared, for example,

---

[9] If a time series consists of only *annual* data, it will include trend, cyclical, and irregular elements, but seasonal variations will *not* be present because they occur *within* a year.

## FIGURE 13-11 A crude seasonal index

| Month (1) | Month's receipts (2) | Seasonal index [(2) ÷ Average month × 100] (3) |
|---|---|---|
| January | $8,400 | 112.0 |
| February | 8,000 | 106.7 |
| March | 7,400 | 98.7 |
| April | 6,500 | 86.7 |
| May | 6,400 | 85.3 |
| June | 6,200 | 82.7 |
| July | 7,000 | 93.3 |
| August | 7,500 | 100.0 |
| September | 7,700 | 102.7 |
| October | 7,900 | 105.3 |
| November | 8,200 | 109.3 |
| December | 8,800 | 117.3 |
| | $90,000 | 1,200.0 |

Average (mean) month's receipts = $90,000/12 = $7,500

Seasonal index for January = $8,400/$7,500 × 100 = 112.0

by Charles Harse, the owner of a combination massage parlor and health spa, if he were to take the trouble to analyze his monthly receipts for the past year as shown in Fig. 13-11.

As you can see from Fig. 13-11, our crude index of seasonal variation is based on average monthly receipts of $7,500. We have merely expressed the *actual* receipts for each month as a *percentage of the average month*. (In good index number tradition, however, we have omitted the use of the percentage sign.) Thus, the December receipts of $8,800 are 17.3 percent *greater than* (or 117.3 percent *of*) the receipts that could be expected in an average month. With an index of 82.7, however, the month of June had receipts that were only 82.7 percent of the average month. You will notice that the total of the index numbers equals 1,200, as it must in order to give us an average index number value of 100.

There are numerous ways to prepare an index of seasonal variation, and our simple example in Fig. 13-11 has several flaws. However, the *ratio-to-moving-average method* that we will use later, although more sophisticated and the most popular method employed, will still produce seasonal index values by dividing the *actual monthly data* by an *average monthly figure*.

**Reasons for Measuring Seasonal Variation**

The *three main reasons for measuring seasonal variation are:*

1  *To understand seasonal patterns.* The seasonal patterns shown in Fig. 13-3 are of obvious interest to airline executives. By having a measure or index of seasonal variation, they can determine, for example, if the decline in passenger traffic in February is more or less than the typical seasonal amount.

**2** *To project existing patterns into the future.* We have seen that secular trends may be projected as an aid in the *long-term* planning and controlling of important variables. Seasonal variation patterns are also used for forecasting, but the emphasis is on *short-term* planning and control. For example, persistent seasonal patterns may be projected in order to plan for (and control) the appropriate utilization of personnel and to maintain the best levels of inventories and cash balances. A short-term forecast based on Fig. 13-3 would have obvious planning and control implications for scheduling flight and maintenance crews and for maintaining and financing inventories of fuel and parts.

**3** *To eliminate the seasonal component.* The cyclical component in a time series, like the trend and seasonal components, is also used in forecasting. Anticipated cyclical movements are frequently used in the preparation of *intermediate-term* forecasts. However, the cyclical element is often obscured by the swings in seasonal activity, and so the seasonal element must be measured and eliminated in order to reveal the cyclical pattern. Data published by the government in charts and tables are often on a seasonally adjusted basis for just this reason (see Fig. 13-12). We will look at a technique to "deseasonalize" data in a later section.

**FIGURE 13-12**

**Seasonally adjusted data (Source: United States Department of Labor, Bureau of Labor Statistics,** *Monthly Labor Review,* **August 1978, p. 4.)**

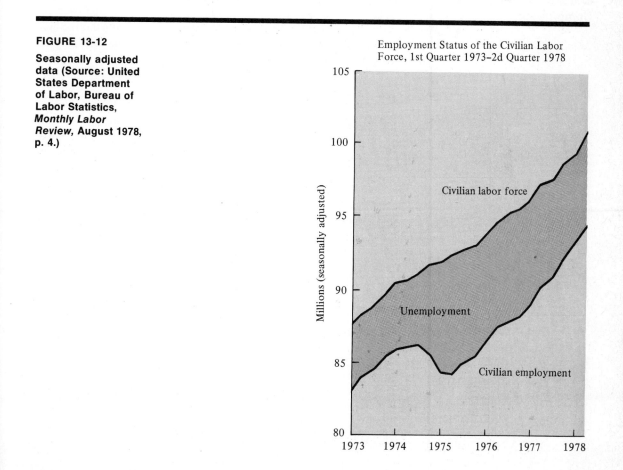

Employment Status of the Civilian Labor Force, 1st Quarter 1973–2d Quarter 1978

CHAPTER 13/FORECASTING TOOLS: TIME-SERIES ANALYSIS

FIGURE 13-13

Production data, Otto S. Guzzling Refining Company (millions of gallons)

| Month | 1976 | 1977 | 1978 | 1979 | 1980 | 1981 |
|---|---|---|---|---|---|---|
| January | 8 | 10 | 11 | 9 | 8 | 12 |
| February | 9 | 12 | 8 | 9 | 10 | 14 |
| March | 9 | 10 | 8 | 13 | 12 | 13 |
| April | 12 | 14 | 16 | 15 | 19 | 18 |
| May | 16 | 15 | 17 | 14 | 22 | 19 |
| June | 13 | 17 | 16 | 13 | 20 | 16 |
| July | 17 | 20 | 23 | 25 | 21 | 23 |
| August | 15 | 14 | 12 | 10 | 18 | 19 |
| September | 10 | 9 | 14 | 14 | 16 | 18 |
| October | 8 | 10 | 13 | 15 | 15 | 19 |
| November | 10 | 10 | 14 | 16 | 12 | 18 |
| December | 14 | 15 | 17 | 15 | 17 | 17 |

**Seasonal Index Computation**

As we mentioned earlier, the *ratio-to-moving average method* of computing a seasonal index that we will use in this section is the most popular approach for measuring the seasonal element in a time series. In order to demonstrate this method, let us use the monthly gasoline production figures (expressed in millions of gallons) of the Otto S. Guzzling Refining Company presented in Fig. 13-13.

The *steps in the ratio-to-moving-average method are listed below*[10] *and demonstrated in Figs. 13-14 and 13-15.*

1 *Compute a 12-month moving total* of the original data, and place this total opposite the *seventh*[11] of the 12 months (see column 4 of Fig. 13-14). You will notice that the total for the 12 months in 1976 (a value of 141) is placed opposite the seventh month of July. The next 12-month moving total of 143 is the sum of the monthly data from February 1976 through January 1977—i.e., the total value of 143 was found by dropping the value of January 1976 and adding the value of January 1977 to the previous total of 141. This procedure is continued until the last month of the last year has been included in the moving total.

2 *Compute a 12-month moving average* (see column 5 in Fig. 13-14) by dividing each of the 12-month totals by 12. Thus, the first 12-month moving average of

---

[10] There are no formidable conceptual difficulties in applying these steps, as you will see; what is formidable, however, is the horrendous amount of clerical effort required to compute a seasonal index if the arithmetic is done by hand. Fortunately, however, computers are ideally suited to massage the mass of data and produce the hundreds of intermediate and final calculations often required.

[11] The exact center of the 12-month period is, of course, midway between the sixth and seventh months, and many texts place the moving total in that position. Then, in order to center moving total figures opposite the data for a particular month, it is necessary to compute an additional 24-month moving total by adding two 12-month moving totals. For the purposes of this text, the extra work required to achieve a minor timing refinement is not necessary.

11.8 in Fig. 13-14 was found by dividing the moving total of 141 by 12, and the second moving average of 11.9 is 143 divided by 12. Again, this procedure is continued until the last average has been found using the last moving total.

**3** *Compute the specific seasonals.* For the month of July 1976 we have the actual production data in column 3 (17 million gallons), and we now have a yearly moving average of 11.8 (million gallons). If you recall, back in Fig. 13-11 we divided the actual data by the average month's value to get a crude seasonal index. Well, this is exactly what we have done in column 6 of Fig. 13-14—we have divided the actual data in July 1976 (17) by the yearly moving average of 11.8 (and multiplied the result by 100) to get a seasonal index value[12] of 144.1. The second index number (called a *specific seasonal*) was found by dividing the actual value of 15 for August 1976 by the corresponding moving average of 11.9. This same procedure is continued until the data are exhausted (or until you are), as shown in Fig. 13-14.

**4** *Arrange the specific seasonals.* Our problem now is *not* that we lack seasonal index values; rather, our problem is that we have *five* specific seasonals for every month of the year. Obviously, we must distill from all these index values a *single* representative or typical index number for each month. We approach this task by preparing Fig. 13-15. (The data in Fig. 13-15 are simply the specific seasonals from column 6 of Fig. 13-14 arranged in a more convenient way.)

**5** *Compute the typical seasonal index values.* To find a typical seasonal index number for each month, we will apply a measure of central tendency to the five specific seasonals for the month. We have chosen to use a modified arithmetic mean—i.e., a measure found by dropping the high and low values as indicated in Fig. 13-15 and computing the mean of the three middle values.[13] Dividing the total of the three middle values for each month by 3 produces the single modified mean index value for each month shown near the bottom of Fig. 13-15.

**6** *Complete the seasonal index.* The modified mean or typical values in Fig. 13-15 are essentially the seasonal index number series that we have been looking for. You will notice, however, that these index values total 1,199.4; you will also remember from Fig. 13-11 that a seasonal index should, in theory, total 1,200. To adjust a total to 1,200,[14] we would merely (a) divide the computed total *into* 1,200 to get a correction factor and (b) multiply the correction factor by each of the monthly index numbers. The *final seasonal index* in Fig. 13-15 has been adjusted in this way.

The purpose of the above steps, of course, was to isolate the effects of seasonal

---

[12] What does the 144.1 mean? It means that gasoline production in July 1976 was 44.1 percent greater than during the average month.

[13] Other texts also use the arithmetic mean or the median as the measure of central tendency. The choice often appears to be rather arbitrary.

[14] As a practical matter, the total of 1,199.4 in our example is close enough to 1,200 that no adjustment is needed. But the adjusting procedure is shown here in case you have to deal with a problem in which your computed total is, say, 1,175 or 1,220.

## FIGURE 13-14

Computation of seasonal index by ratio-to-moving-average method

| Year (1) | Month (2) | Production data (millions of gallons) (3) | 12-month moving total (4) | 12-month moving average [(4) ÷ 12] (5) | Specific seasonals [(3) ÷ (5) × 100] (6) |
|---|---|---|---|---|---|
| 1976 | January | 8 | | | |
| | February | 9 | | | |
| | March | 9 | | | |
| | April | 12 | | | |
| | May | 16 | | | |
| | June | 13 | | | |
| | July | 17 | 141 | 11.8 | 144.1 |
| | August | 15 | 143 | 11.9 | 126.1 |
| | September | 10 | 146 | 12.2 | 82.0 |
| | October | 8 | 147 | 12.2 | 65.6 |
| | November | 10 | 149 | 12.4 | 80.6 |
| | December | 14 | 148 | 12.3 | 113.8 |
| 1977 | January | 10 | 152 | 12.6 | 79.4 |
| | February | 12 | 155 | 12.9 | 93.0 |
| | March | 10 | 154 | 12.8 | 78.1 |
| | April | 14 | 153 | 12.8 | 109.4 |
| | May | 15 | 155 | 12.9 | 116.3 |
| | June | 17 | 155 | 12.9 | 131.8 |
| | July | 20 | 156 | 13.0 | 153.8 |
| | August | 14 | 157 | 13.1 | 106.9 |
| | September | 9 | 153 | 12.8 | 70.3 |
| | October | 10 | 151 | 12.6 | 79.4 |
| | November | 10 | 153 | 12.8 | 78.1 |
| | December | 15 | 155 | 12.9 | 116.3 |
| 1978 | January | 11 | 154 | 12.8 | 85.9 |
| | February | 8 | 157 | 13.1 | 61.1 |
| | March | 8 | 155 | 12.9 | 62.0 |
| | April | 16 | 160 | 13.3 | 120.3 |
| | May | 17 | 163 | 13.6 | 125.0 |
| | June | 16 | 167 | 13.9 | 115.1 |
| | July | 23 | 169 | 14.1 | 163.1 |
| | August | 12 | 167 | 13.9 | 86.3 |
| | September | 14 | 168 | 14.0 | 100.0 |
| | October | 13 | 173 | 14.4 | 90.3 |
| | November | 14 | 172 | 14.3 | 97.9 |
| | December | 17 | 169 | 14.1 | 120.6 |
| 1979 | January | 9 | 166 | 13.8 | 65.2 |
| | February | 9 | 168 | 14.0 | 64.3 |
| | March | 13 | 166 | 13.8 | 94.2 |
| | April | 15 | 166 | 13.8 | 108.7 |
| | May | 14 | 168 | 14.0 | 100.0 |
| | June | 13 | 170 | 14.2 | 91.6 |
| | July | 25 | 168 | 14.0 | 178.6 |
| | August | 10 | 167 | 13.9 | 71.9 |

## FIGURE 13-14

Computation of seasonal index by ratio-to-moving-average method (continued)

| Year (1) | Month (2) | Production data (millions of gallons) (3) | 12-month moving total (4) | 12-month moving average [(4) ÷ 12] (5) | Specific seasonals [(3) ÷ (5) × 100] (6) |
|---|---|---|---|---|---|
| | September | 14 | 168 | 14.0 | 100.0 |
| | October | 15 | 167 | 13.9 | 107.9 |
| | November | 16 | 171 | 14.2 | 112.7 |
| | December | 15 | 179 | 14.9 | 100.7 |
| 1980 | January | 8 | 186 | 15.5 | 51.6 |
| | February | 10 | 182 | 15.2 | 65.8 |
| | March | 12 | 190 | 15.8 | 75.9 |
| | April | 19 | 192 | 16.0 | 118.8 |
| | May | 22 | 192 | 16.0 | 137.5 |
| | June | 20 | 188 | 15.6 | 128.2 |
| | July | 21 | 190 | 15.8 | 132.9 |
| | August | 18 | 194 | 16.2 | 111.1 |
| | September | 16 | 198 | 16.5 | 97.0 |
| | October | 15 | 199 | 16.6 | 90.4 |
| | November | 12 | 198 | 16.5 | 72.7 |
| | December | 17 | 195 | 16.2 | 104.9 |
| 1981 | January | 12 | 191 | 15.9 | 75.5 |
| | February | 14 | 193 | 16.1 | 87.0 |
| | March | 13 | 194 | 16.2 | 80.2 |
| | April | 18 | 196 | 16.3 | 110.4 |
| | May | 19 | 200 | 16.6 | 114.5 |
| | June | 16 | 206 | 17.2 | 93.0 |
| | July | 23 | | | |
| | August | 19 | | | |
| | September | 18 | | | |
| | October | 19 | | | |
| | November | 18 | | | |
| | December | 17 | | | |

SOURCE: Fig. 13-13.

variation from the other time-series elements present in the original data. How was this done? Very briefly, the original data in column 3 of Fig. 13-14 was made up of trend, cyclical, seasonal, and irregular movements. The *rationale of the ratio-to-moving-average method is* (1) that the *12-month moving average* in column 5 of Fig. 13-14 is an *approximate estimate of the combined influences of the trend and cyclical components* (because the process of yearly averaging tends to smooth out the seasonal element) and (2) that by dividing the actual data in column 3 by the moving average, the net result in column 6 of Fig. 13-14 is that the trend and cyclical factors have been removed and the remaining specific seasonals in percentage form are the fluctuations occurring because of seasonal and irregular factors. In other words,

**FIGURE 13-15**

Arranging specific seasonals and computing seasonal index

| Year | Jan. | Feb. | Mar. | Apr. | May | June | July | Aug. | Sept. | Oct. | Nov. | Dec. | Total |
|---|---|---|---|---|---|---|---|---|---|---|---|---|---|
| 1976 | | | | | | | 144.1 | ~~126.1~~ | 82.0 | ~~65.6~~ | 80.6 | 113.8 | |
| 1977 | 79.4 | ~~93.0~~ | 78.1 | 109.4 | 116.3 | ~~131.8~~ | 153.8 | 106.9 | ~~70.5~~ | 79.4 | 78.1 | 116.3 | |
| 1978 | ~~85.9~~ | ~~61.1~~ | ~~62.0~~ | ~~120.3~~ | ~~100.0~~ | 115.1 | 163.1 | 86.3 | ~~100.0~~ | 90.3 | 97.9 | ~~120.6~~ | |
| 1979 | 65.2 | 64.3 | 94.2 | ~~108.7~~ | ~~137.5~~ | ~~91.6~~ | ~~178.6~~ | 71.9 | 100.0 | ~~107.9~~ | ~~112.7~~ | ~~100.7~~ | |
| 1980 | ~~51.6~~ | 65.8 | 75.9 | 118.8 | 114.5 | 128.2 | ~~132.9~~ | 111.1 | 97.0 | 90.4 | ~~72.7~~ | 104.9 | |
| 1981 | 75.5 | 87.0 | 80.2 | 110.4 | | 93.0 | | | | | | | |
| Total of middle three | 220.1 | 217.1 | 234.2 | 338.6 | 355.8 | 336.3 | 461.0 | 304.3 | 279.0 | 260.1 | 256.6 | 335.0 | Total |
| Modified mean | 73.4 | 72.4 | 78.1 | 112.9 | 118.6 | 112.1 | 153.6 | 101.4 | 93.0 | 86.7 | 85.6 | 111.6 | 1,199.4 |
| Seasonal index* | 73.4 | 72.4 | 78.1 | 113.0 | 118.7 | 112.2 | 153.7 | 101.5 | 93.0 | 86.7 | 85.6 | 111.7 | 1,200.0 |

* Correction factor = 1200.0/1199.4 = 1.0005002.

Column 3, Fig. 13-14

$$\frac{\overbrace{\bar{X} \times S \times \bar{C} \times I}}{\underbrace{\bar{X} \times \bar{C}}} = S \times I\} \text{ Column 6, Fig. 13-14}$$

Column 5, Fig. 13-14

Finally, the irregular fluctuations are removed in Fig. 13-15 through the process of eliminating high and low specific seasonal values.

**Uses for the Seasonal Index**

As we saw earlier, the seasonal index is used for *forecasting* and *deseasonalizing* purposes. Sales volume, production schedules, personnel and inventory requirements, financial needs—all these variables are typically subject to short-term forecasts for planning and control purposes. A seasonal index may be used in preparing such forecasts. For example, suppose that the *expected trend value* for sales of a firm next February is $17,000—i.e., $17,000 is the expected sales based on the past trend pattern.[15] Now let us also assume that the firm's February seasonal index value is 90.2. Thus, an analyst would need to adjust the projected trend value by multiplying it by the seasonal index for February. The result would be a *seasonally adjusted forecast* for February of $15,334 ($17,000 × .902).[16]

Original time-series data may also need to be seasonally adjusted or *deseasonalized* so that the effects of the seasonal component can be eliminated in order to examine, for example, the cyclical pattern. Deseasonalizing is an easy operation: You simply divide the original data by the appropriate seasonal index values. Figure 13-16 shows the seasonally adjusted monthly production figures for the Otto S. Guzzling Refining Company for 1981. (You will notice that the data in column 2 of Fig. 13-16 come from Fig. 13-13 and that the seasonal index in column 3 is the one computed in Fig. 13-15.) The deseasonalized production values in column 4 of Fig. 13-16 contain trend, cyclical, and irregular components. In other words,

Original data
(column 2, Fig. 13-16)

$$\frac{\overbrace{T \times S \times C \times I}}{\underbrace{S}} = T \times C \times I\} \text{ Column 4, Fig. 13-16}$$

Seasonal index
(column 3, Fig. 13-16)

---

[15] Although we need not go into the details here, it is not difficult to modify a *yearly* trend equation such as we computed earlier in the chapter into one which can be used to project *monthly* trend values.

[16] With an estimate of the *cyclical* element, the analyst could further adjust the February forecast. For example, if the cyclical factor is expected to be +4 percent in February, the forecast would be $15,947 ($17,000 × .902 × 1.04). Of course, we are not so naïve as to expect sales in February to be *exactly* $15,947. (After all, there are irregular movements possible, and our trend and seasonal figures are only approximations. But a forecast of $15,947 may be better than no explicit forecast at all!)

**FIGURE 13-16**

Deseasonalized production data, 1981, Otto S. Guzzling Refining
Company (millions of gallons)

| Month (1) | Unadjusted pro- duction data, 1981 (2) | Seasonal index (3) | Seasonally adjusted production [(2) ÷ (3) × 100] (4) |
|---|---|---|---|
| January | 12 | 73.4 | 16.3 |
| February | 14 | 72.4 | 19.3 |
| March | 13 | 78.1 | 16.6 |
| April | 18 | 113.0 | 15.9 |
| May | 19 | 118.7 | 16.0 |
| June | 16 | 112.2 | 14.3 |
| July | 23 | 153.7 | 15.0 |
| August | 19 | 101.5 | 18.7 |
| September | 18 | 93.0 | 19.4 |
| October | 19 | 86.7 | 21.9 |
| November | 18 | 85.6 | 21.0 |
| December | 17 | 111.7 | 15.2 |

**Self-testing Review 13-2**

The following data represent monthly receipts (in thousands of dollars) of the
Natalie A. Tyred fashion model school:

| Year | Month | Receipts | Year | Month | Receipts |
|---|---|---|---|---|---|
| 1979 | January | $10 | 1981 | January | $10 |
| | February | 12 | | February | 12 |
| | March | 14 | | March | 15 |
| | April | 15 | | April | 16 |
| | May | 15 | | May | 18 |
| | June | 8 | | June | 7 |
| | July | 9 | | July | 9 |
| | August | 10 | | August | 10 |
| | September | 12 | | September | 15 |
| | October | 13 | | October | 14 |
| | November | 15 | | November | 16 |
| | December | 18 | | | |
| | | | | | |
| 1980 | January | 9 | | | |
| | February | 13 | | | |
| | March | 15 | | | |
| | April | 18 | | | |
| | May | 17 | | | |
| | June | 6 | | | |
| | July | 11 | | | |
| | August | 10 | | | |
| | September | 14 | | | |
| | October | 13 | | | |
| | November | 16 | | | |
| | December | 17 | | | |

1  Compute a seasonal index by the ratio-to-moving-average method. (Since the data are limited, you must use the arithmetic mean in place of the modified mean to compute the typical seasonal index values.)

2  Use the seasonal index computed in problem **1** above to deseasonalize the following monthly receipts (in thousands of dollars) for 1982 of the Natalie A. Tyred organization:

| Month | Unadjusted receipts |
|-------|---------------------|
| January | $11 |
| February | 13 |
| March | 14 |
| April | 15 |
| May | 19 |
| June | 8 |
| July | 12 |
| August | 14 |
| September | 14 |
| October | 16 |
| November | 17 |
| December | 21 |

## IDENTIFYING CYCLICAL AND IRREGULAR COMPONENTS

In earlier sections of this chapter we referred to a time-series model that describes the relationship between the time-series components and the original data ($Y$) as follows:

$$Y = T \times S \times C \times I$$

This time-series model is frequently referred to as the *classical model,* and it may be used to explain the procedure for identifying the cyclical and irregular components.

Once trend values and a seasonal index have been prepared from the original data using the methods that have now been presented in this chapter, it is possible to cancel out and thus eliminate these elements in order to obtain a theoretical series in which only the cyclical and irregular components are at work.[17] That is, $(T \times S \times C \times I) \div T = S \times C \times I$, and $(S \times C \times I) \div S = C \times I$. And once the trend and seasonal elements have been eliminated, the usual procedure is to then cancel out and thus *eliminate the irregular movements* by computing a moving average. The remaining or *residual* variations are considered to be the *cyclical forces* at work. Of course, it is desirable to try to measure the cyclical element because of the importance of cyclical factors in short- and intermediate-term forecasts.

---

[17] If the original data are expressed in yearly values (or if the data are presented on a seasonally adjusted basis), the seasonal component either does not exist or has already been eliminated by deseasonalization. In this case, the procedure to identify the $C$ and $I$ movements would be the same, except that the adjustment to eliminate the seasonal element would obviously not be needed.

**FIGURE 13-17**

Worksheet format for identifying the cyclical component

| Month (1) | Original data ($T \times S \times C \times I$) (2) | Trend value ($T$) (3) | Percent of trend ($S \times C \times I$) [(2) ÷ (3)] (4) | Seasonal index ($S$) (5) | Cyclical irregular percentages ($C \times I$) [(4) ÷ (5)] (6) | Moving total of (6) percentages (7) | Cyclical percentages ($C$) [(7) ÷ number of months (or quarters) used in moving total] (8) |
|---|---|---|---|---|---|---|---|
| | | | | | | | |

The calculations required to identify the cyclical and irregular forces are not difficult, but they are tedious and better left to more advanced texts. A typical work sheet to perform the steps just described, however, might be set up as shown in Fig. 13-17.

## TIME-SERIES ANALYSIS IN FORECASTING: A SUMMARY

There are many procedures used to forecast the future, but essentially these procedures are all based on a very limited number of basic assumptions. *One* such assumption is that past patterns will *persist* into the future; *another* is that measurable past data fluctuations will *recur regularly* and can thus be projected into the future. The following *uses of time-series analysis in forecasting,* as we have seen, are based on both of these assumptions:

1 *For long-term forecasting.* In making forecasts several years into the future, the analysis, and then projection, of secular trend is an important element. We have seen how it is possible to use a trend equation to make such projections. A long-term trend forecast often does not take cyclical forces into account, and the seasonal element is of no concern in the projection of annual data.

2 *For intermediate-term forecasting.* The cyclical component may be incorporated into forecasts of shorter duration by multiplying the projected trend value by an estimate of the expected percentage change in the data as a result of cyclical forces. That is, projected trend times the cyclical percentage equals the forecast value. (Of course, the assumptions that relationships or patterns which existed in the past will persist into the future with recurring regularity are very shaky when applied to the cyclical component because successive cycles vary widely in timing, pattern, and percentage changes.)

3 *For short-term forecasting.* We have seen how a seasonal index may be used with a monthly trend projection to produce a short-term seasonally adjusted forecast. And in footnote 16, we saw how a cyclical estimate could also be incorporated into the short-term forecast. In short-term forecasts, as in longer-term projections, an attempt to predict irregular movements is usually *not* made.

## PROBLEMS IN THE USE OF TIME-SERIES ANALYSIS

Economists state their GNP growth projections to the nearest tenth of a percentage point to prove that they have a sense of humor.

—E. R. Fiedler

Give them a number or give them a date, but never both.

—E. R. Fiedler

It is surprising that one soothsayer can look at another without smiling.

—Cicero

The fact that a particular forecast made with the aid of time-series analysis techniques may prove to be incorrect does not discredit the use of the techniques discussed in this chapter. After all, the incorrect forecast might have been much closer to the actual results than a hunch.[18] If a time-series analysis leads to forecasts that reduce the *avoidable* risk, it will have been worthwhile. There are, however, potential problem areas associated with the use of time-series analysis. And these problem areas lead to questions that cannot be safely ignored. A few of the *precautionary questions that should be considered* in a particular situation are:

1   *How appropriate is the classical model?* A model is an abstract representation of reality. In the classical model, each component represents the theoretical effects of a myriad of causal factors that have been grouped together. Obviously, when we create a model and apply it to the analysis of actual data, we can expect results that are proportional to the accuracy and sophistication of the model itself. Unfortunately, our time-series model is not too sophisticated. It assumes, for example, that the four components are *independent of each other,* and this hardly conforms to reality in many cases. (There are many people who would probably alter their seasonal Christmas spending habits during a cyclical recession.)

2   *How valid are our assumptions about persistence and regularity?* If there are *no persuasive independent reasons* to support these assumptions, the rather mechanical mathematical procedures that produce a trend equation and a seasonal index can lead to a very precise-looking forecast that may be completely out of touch with reality. The assumptions must always be kept in mind, and the analyst must adjust computations in light of the subjective and qualitative factors that are almost always present.

3   *How reliable are the input data?* Lack of data is often a problem, and when adequate data are available, they may not be strictly comparable. During the rather long intervals of time over which data should be gathered before usable patterns can emerge, a number of things can happen, for example, to change the quality of a measured variable. It is not inconceivable that the quality of a successful product could deteriorate to the point that the past trend could be significantly altered in a short period of time.

The above questions are merely a few of the ones that the analyst should consider in using time-series analysis techniques. Though the limitations and pitfalls are many, time-series analysis can be a very useful tool when used to provide an initial and approximate forecast.

**SUMMARY**

Planning and decision making involve expectations of what the future will bring, and so administrators are required to make forecasts. One forecasting tool used by managers is time-series analysis. A time series is worth studying because (1) it may

---

[18] An alternative forecasting approach would be for you to sit in a trance on a tripod above a chasm that emitted noxious vapors and make oracular utterances for someone to record. This was the approach used by the Greeks in ancient Delphi.

enhance understanding of past and current patterns of change, and (2) it may provide clues about future patterns to aid in forecasting.

The four components which may be identified in a time series are secular trend, seasonal variation, cyclical fluctuations, and irregular movements. A classical time-series model is used to describe the relationship between these components.

The first component studied in this chapter, secular trend, is examined in order to describe historical patterns in a series and to project persistent patterns into the future. Once the trend element is identified, it may be removed from the original data to reveal the movement of other components. A linear trend equation ($Y_t = a + bx$) is discussed and computed in the chapter. By using this equation, it is possible to project a trend line into the future.

The second component, seasonal variation, is measured by constructing an index of seasonal variation. A ratio-to-moving-average method was employed in the chapter to prepare such an index. The reasons for measuring seasonal variation are (1) to understand seasonal patterns, (2) to project existing patterns into the future, and (3) to eliminate the seasonal component from the time series by a process called deseasonalization.

Once trend and seasonal elements have been measured, it is then possible to eliminate these elements in order to obtain a theoretical series in which only the cyclical and irregular components are at work. The purpose of measuring trend, seasonal, and cyclical components, as mentioned above, is often for forecasting. However, there are limitations and pitfalls in the uncritical application of time-series computations for the purpose of preparing forecasts.

## Important Terms and Concepts

1  Forecast
2  Time-series analysis
3  Time series
4  Secular trend
5  Seasonal variation
6  Cyclical fluctuations
7  Irregular movements
8  $Y = T \times S \times C \times I$
9  Trend analysis
10  $Y$ intercept
11  Slope of the line
12  $Y_t = a + bx$
13  $\Sigma(Y - Y_t) = 0$
14  $\Sigma(Y - Y_t)^2 =$ least value

15  Trend projection
16  Index of seasonal variation
17  Ratio-to-moving-average method
18  Specific seasonals
19  Deseasonalizing
20  Seasonally adjusted forecast
21  Classical model
22  Residual variations
23  Long-term forecasting
24  Short-term forecasting
25  Assumption of persistence
26  Method of least squares
27  Linear trend equation

## Problems

1  The Icehole Swimming Pool Company of Fairbanks, Alaska, is planning an expansion program. The company needs a forecast of annual sales for each of

the next 5 business years in order to plan this expansion. An analysis of the company's records indicated the following:

| Year | Annual sales (thousands) |
|------|------|
| 1972 | 110 |
| 1973 | 125 |
| 1974 | 135 |
| 1975 | 150 |
| 1976 | 170 |
| 1977 | 185 |
| 1978 | 196 |
| 1979 | 216 |
| 1980 | 230 |

**a** Construct a time-series graph with the data given above.

**b** Compute the secular trend equation with the least-squares method. Plot this equation on the time-series graph prepared in problem **1a**.

**c** Compute the needed 5-year forecast with the equation found in problem **1b**.

**2** Radio station WINO of Sterling, Illinois, is concerned about an apparent secular trend decline in the number of persons listening to AM radio in its market area. FM listenership is apparently on a secular upswing. The data given below were prepared by a marketing research agency:

| Year | AM radio average audience size (hundreds) | FM radio average audience size (hundreds) |
|------|------|------|
| 1972 | 31 | 3 |
| 1973 | 32 | 3 |
| 1974 | 33 | 6 |
| 1975 | 30 | 7 |
| 1976 | 29 | 10 |
| 1977 | 30 | 11 |
| 1978 | 28 | 14 |
| 1979 | 26 | 14 |
| 1980 | 24 | 17 |

**a** Construct a time-series graph with the data given above for both the AM and FM audience size series.

**b** Compute the secular trend equation with the least-squares method, for both AM and FM.

**c** Forecast the total audience size for both AM and FM in 1982.

**3** The security chief of Horace Cints University is making plans to assign personnel and schedule vacations next year in the traffic division. Official records indicate the total issuance of traffic tickets over the past 3 years:

|  | Traffic tickets issued | | |
| Month | 1979 | 1980 | 1981 |
| --- | --- | --- | --- |
| January | 100 | 90 | 110 |
| February | 90 | 110 | 120 |
| March | 80 | 90 | 90 |
| April | 90 | 110 | 100 |
| May | 110 | 130 | 140 |
| June | 130 | 120 | 140 |
| July | 140 | 150 | 140 |
| August | 160 | 170 | 160 |
| September | 200 | 200 | 220 |
| October | 210 | 210 | 220 |
| November | 180 | 200 | 240 |
| December | 240 | 220 | 280 |

   **a** Determine seasonal indexes by the ratio-to-moving-average method. (You must use the arithmetic mean to average the specific seasonals.)

   **b** Comment on how the seasonal indexes might be used by the chief.

**4** The chief assistant to the assistant chief in charge of university reports must give the trustees of Horace Cints a report on traffic tickets issued during the first 6 months of 1982. Actual issuances are recorded below:

| Month | Tickets issued |
| --- | --- |
| January | 120 |
| February | 120 |
| March | 100 |
| April | 90 |
| May | 130 |
| June | 150 |

   **a** Deseasonalize the ticket issuances with the seasonal index values obtained in problem **3**.

   **b** Comment on the apparent increase or decrease in tickets issued thus far in 1982.

**Topics for Discussion**

**1** "All administrators must make forecasts." Comment on this statement.

**2** Why should we study time series?

**3** Identify and discuss the components found in a time series.

**4** What is the classical time-series model, and how may it be used?

**5** Compare the reasons for measuring trend with the reasons for measuring seasonal variation.

**6** Why are the assumptions of persistence and regularity important in forecasting?

7   Apply the straight-line equation to a firm whose *total cost* curve is a straight line for a particular product.[19]

8   In economics, the consumption function is often drawn as a straight line. Explain why the slope of the consumption function is therefore the marginal propensity to consume.[19]

9   Explain the assumptions made when a secular trend is projected into the future.

10  A seasonal index for January is 112.6. What does this number mean?

11  Discuss the steps in the ratio-to-moving-average method of computing a seasonal index.

12  Explain how a short-term forecast could be adjusted to take the seasonal element into account.

13  Why is the method of identifying cyclical fluctuations sometimes referred to as the residual method?

14  Summarize how time-series analysis may be used in forecasting.

15  What problems are there in the use of time-series analysis?

**Answers to Self-testing Review Questions**

13-1

1 a   $a = 188/11 = 17.09$
        $b = 226/110 = 2.05$
        $Y_t = 17.09 + 2.05x$

  b   $Y_t = 17.09 + 2.05(7)$
        $= 17.09 + 14.35$
        $= 31.44$

  c   If past production patterns persist, a tentative forecast of saddle production in 1983 would indicate that approximately 3,144 saddles would be produced.

2   Since the origin is the middle of 1979, and since $x$ is in one year units, the value of $x$ in 1983 is 4. Thus,
        $Y_t$ 1983 $= 10 + 3(4)$
        $= 22$

---

[19] This is an optional and tough question. Don't worry if you are stumped by it.

**13-2**

| Year | Specific seasonals | | | | | | | | | | | |
|---|---|---|---|---|---|---|---|---|---|---|---|---|
| | Jan. | Feb. | Mar. | Apr. | May | June | July | Aug. | Sept. | Oct. | Nov. | Dec. |
| 1979 | | | | | | | 71.4 | 80.0 | 95.2 | 103.2 | 116.3 | 137.4 |
| 1980 | 69.8 | 99.2 | 114.5 | 136.4 | 128.8 | 45.1 | 83.3 | 75.2 | 106.1 | 98.5 | 122.1 | 128.8 |
| 1981 | 75.8 | 91.6 | 114.5 | 121.2 | 136.4 | 53.0 | | | | | | |
| Mean | 72.8 | 95.4 | 114.5 | 128.8 | 132.6 | 49.0 | 77.4 | 77.6 | 100.6 | 100.8 | 119.2 | 133.1 |
| Seasonal index* | 72.7 | 95.3 | 114.3 | 128.6 | 132.4 | 48.9 | 77.3 | 77.5 | 100.4 | 100.6 | 119.0 | 132.9 |

* Correction factor = 1200.0/1201.8 = .9985

**2**

| Month | Unadjusted receipts | Seasonal index | Seasonally adjusted receipts |
|---|---|---|---|
| January | $11 | 72.7 | $15.1 |
| February | 13 | 95.3 | 13.6 |
| March | 14 | 114.3 | 12.2 |
| April | 15 | 128.6 | 11.7 |
| May | 19 | 132.4 | 14.4 |
| June | 8 | 48.9 | 16.4 |
| July | 12 | 77.3 | 15.5 |
| August | 14 | 77.5 | 18.1 |
| September | 14 | 100.4 | 13.9 |
| October | 16 | 100.6 | 15.9 |
| November | 17 | 119.0 | 14.3 |
| December | 21 | 132.9 | 15.8 |

# CHAPTER 14

# FORECASTING TOOLS: SIMPLE LINEAR REGRESSION AND CORRELATION

**LEARNING OBJECTIVES**

After studying this chapter, working the problems, and answering the discussion questions, you should be able to:

☞ Explain the purposes of regression and correlation analysis.

☞ Compute the regression equation and the standard error of estimate and then use these measures to prepare interval estimates of the dependent variable for forecasting purposes.

☞ Compute (and explain the meaning of) the coefficients of determination and correlation.

☞ Point out several errors that may be made in the use of regression and correlation techniques.

**CHAPTER OUTLINE**

As we have seen, in order to make a decision it is often necessary to prepare a forecast. In order to prepare a budget, it is necessary to predict revenues. A university must predict enrollment before making up class schedules. These and other decisions can be made easier if a relationship can be established between the variable to be predicted and some other variable that is either known or is significantly easier to anticipate.

For our purposes, the term "relationship" means that changes in two (or more) variables are *associated with each other*. For example, we might find a high degree of relationship between the consumption of fuel oil and the number of cold days during the winter, or between gasoline sales and the number of registered motor vehicles. That is, we might logically expect a change in the number of cold days to be accompanied by a change in the consumption of fuel oil, or a change in the number of registered automobiles to be accompanied by a change in the demand for gasoline.

Our purpose in this chapter is to study some statistical techniques for measuring and evaluating the relationship between two variables. Therefore, after some *introductory concepts* have been presented, we will devote the remainder of the chapter to the topics of *regression analysis, the standard error of estimate, correlation analysis,* and the *common errors and limitations* involved in the use and interpretation of regression and correlation measures.

## INTRODUCTORY CONCEPTS
### Regression and Correlation Analysis: A Preview

The tools of *regression analysis* and *correlation analysis* have been developed to study and measure the statistical relationship that exists between two or more variables. (Although we shall only consider the relationship between *two* variables —i.e., we will only be concerned with *simple* regression and correlation—the measures introduced in this chapter may also be applied to *three or more* variables—i.e., to *multiple* regression and correlation situations.)

In regression analysis, an *estimating equation* is developed to describe the pattern or functional nature of the relationship that exists between the variables. As the name implies, an analyst prepares an estimating (or *regression*) equation to make estimates of values of one variable from given values of the other. The variable to be estimated is called the *dependent variable* and is customarily plotted on the vertical (or $Y$) axis of a chart. (The dependent variable is therefore identified by the symbol $Y$.) The variable that presumably exerts an influence on or explains variations in the dependent variable is customarily plotted on the horizontal (or $X$) axis, is termed the *independent variable,* and is identified by the symbol $X$. Thus, *if* Hiram N. Hess, the personnel manager of the Tackey Toy Manufacturing Company (widely known for its motto "The little monsters deserve Tackey quality"), finds that there is a close and logical relationship between the productive output of employees in a certain department of the company and their earlier performance on an aptitude test, *if* he computes an estimating equation (by a method to be presented in a few pages), and *if* he has the aptitude test score of a potential employee, *then* he may use his estimating equation to forecast the future output (the dependent variable) of the job applicant on the basis of the aptitude test results (the independent variable). In addition to the techniques used to make estimates of the dependent variable, regression analysis also includes a method of measuring the *dependability* of these estimates.

In correlation analysis, the purpose is to measure the *degree* or *closeness* of the relationship between the variables. In other words, regression analysis asks "What is the pattern of the existing relationship?" and correlation analysis asks "How close is the relationship described in the regression equation?" Although it is possible to be concerned with only regression analysis, or with only an analysis of correlation, the two are typically considered together in business and economic applications. In fact, the term "correlation analysis" is sometimes used to include *both* the regression and correlation elements described here.

To summarize, in the following pages we will concentrate on three main topics (the first two of which are a part of regression analysis):

1   The computation of the regression or estimating equation and its use in providing an estimate of the value of the dependent variable ($Y$) from a given value of the independent variable ($X$).

2   The computation of a measure that will indicate the possible error involved in using the estimating equation as a basis for forecasting.

3   The preparation of measures that will indicate the closeness of the association or correlation that exists between the variables.

## A Logical Relationship: The First Step

**Your sales last year just paralleled the sales of rum cokes in Rio de Janeiro, as modified by the sum of the last digits of all new telephone numbers in Toronto. So, why bother with surveys of your own market? Just send away for the data from Canada and Brazil.[1]**

**— Lydia Strong**

Two series may vary together for several reasons. And analysts may correctly assume a causal relationship in their interpretation of correlation measures. But just the fact that two variables are associated in a statistical sense does not guarantee the existence of a causal relationship. In other words, the existence of a causal relationship usually does imply correlation, but, as the above quote adequately shows, *statistical correlation alone does not in any way prove causality.* (If you believe that it does, you should indeed send away for the data from Canada and Brazil!)

---

[1] From "What's Ahead? The Gentle Art of Business Forecasting," *Management Review,* September 1956.

Causal relationships may fit into *cause-and-effect* or *common cause* categories. A cause-and-effect relationship exists if a change in one variable causes a change in the other variable. For example, a change in the temperature of a chemical process may cause a change in the yield of the process, and a change in the level of output may cause a change in total production cost. Alternatively, two series may vary together because of a common cause acting on both series in the same way. One could probably find a close relationship between jewelry sales and phonograph record sales, but one of these series is not the cause and the other the effect. Rather, changes in both series are probably a result of changes in consumer income.

Of course, as we saw in the quote at the beginning of this section, some relationships are purely accidental. So if a relationship were found between furniture sales in the United States and the average temperature in Tanzania, for example, it would be a meaningless exercise to analyze the data. Relationships such as this one are known as *spurious correlations*.

Probably the first step in the study of regression and correlation, therefore, is to determine if a logical relationship may exist between the variables to support further analysis. Unfortunately, presenting a summary of many types of variables, and the forces at work on those variables, is beyond the scope of this book; fortunately, however, such a summary is to be found in the other economics, accounting, finance, marketing, and management courses that you have taken or will take. It is only through the use of reason and judgment (along with the application of knowledge about the variables and the forces at work) that an analyst may assume causality. Yet without this assumption, there is not much point in proceeding with regression and correlation analyses.[2]

## REGRESSION ANALYSIS

As you know, our main purpose in this section is to compute an estimating or regression equation that describes the relationship between the variables. Our discussion will be limited to an analysis of *simple linear regression,* that is, to the case in which the relationship between only two variables can be adequately described by a straight line.

## The Scatter Diagram

After it has been determined that a logical relationship may exist between variables to support further analysis, perhaps the next step is to use a chart to plot the available data. This chart—called a *scatter diagram*—shows plotted points, and each point represents an item for which we have a value for both the dependent variable and the independent variable. *The scatter diagram serves two purposes:* (1) It helps determine if there is a useful relationship between the two variables, and (2) it helps to determine the type of equation to use to describe the relationship.

---

[2] Beginning students sometimes go to one of two extremes: Either they fail to realize the importance of determining if a logical relationship exists, mechanically apply statistical procedures to the data, and arrive at possibly spurious correlations to which they erroneously assign interpretations of cause and effect, or they think that since one cannot prove causality from correlation, it is necessary to conclude that there is no connection at all between correlation and causality.

**FIGURE 14-1**

Output and aptitude test results of eight employees
of Tackey Toy Manufacturing Company

| Employee | Output (Y) (dozens of units) | Aptitude test results (X) |
|---|---|---|
| A | 30 | 6 |
| B | 49 | 9 |
| C | 18 | 3 |
| D | 42 | 8 |
| E | 39 | 7 |
| F | 25 | 5 |
| G | 41 | 8 |
| H | 52 | 10 |

We can illustrate the purposes of a scatter diagram by using the data in Fig. 14-1. This figure gives us the output for a time period in dozens of units (the dependent variable) and the aptitude test results (the independent variable) for eight employees in a department of the Tackey Toy Manufacturing Company. If the aptitude test does what it is supposed to do, it is reasonable to assume that employees with higher aptitude scores will be among the higher producers.[3]

The data for an employee represent *one* point on the scatter diagram shown in Fig. 14-2. (The points representing employees C and F have been labeled to show you how the pairs of observations for the employees are used to prepare the points on the chart.) As you will notice in Fig. 14-2, the eight points form a path that can be described by a *straight line,* and a high degree of relationship is indicated by the fact that the points are all close to this straight-line path. You will also notice that there is a *positive* (or *direct*) relationship between the variables—i.e., as aptitude test results *increase,* output *also increases.* Of course, it would be quite possible for variables to have a *negative* (or *inverse*) relationship (as the X value increases, the Y value decreases). Figure 14-3 shows some possible scatter diagram forms. You will notice that in Fig. 14-3*a* and *b* there is a positive linear relationship. Relationships, of course, need not be linear, as shown in Fig. 14-3*d* to *f*.[4] In Fig. 14-3*f*, the variables might be family income and age of the head of the household. (Income tends to rise for a period and then fall off when retirement age is reached.) Finally, it is possible that a scatter diagram such as the one in Fig. 14-3*g* might show *no relationship* at all between the variables.

---

[3] Obviously, our group of eight hypothetical employees would probably represent a small sample from which only very tentative conclusions could be drawn. We have chosen to keep the number of employees small and the data simple, however, in order to minimize the amount of computations necessary in later sections.

[4] The equations necessary to describe these curvilinear relationships differ from the linear one we will introduce in the next section.

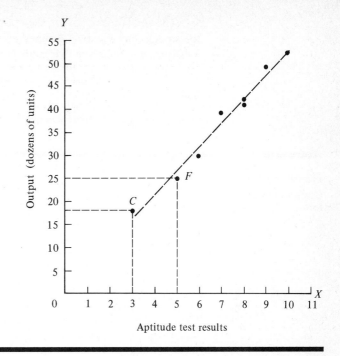

**FIGURE 14-2**

**Scatter diagram (Source: Fig. 14-1)**

## THE LINEAR REGRESSION EQUATION

The straight line in the scatter diagram in Fig. 14-2 (and the straight lines in Fig. 14-3) that describes the relationship between the variables is called a *regression* (or *estimating*) line. And, as you know, the equation used to fit the regression line to the scatter diagram data is called the regression or estimating equation.[5] If you remember the discussion in the last chapter about linear trend computation, you should have little difficulty here with the regression equation. Why? Because the general straight-line equation presented there is also used in fitting the regression line to the data. That is, the *method of least squares* will once again be used to fit a line to the observed data. Therefore, the formula for the regression equation is:

$$Y_c = a + bX \qquad (14\text{-}1)$$

where   $a = Y$ intercept (the value of $Y_c$ when $X = 0$)

$b =$ slope of the regression line (the increase or decrease in $Y_c$ for each change of one unit of $X$)

$X =$ a given value of the independent variable

$Y_c =$ a computed value of the dependent variable

---

[5] The term "regression" was used in 1877 by Sir Francis Galton in connection with a study he had made of human height. Since his study showed that the height of the children of tall parents tended to regress (or move back) toward the average height of the population, Galton called the line describing this relationship a "line of regression." The word "regression" has thus stuck with us, even though words such as "estimating" or "predictive" are probably more appropriate.

The similarities between the regression line and the trend line do not end with the straight-line equation. The regression line (like the trend line and the arithmetic mean) has the following two mathematical properties:

$$\Sigma(Y - Y_c) = 0$$

and
$$\Sigma(Y - Y_c)^2 = \text{a minimum or least value}$$

In other words, the regression line will be fitted to the data in the scatter diagram

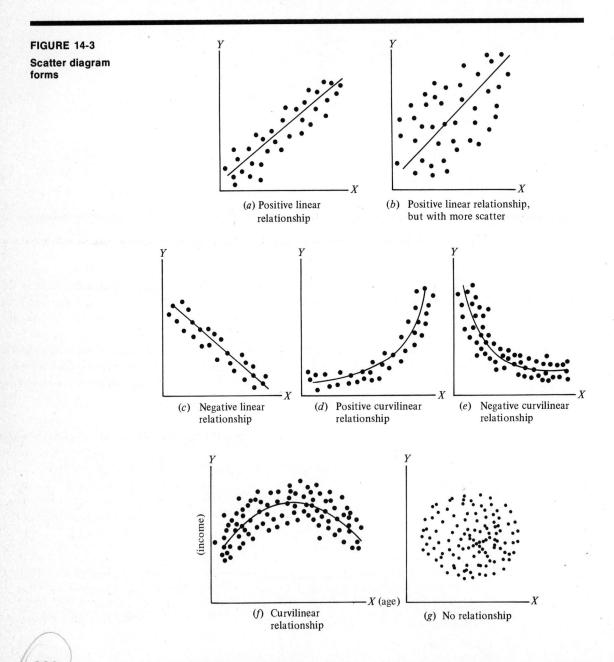

**FIGURE 14-3**

**Scatter diagram forms**

(a) Positive linear relationship

(b) Positive linear relationship, but with more scatter

(c) Negative linear relationship

(d) Positive curvilinear relationship

(e) Negative curvilinear relationship

(f) Curvilinear relationship

(g) No relationship

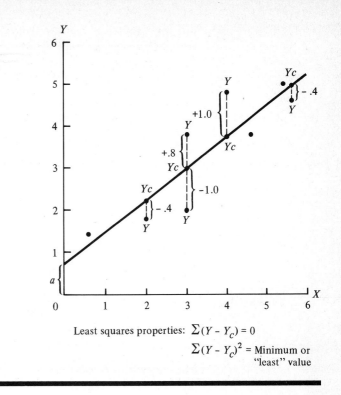

Least squares properties: $\Sigma(Y - Y_c) = 0$

$\Sigma(Y - Y_c)^2 =$ Minimum or "least" value

**FIGURE 14-4**

in such a way that the *positive* deviations of the scatter points *above* the line in the diagram will cancel out the *negative* deviations of the scatter points below the line, and the resulting sum will be zero (see Fig. 14-4).

For our purposes, the computations for regression and correlation analysis may be facilitated if the formulas are stated in terms of deviations from the *means of the X and Y variables, that is, in terms of deviations from $\bar{X}$ and $\bar{Y}$*. Therefore, using the following symbols:

$$x = (X - \bar{X})$$
$$y = (Y - \bar{Y})$$

and
$$xy = (X - \bar{X})(Y - \bar{Y})$$

the values of $a$ and $b$ in the regression equation may be computed with the following formulas:

$$b = \frac{\Sigma(xy)}{\Sigma(x^2)} \qquad (14\text{-}2)$$

$$a = \bar{Y} - b\bar{X} \qquad (14\text{-}3)$$

We are now ready to compute the regression equation for the data presented in Fig. 14-1. A work sheet for computing the values required to solve for $b$ and $a$ [using formulas (14-2) and (14-3)] is given in Fig. 14-5. The values of $b$ and $a$ are computed as follows:

**FIGURE 14-5**

| Employee | Output in dozens of units (Y) | Test results (X) | y (Y − Ȳ) | x (X − X̄) | xy | x² | y² |
|---|---|---|---|---|---|---|---|
| A | 30 | 6 | −7 | −1 | 7 | 1 | 49 |
| B | 49 | 9 | 12 | 2 | 24 | 4 | 144 |
| C | 18 | 3 | −19 | −4 | 76 | 16 | 361 |
| D | 42 | 8 | 5 | 1 | 5 | 1 | 25 |
| E | 39 | 7 | 2 | 0 | 0 | 0 | 4 |
| F | 25 | 5 | −12 | −2 | 24 | 4 | 144 |
| G | 41 | 8 | 4 | 1 | 4 | 1 | 16 |
| H | 52 | 10 | 15 | 3 | 45 | 9 | 225 |
| | 296 | 56 | 0 | 0 | 185 | 36 | 968 |

$$\bar{Y} = \frac{\Sigma Y}{N} = \frac{296}{8} = 37; \quad \bar{X} = \frac{\Sigma X}{N} = \frac{56}{8} = 7$$

$$b = \frac{\Sigma(xy)}{\Sigma(x^2)} = \frac{185}{36} = 5.138 \text{ or } 5.14$$

$$a = \bar{Y} - b\bar{X} = 37 - (5.14)(7) = 1.02$$

Therefore, the regression equation that describes the relationship between the output of Tackey Toy employees and their aptitude test results is:

$$Y_c = 1.02 + 5.14X$$

## Using the Regression Equation in Forecasting

There was once a young manager named Hess
Whose forecasts were always a mess.
So his boss did appear,
And in voice loud and clear,
Said, "Hess, son, try regression, or consider another career!"
— One of the authors of this text, who, for obvious reasons,
declines to be identified

The primary use of the regression equation is to estimate values of the dependent variable given values of the independent variable. Suppose, for example, that the unfortunate Hiram Hess, personnel manager for Tackey Toys, is considering hiring an applicant who scored a 4 on the aptitude test. The supervisor of the department wants someone hired who can produce at least 30 dozen units. Of course, it is not possible to tell exactly what the applicant's future production might be, but Hiram can use the equation computed in the preceding section to arrive at an estimate or forecast of the applicant's future output. How? By simply substituting 4 for $X$ in the regression equation. The estimate would be computed as follows:

$$Y_c = 1.02 + 5.14(4)$$
$$= 1.02 + 20.56$$
$$= 21.58 \text{ dozen units of output}$$

This prediction is shown graphically in Fig. 14-6.

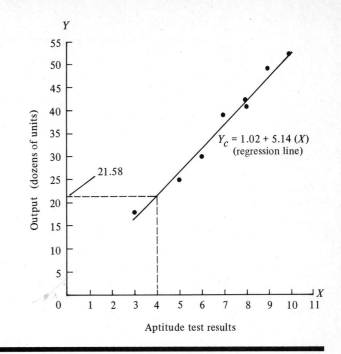

**FIGURE 14-6**

Aptitude test results

---

**Self-testing Review 14-1**

The questions in this review are based on the following data:

| Y | X |
|-----|-----|
| 255 | 5 |
| 100 | 2 |
| 307 | 6 |
| 150 | 3 |

**1** What are the values of
   **a** $\Sigma(xy)$?
   **b** $\Sigma(x^2)$? *25*

**2** What is the regression equation for these data?

**3** What is the estimate of Y when X is 4?

---

**THE STANDARD ERROR OF ESTIMATE**

When an estimate is made from a regression equation, this question naturally arises: "How dependable is the estimate?" Obviously, an important factor in determining dependability is the closeness of the relationship between the variables. When the points in a scatter diagram are closely spaced around the regression line, as they are in Fig. 14-7a—i.e., when there is a relatively small amount of scatter—it is logical to assume that an estimate based on that relationship would probably be more dependable than an estimate based on a regression line such as that shown in Fig. 14-7b, in which the scatter or variation is much greater.[6] Therefore, if we had a measure of

---

[6] It is assumed that both parts of Fig. 14-7 have the same scales for the variables and the same regression line.

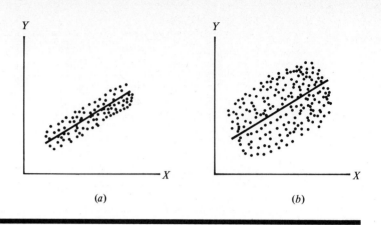

**FIGURE 14-7**

**Varying degrees of spread or scatter**

(a)                    (b)

the extent of the spread or scatter of the points around the regression line, we would be in a better position to judge the dependability of estimates made using the line. (You just know we are leading up to something here, don't you?)

   You will not be surprised to learn that we *do have a measure* that indicates the extent of the spread, scatter, or dispersion of the points about the regression line. From an estimating standpoint, *the smaller this measure is, the more dependable the prediction is likely to be.* (The numerical value of this measure for Fig. 14-7a is smaller than the value for Fig. 14-7b because the dispersion is smaller in Fig. 14-7a.) The name of this measure is the *standard error of estimate* (the symbol is $Sy \cdot x$), and, as the name implies, it is used to qualify the estimate made with the regression equation by indicating the extent of the possible variation (or error) that may be present.

**Computation of Sy · x**

One formula that may be used to compute the standard error of estimate bears a striking resemblance to the formula used to compute the standard deviation for ungrouped data. The primary difference between the two formulas lies in the fact that the standard deviation is measured from the mean, while the standard error of estimate is measured from the regression line. (Both the mean and the regression line, of course, indicate central tendency.) This formula for the standard error of estimate is:[7]

$$Sy \cdot x = \sqrt{\frac{\Sigma(Y - Y_c)^2}{n - 2}} \qquad (14\text{-}4)$$

In using this formula, we must compute a value for $Y_c$ for each value of $X$ by plugging each $X$ value into the regression equation. We must then compute the difference between these $Y_c$ values and the corresponding observed values of $Y$. A work sheet

---

[7] The $n - 2$ in the denominator is used in this case because it is assumed that we are dealing with sample data. A rather detailed explanation could be given of why $n - 2$ is appropriate, but we will leave this explanation to other texts.

**FIGURE 14-8**

| Employee | Output (Y) | Test results (X) | $Y_c$ | $(Y - Y_c)$ | $(Y - Y_c)^2$ |
|----------|------------|------------------|-------|-------------|----------------|
| A | 30 | 6 | 31.86 | −1.86 | 3.4596 |
| B | 49 | 9 | 47.28 | 1.72 | 2.9584 |
| C | 18 | 3 | 16.44 | 1.56 | 2.4336 |
| D | 42 | 8 | 42.14 | −0.14 | 0.0196 |
| E | 39 | 7 | 37.00 | 2.00 | 4.0000 |
| F | 25 | 5 | 26.72 | −1.72 | 2.9584 |
| G | 41 | 8 | 42.14 | −1.14 | 1.2996 |
| H | 52 | 10 | 52.42 | − .42 | .1764 |
|   | 296* |   | 296.00* | 0.0 | 17.3056 |

* Note that the sum of Y and $Y_c$ are equal. This must always be true if $\Sigma(Y - Y_c) = 0$.

for calculating the standard error of estimate for our Tackey Toy example is presented in Fig. 14-8. The standard error of estimate is calculated as follows:

$$Sy \cdot x = \sqrt{\frac{\Sigma(Y - Y_c)^2}{n - 2}}$$

$$= \sqrt{\frac{17.3056}{6}}$$

$$= \sqrt{2.88}$$

$$= 1.69 \text{ dozens of units of output (The value of } Sy \cdot x \text{ will always be expressed in the units of the } Y \text{ variable.)}$$

Although it is helpful to use formula (14-4) to explain the nature of the standard error of estimate, a much easier formula to apply is:

$$Sy \cdot x = \sqrt{\frac{\Sigma(y^2) - (b)(\Sigma xy)}{n - 2}} \qquad (14\text{-}5)$$

As you will notice, all the values needed for this formula are available from Fig. 14-5, which was used to prepare the regression equation. Using the values from Fig. 14-5:[8]

$$Sy \cdot x = \sqrt{\frac{968 - (5.14)(185)}{8 - 2}}$$

$$= \sqrt{\frac{17.1}{6}}$$

$$= \sqrt{2.85}$$

$$= 1.69 \text{ dozens of units of output}$$

---

[8] The slight difference between the values obtained using formulas (14-4) and (14-5) is due to rounding. For more lengthy problems than our example, the values needed for computing the regression equation and $Sy \cdot x$ may be more easily obtained with the following formulas: $\Sigma(xy) = \Sigma(XY - \bar{X}\Sigma Y)$, $\Sigma(x^2) = \Sigma(X^2) - \bar{X}\Sigma X$, and $\Sigma(y^2) = \Sigma(Y^2) - \bar{Y}\Sigma Y$.

Just as there is a similarity between the computation of the standard deviation and the computation of the standard error of estimate, so too is there a similarity in their interpretation. We know that the standard deviation is a measure of spread or dispersion about the mean, and the standard error of estimate is a measure of scatter or dispersion about the regression line. And we have seen in many places (perhaps in too many places!) that in a normal distribution (1) the middle 68.3 percent of the distribution values lie within a range of one standard deviation above and below the mean, (2) the middle 95.4 percent of the values lie within $\pm 2.00$ standard deviations, and (3) the middle 99.7 percent of the values lie within $\pm 3.00$ standard deviations. Therefore, *if* the observed $Y$ values are normally distributed about the regression line, approximately 68 percent of the points in the scatter diagram will fall within a range of one standard error of estimate above and below the regression line (see Fig. 14-9). This range of $\pm 1 Sy \cdot x$ is represented by the *nearest* dashed lines to either side of the regression line in Fig. 14-9; the range of $\pm 3 Sy \cdot x$ in Fig. 14-9 accounts for virtually all the points.

The normal curves in Fig. 14-9 have been added to help you better understand the relationship of the standard error of estimate to the regression line. If, in Fig. 14-9, a value of $X_1$ were plugged into the regression equation, an estimate of $Y_1$ would be obtained ($Y_1 = a + bX_1$). It is important to note, however, that *the value of $Y_1$ is a point estimate*, and considering the scatter about the regression line in Fig. 14-9, it is unlikely that this estimate ($Y_1$) will be exact. The dependability of this point estimate depends on the size of the standard error of estimate—as we have already seen, the smaller the standard error, the closer the point estimate is likely to be to the ultimate value of the dependent variable. With a knowledge of the standard error of estimate, however, we are not limited to the use of a point estimate. *Rather, an interval estimate may be computed, and a probability may be assigned to this interval estimate.* Thus, an interval estimate could be computed using this form:

$$Y_c \pm Z(Sy \cdot x)$$

Or, as shown in Fig. 14-9, the interval estimate with a predictive probability of 95.4 percent is $Y_1 \pm 2(Sy \cdot x)$.

Let's see now how we can apply the above concepts to our Tackey Toy example problem. When we last left Hiram Hess, he had obtained a *point estimate of 21.58* dozen units of output for an applicant who had an aptitude test score of 4. And we have now seen that the standard error of estimate for this problem is 1.69 dozen units of output. Assuming that the points in our example are normally distributed about the regression line,[9] we can prepare a 95 percent *prediction interval* for Hiram as follows:

$$
\begin{aligned}
\text{Interval} &= Y_c \pm Z(Sy \cdot x) \\
&= 21.58 \pm 1.96(1.69) \\
&= 21.58 \pm 3.31 \\
&= 18.27 \text{ to } 24.89
\end{aligned}
$$

---

[9] In a practical situation with a sample of the size we have in our example problem (only 8), it would be more appropriate to use a *t*-distribution value along with a correction factor to compute the prediction interval. For the purposes of this text, however, we need not go into these additional details.

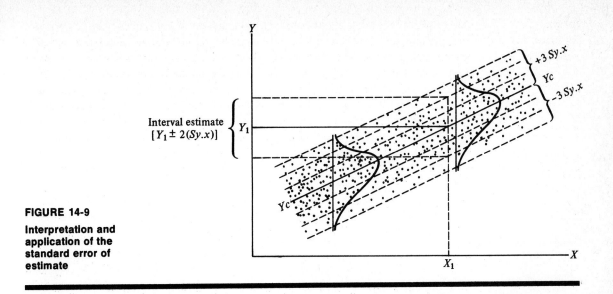

**FIGURE 14-9**

**Interpretation and application of the standard error of estimate**

Interval estimate $[Y_1 \pm 2(Sy.x)]$

This means that Hiram can predict that 95 percent of the time applicants with a score of 4 on the test will have an output between 18.27 and 24.89 dozen units. Since the supervisor is looking for someone who can produce 30 dozen units, this applicant is unlikely to be suited for that particular position.

**Self-testing Review 14-2**

Use the data in Self-testing Review 14-1 to answer the following questions:

**1** What is the value of $\Sigma(y^2)$?

**2** What is the standard error of estimate?

**3** Assuming that there is a normal distribution, what would be the 68 percent prediction interval for the estimate of $Y$ computed in problem **3** of Self-testing Review 14-1?

**CORRELATION ANALYSIS**

In addition to an equation that describes the relationship between the two variables, we need measures that will indicate the closeness of the association or correlation that exists between the variables. In this section we will briefly examine two of these correlation measures: the *coefficient of determination* and the *coefficient of correlation*.

**Coefficient of Determination**

Let us assure you at the outset that we do not plan to go into great detail in discussing the coefficient of determination (its symbol is $r^2$). However, before defining this measure it is desirable to consider the several terms and concepts illustrated in Fig. 14-10. If the mean of the $Y$ variable ($\bar{Y}$) alone had been used to estimate the dependent variable, we would, of course, expect to find quite a bit of possible deviation between our estimate and the value of $Y$. A single point ($Y$) in Fig. 14-10 has been used to indicate the considerable *total deviation* that exists in this case between $Y$ and $\bar{Y}$. But when the regression line is used as the basis for estimating the dependent variable, we can expect to have a closer estimate of $Y$. As indicated in Fig. 14-10, our regression line is indeed closer to most of the points in the scatter

CHAPTER 14/FORECASTING TOOLS: SIMPLE LINEAR REGRESSION AND CORRELATION

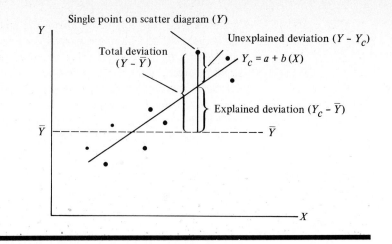

**FIGURE 14-10**

Conceptual representation of total, explained, and unexplained deviations

diagram. Thus, in the case of the single point $(Y)$ in Fig. 14-10, the regression line explains or accounts for part of the deviation between $Y$ and $\bar{Y}$—i.e., the explained deviation is $Y_c - \bar{Y}$. Unfortunately, the regression line does not account for all the deviation, since the distance between $Y$ and $Y_c$ is still unexplained.

In other words, we have the following situation in Fig. 14-10:

Total deviation = explained deviation + unexplained deviation
$$Y - \bar{Y} \quad = \quad (Y_c - \bar{Y}) \quad + \quad (Y - Y_c)$$

And if we were to consider the total variation in an entire scatter diagram, we would find the following situation:

Total variation = explained variation + unexplained variation
$$\Sigma(Y - \bar{Y})^2 \quad = \quad \Sigma(Y_c - \bar{Y})^2 \quad + \quad \Sigma(Y - Y_c)^2$$

Now that we have dazzled you with all these thoughts, we can define $r^2$. The *coefficient of determination is a measure of the portion of the total variance in the Y variable that is explained or accounted for by the introduction of the X variable* (and thus the regression line). That is:

$$r^2 = \frac{\text{explained variation}}{\text{total variation}}$$

where the total variation of $\Sigma(Y - \bar{Y})^2$ (we have already seen in Fig. 14-5) is $\Sigma(y^2)$, and where the explained variation of $\Sigma(Y_c - \bar{Y})^2$ is $b(\Sigma xy)$—a fact that could be demonstrated with numerous mathematical manipulations, but we will spare you that aggravation.

Thus,

$$r^2 = \frac{b(\Sigma xy)}{\Sigma(y^2)} \tag{14-6}$$

To compute the coefficient of determination for our Tackey Toy problem, we need only refer to the regression equation and Fig. 14-5 to get the necessary data:

$$r^2 = \frac{b(\Sigma xy)}{\Sigma(y^2)} = \frac{(5.14)(185)}{968} = \frac{950.9}{968} = .982$$

What does the $r^2$ value of .982 mean? Congratulations on yet another incisive question. It means that 98.2 percent of the variation in the $Y$ variable is explained or accounted for by variation in the $X$ variable. Or, in our example, we can conclude that 98.2 percent of the variation in output is explained by variation in test score results. Since it is obvious that the value of $r^2$ cannot exceed 1.00—after all, you can't explain more than 100 percent of the variation in $Y$!—the value of .982 is quite high. Such a high value is, of course, desirable for forecasting purposes because the higher the value of $r^2$, the smaller the value of $Sy \cdot x$. (Can you figure out why this is true?)

## Coefficient of Correlation

The coefficient of correlation ($r$) is simply the square root of the coefficient of determination ($r^2$). Thus, for our Tackey Toy problem, the coefficient of correlation is .991:

$$r = \sqrt{.982} = .991$$

The coefficient of correlation is not as useful as the coefficient of determination, since it is an abstract decimal and is not subject to precise interpretation. (As the square root of a percentage, it cannot itself be interpreted in percentage terms.) About all that $r$ does is to provide a scale against which the closeness of the relationship between $X$ and $Y$ can be measured. In other words, the value of $r$ will be on a scale between zero and $\pm 1.00$. When $r$ is zero, there is *no* correlation, and when $r = +1.00$ *or* $-1.00$, there is *perfect* correlation. (The algebraic sign of $r$ will always be the same as that of $b$ in the regression equation.) Thus, the closer $r$ is to its limit of $\pm 1.00$, the better the correlation, and the smaller the value of $r$, the poorer the relationship between the variables.

## A Graphical Summary

Let us now graphically summarize, by means of scatter diagrams, some of the relationships discussed in this chapter:

1   In Fig. 14-11a, we have an example of *perfect positive correlation,* with all points in the diagram falling on the regression line. Therefore, $r = +100$, $r^2 = 1.00$, and $Sy \cdot x = 0$ because there is an absence of spread or scatter about the line.

2   In Fig. 14-11b, we have an example of *perfect negative correlation.* The values of various measures are as indicated.

3   In Fig. 14-11c and d, there is positive correlation, but the values of $r$ and $r^2$ are less in d than in c. Assuming the same scales for $X$ and $Y$ and the same regression line, the value of $Sy \cdot x$ would be greater in d than in c.

4   In Fig. 14-11e, there is no correlation. The regression line would simply be a horizontal line drawn at $\bar{Y}$ with no slope. Both $r$ and $r^2$ would be zero, and $Sy \cdot x$ would equal the standard deviation of the $Y$ variable (this is the upper limit for $Sy \cdot x$).

## Self-testing Review 14-3

Using the answers obtained in Self-testing Reviews 14-1 and 14-2, perform the following:

1   Compute the explained variation.

**341**

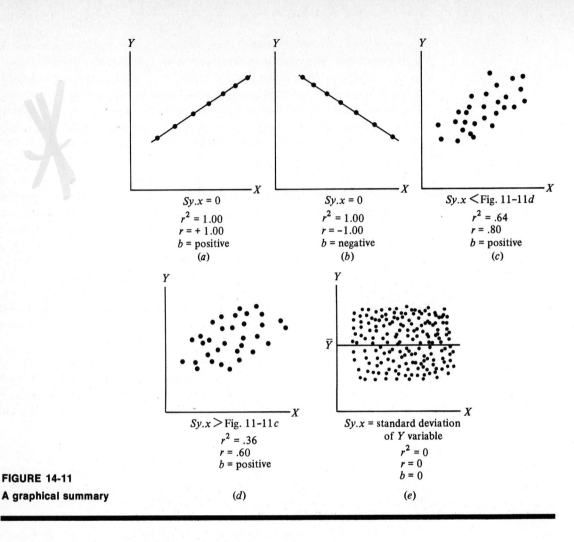

$Sy.x = 0$
$r^2 = 1.00$
$r = +1.00$
$b$ = positive
(a)

$Sy.x = 0$
$r^2 = 1.00$
$r = -1.00$
$b$ = negative
(b)

$Sy.x <$ Fig. 11–11$d$
$r^2 = .64$
$r = .80$
$b$ = positive
(c)

$Sy.x >$ Fig. 11–11$c$
$r^2 = .36$
$r = .60$
$b$ = positive
(d)

$Sy.x$ = standard deviation
of $Y$ variable
$r^2 = 0$
$r = 0$
$b = 0$
(e)

**FIGURE 14-11**

**A graphical summary**

**2** Compute the coefficient of determination.

**3** Interpret the coefficient of determination.

**4** Compute the coefficient of correlation.

**COMMON ERRORS AND LIMITATIONS**

As is true with all statistical methods, regression and correlation analysis is subject to misuses and misinterpretations that can be quite serious. *Some of the more common mistakes are briefly summarized below.*

**1** *Correlation analysis is sometimes used to prove the existence of a cause-and-effect relationship.* The coefficient of determination tells us nothing about the type of relationship between the two variables. Rather, it indicates the proportion of variation that is explained *if* there is a causal relationship.

**2** *The coefficient of correlation is sometimes interpreted as a percentage.* This can be a very serious mistake. For example, if a coefficient of correlation of .7

is interpreted as meaning that 70 percent of the variation in $Y$ is explained, this is significantly above the 49 percent that actually is explained.

3   *The coefficient of determination is also subject to misinterpretation.* It is sometimes interpreted as the percentage of the variation in the dependent variable *caused* by the independent variable. This is simply nonsense. It should always be remembered that it is the variation in the dependent variable that is being explained or accounted for (but not necessarily caused) by the $X$ variable.

4   *When making estimates from the regression equation, it is incorrect to make estimates beyond the range of the original observations.* There is no way of knowing what the nature of the regression equation would be if we encountered values of the dependent variable larger or smaller than those we have observed. For example, in the Tackey Toy illustration used in this chapter it would be ridiculous to assume that an employee's output would increase indefinitely as his or her test results increased. There would obviously be some upper limit to aptitude test scores, and regardless of aptitude, the speed of productive equipment and physical endurance would also set a limit to productive output. Since the largest value of $X$ we observed was 10, we could place no reliance on estimates we might make of the output of employees who might have test scores of 11 or 12.

5   *When time-series data are correlated, the coefficient of determination cannot be interpreted as precisely as we have indicated above.* This is because the pairs of observations are not independent, a fact which tends to inflate the value of $r^2$. There are methods for dealing with this problem, but they are beyond the scope of this book.

## SUMMARY

Regression and correlation analysis is used extensively in business, economics, psychology, and science. It has been used, for example, to forecast sales, to predict the academic success of students, and to estimate the quantity of a product that can be sold at a particular price.

In regression analysis, an estimating or regression equation is developed to describe the pattern or functional nature of the relationship that exists between variables. In addition, regression analysis includes the computation and application of a standard error of estimate that will indicate the possible error involved in using the estimating equation as a basis for forecasting.

In correlation analysis, we are concerned with preparing measures (the coefficient of determination and the coefficient of correlation) that will indicate the closeness of the association or correlation that exists between the variables.

The first step in the study of regression and correlation is probably to determine if a logical relationship exists between the variables to support further analysis. Scatter diagrams may then be used (1) to help show if there is a useful relationship between the variables and (2) to help determine the type of equation (linear or curvilinear) to use to describe the relationship. In this chapter we have illustrated the techniques used to compute a simple linear regression equation, a standard error of estimate, and the measures of $r^2$ and $r$. In applying these measures, however, you should remember their limitations and avoid the common errors of interpretation described in the preceding section.

1  Relationship
2  Simple linear regression and correlation
3  Multiple regression and correlation
4  Regression equation
5  Estimating equation
6  Dependent variable
7  Independent variable
8  Causal relationships
9  Spurious correlations
10  Scatter diagram
11  Regression analysis
12  Positive correlation
13  Negative correlation
14  Curvilinear correlation
15  Method of least squares
16  $Y_c = a + bX$
17  Standard error of estimate
18  $Sy \cdot x = \sqrt{\dfrac{\Sigma(Y - Y_c)^2}{n - 2}}$
19  Point estimate
20  Prediction interval
21  Coefficient of determination
22  $r^2 = \dfrac{\text{explained variation}}{\text{total variation}}$
23  $r^2 = \dfrac{b(\Sigma xy)}{\Sigma(y^2)}$
24  Coefficient of correlation
25  Perfect correlation

**Problems**

1  A United States military organization wanted to know if there was a relationship between the number of recruiting offices located in particular cities and the total number of persons enlisting in those cities. Total enlistment data were obtained for 1 year from 10 cities and are given below:

| City | Total enlistment in each city for 1 year | Number of recruiting offices in each city |
|------|------|------|
| Austin | 20 | 1 |
| Pittsburgh | 40 | 2 |
| Chicago | 60 | 4 |
| Los Angeles | 60 | 3 |
| Denver | 80 | 5 |
| Atlanta | 100 | 4 |
| Cleveland | 80 | 5 |
| Louisville | 50 | 2 |
| New Orleans | 110 | 5 |
| Kansas City | 30 | 1 |

a  Construct a scatter diagram.
b  Compute a least-squares regression equation, and fit it to the scatter diagram.
c  Compute the standard error of estimate.
d  Develop 95.4 percent prediction belts about the regression line in the scatter diagram.
e  Compute the coefficients of determination and correlation, and explain their meaning.
f  Estimate with a 95.4 probability level the number of enlistments the organization could expect in a city with three recruiting offices.

**g** What other factors might influence enlistment other than the number of recruiting offices located in a particular city?

**2** Gus Gas operates several Slochoke Service Stations on major traffic arteries in Aledo, Texas. Gus wants to know if there is any relationship between the total gallons of gasoline pumped at each station last year and the average number of cars that traveled each day on the streets where his service stations are located. Gus obtained the following data from the Texas Highway Department and Slochoke Company records:

| Service station location | Total gallons of gasoline pumped (thousands) | Average daily traffic count at location (hundreds) |
|---|---|---|
| E. Regular Street | 100 | 3 |
| S. Lead Street | 112 | 4 |
| N. Main Street | 150 | 5 |
| Highway 606 | 210 | 7 |
| Baker Boulevard | 60 | 2 |
| E. High Street | 85 | 3 |
| Country Club Road | 77 | 2 |

**a** Construct a scatter diagram.
**b** Compute a least-squares regression equation, and fit it to the scatter diagram.
**c** Compute the standard error of estimate.
**d** Develop 95.4 percent prediction belts about the regression line fitted to the scatter diagram.
**e** Compute the coefficient of determination, and explain its meaning.
**f** Estimate a 68 percent prediction interval for the total gallons that would be expected if a new station were located on Smith Road. The average daily traffic count on Smith Road is 275 vehicles.

**3** The head nurse of Mortality Memorial Hospital is interested in determining whether a relationship exists between the number of days a patient has been in the hospital and the number of times a patient requests a nurse during a 24-hour period. The head nurse recorded the following data:

| Patient | Number of days in Mortality Memorial | Total number of requests for nurses |
|---|---|---|
| Mr. Chill Blains | 2 | 2 |
| Mr. Leow B. Count | 4 | 3 |
| Ms. Hedda (Dizzy) Spell | 5 | 3 |
| Ms. Cara Reck | 6 | 4 |
| Mr. A. Penn Dix | 3 | 2 |
| Ms. Anne S. Teesha | 8 | 6 |
| Mr. I. M. Hurting | 15 | 10 |
| Mr. Hy P. Dermik | 7 | 5 |
| Ms. Ivy Pellagra | 15 | 11 |
| Ms. Hada Bortion | 2 | 1 |

**a** Construct a scatter diagram.
**b** Compute a least-squares regression equation, and fit it to the scatter diagram.
**c** Compute the standard error of estimate.
**d** Develop 95.4 percent prediction belts about the regression line fitted to the scatter diagram.
**e** Compute the coefficient of determination, and explain its meaning.

**4** The Rip-Off Vending Machine Company operates coffee vending machines in several office buildings. The company wants to study the relationship, if any, which exists between the number of cups of coffee sold per day and the number of persons working in each office. Data for this study were collected by the company and are presented below:

| Location of coffee machine | Number of cups of coffee sold at this location | Number of persons working at this location |
|---|---|---|
| 1 | 10 | 5 |
| 2 | 20 | 6 |
| 3 | 30 | 14 |
| 4 | 40 | 19 |
| 5 | 30 | 15 |
| 6 | 20 | 11 |
| 7 | 40 | 18 |
| 8 | 40 | 22 |
| 9 | 50 | 26 |
| 10 | 10 | 4 |

**a** Construct a scatter diagram.
**b** Compute a least-squares regression equation, and fit it to the scatter diagram.
**c** Compute the standard error of estimate.
**d** Develop 95.4 percent prediction belts about the regression line fitted to the scatter diagram.
**e** Compute the coefficients of determination and correlation, and explain their meaning.

**Topics for Discussion**

**1** What is the purpose of regression and correlation analysis?
**2** What type of relationship exists between the following pairs of variables?
   **a** Gasoline sales and the number of automobiles.
   **b** Gasoline sales and motor oil sales.
   **c** Gasoline sales and the number of books in the New York Public Library.
**3** What is the purpose of the scatter diagram?
**4** Discuss the use of the regression equation in forecasting.
**5** How is the standard error of estimate used when making predictions from the regression equation?
**6** What is meant by "simple linear regression analysis"?

**7** What is the interpretation of the coefficient of determination?

**8** Discuss the common errors made in interpreting the results of correlation analyses.

**9** What is the relationship between the values of $r^2$ and $Sy \cdot x$?

**10** "The standard error of estimate is similar in many respects to the standard deviation." Discuss this statement.

**Answers to Self-testing Review Questions**

**14-1**

| $Y$ | $X$ | $y$ <br> $(Y - \bar{Y})$ | $x$ <br> $(X - \bar{X})$ | $xy$ | $x^2$ | $y^2$ |
|---|---|---|---|---|---|---|
| 255 | 5 | 52 | 1 | 52 | 1 | 2,704 |
| 100 | 2 | −103 | −2 | 206 | 4 | 10,609 |
| 307 | 6 | 104 | 2 | 208 | 4 | 10,816 |
| 150 | 3 | − 53 | −1 | 53 | 1 | 2,809 |
| 812 | 16 | | | 519 | 10 | 26,938 |

$$\bar{Y} = 203 \qquad \bar{X} = 4$$

**1 a** $\Sigma(xy) = 519$
 **b** $\Sigma(x^2) = 10$

**2** $Y_c = a + bX$

$$b = \frac{\Sigma(xy)}{\Sigma(x^2)} = \frac{519}{10} = 51.9$$

$$a = \bar{Y} - b\bar{X} = 203 - 51.9(4) = -4.6$$

**3** $Y_c = -4.6 + 51.9(4)$
$$= -4.6 + 207.6$$
$$= 203$$

**14-2**

**1** 26,938

**2** $Sy \cdot x = \sqrt{\dfrac{\Sigma(y^2) - (b)(\Sigma xy)}{n - 2}} = \sqrt{\dfrac{26,938 - (51.9)(519)}{2}} = \sqrt{\dfrac{1.9}{2}} = .97$

**3** $203 \pm 1.00(.97)$
202.03 to 203.97

**14-3**

**1** Explained variation $= b(\Sigma xy) = (51.9)(519) = 26,936.1$

**2** $r^2 = \dfrac{26,936.1}{26,938.0} = .999$

**3** We can say that 99.9 percent of the variation in $Y$ is explained or accounted for by the variation in $X$.

**4** $r = \sqrt{.999} = .999$

PART A

# CONCLUDING TOPICS

The following comments appear on a poster in the window of a curio shop on Bourbon Street in New Orleans:

> *The objective of all our dedicated employees should be to analyze all situations, to sort out important problems prior to their occurrence, and to have answers to all these problems when they are called upon. . . . However, when you're up to your [anatomical reference deleted] in alligators it's difficult to remember that your initial objective was to drain the swamp.*

Perhaps as you struggled to master the material found in the preceding chapters there were times when you felt bogged down by details; perhaps you wondered why you were expected to learn everything about statistics in a single course; and perhaps you felt in complete sympathy with the author of the Bourbon Street poster. But you will be chagrined (or relieved) to learn that in this book we have only outlined a few of the basic topics in the areas of statistics and quantitative methods.

Although we cannot cover in any great detail many of the important methods of quantitative analysis that have been ignored up to this point, we also cannot conclude this book without some discussion of the uses and limitations of several additional quantitative tools available to the statistician and decision maker.

In Chapters 7 through 10, inferences were made and decisions were reached through the use of estimation and hypothesis-testing techniques. But these earlier techniques are generally valid only if the assumptions that were made about the shapes of various sampling and/or population distributions are correct. However, there is an important group of statistical procedures which do not require that assumptions be made about the shape of a distribution and which are thus referred to as distribution-free procedures. These *nonparametric methods*, in other words, enable the decision maker to test hypotheses without the need to impose certain restrictive assumptions. Although it would certainly have been appropriate to include these nonparametric tests in the chapters of Part 2, we have chosen to place the material in this final part.

In addition, a few other advanced quantitative tools are briefly described in the final chapter. Thus, the chapters included in Part 4 are:

# CHAPTER 15

# NONPARAMETRIC STATISTICAL METHODS

**LEARNING OBJECTIVES**

After studying this chapter, working the problems, and answering the discussion questions, you should be able to:

☞ Identify situations that call for the use of particular nonparametric methods.

☞ Apply sign test procedures for both small and large samples.

☞ Apply the Wilcoxon signed rank test in small-sample situations.

☞ Use the Mann-Whitney test to determine if two small independent random samples are taken from identical populations.

☞ Determine if randomness exists (or if there is an underlying pattern) in a small sequence of sample data through the use of the runs test for randomness.

☞ Compute the Spearman rank correlation coefficient and then test this measure for significance.

**CHAPTER OUTLINE**

GENERAL CONCEPTS
  Use of Nonparametric Techniques
  Chapter Limitations

THE SIGN TEST
  Sign Test Procedure with Small
    Samples
  Sign Test Procedure with Large
    Samples

Self-testing Review 15-1

THE WILCOXON SIGNED RANK TEST
  Wilcoxon Signed Rank Test Procedure
  Self-testing Review 15-2

THE MANN-WHITNEY TEST
  The Mann-Whitney Test Procedure
  Self-testing Review 15-3

**GENERAL CONCEPTS**

A short time ago, some of your skeptical classmates (maybe even you) might have remarked that much of the material covered in Chapters 7, 8, 9, and 10 may not always be relevant because a normal probability distribution cannot always be assumed in a real-world situation. This skepticism, of course, may be justified in certain situations! The validity of the inferences made in those earlier chapters depended upon the accuracy of the assumptions[1] made about such things as (1) the shape of the various sampling distributions of sample statistics and/or the shape of the population distribution and (2) the relationship of these probability distributions to the underlying population parameters. (For those of you who are getting an anxiety attack, a quick review of the Central Limit Theorem in Chapter 6 may be in order.)

In this chapter, however, we will be concerned with *nonparametric statistics*— i.e., with *statistics which do not require that assumptions be made about the shape of a distribution and which are thus distribution-free statistics*. With nonparametric statistics, inferences may be made *regardless* of the shape of the population distribution; with the parametric statistics of the earlier chapters, inferences were valid only if certain restrictive assumptions were true.

It may surprise you to learn that you have already worked with nonparametric techniques. The chi-square methods in Chapter 11 were nonparametric in nature. You may recall that the chi-square procedure was used to compare *observed* (sample) frequencies with *expected* (hypothesized) frequencies and that the expected frequencies were not necessarily restricted to any particular type of distribution.

**Use of Nonparametric Techniques**

When should nonparametric methods be used? *They should be employed in any of the following situations:*

1  When the sample size is so small that the sampling distributions of the statistics do not approximate the normal distribution, and when no assumption can be made about the shape of the population distribution from which the sample is drawn.

---

[1] A statistician and the statistician's spouse found themselves marooned on a remote island. When the spouse asked how they would escape from the island and get to civilization, the statistician replied, "Assuming we have a boat. . . ."

**2** When *rank* or *ordinal* data are used. (Ordinal data only tell us whether one item is higher than, lower than, or equal to another item; they do not tell us the size of the difference.)

**3** When *nominal* data are used. (Nominal data are simply data where names such as "male" or "female" are assigned to items and there is no implication in the names that one item is higher or lower than another.)

## Chapter Limitations

Unfortunately, the subject of nonparametric statistics cannot be covered in its entirety here because of the page constraints of this book (and perhaps because of your dwindling patience as the end of the term approaches). However, this chapter will deal with a few of the more widely used nonparametric techniques.[2] They are the (1) *sign test*, (2) *Wilcoxon signed rank test*, (3) *Mann-Whitney test*, (4) *runs test for randomness*, and (5) *Spearman rank correlation coefficient*. Although most of these techniques may be discussed in the context of both large- and small-sample situations, we will limit our attention primarily to the small-sample cases.

## THE SIGN TEST

When you have paired ordinal measurements taken from the same subjects or matched subjects, and when you are simply interested in whether or not there are real differences regardless of the size of the differences, the sign test should be employed. *The sign test is based on the signs, negative or positive, of the differences between pairs of ordinal data.* Essentially, this test considers only the direction of the differences and not the magnitude of the differences.

## Sign Test Procedure with Small Samples

Let's go through an example. Chicken Out, a national fast-food chain, has developed a new formula for the batter used in coating its chicken, and the marketing department simply wishes to find out whether the new batter is tastier than the original batter. At the present stage of product development, the department is not interested in the degree of taste improvement.

Ten consumers are randomly selected for a taste test. Each consumer first tastes a piece of chicken with the original batter and rates the taste on a scale of 1 to 10, where 1 is very poor and 10 is very good. Then, the same consumer munches a piece of chicken fried with the new batter and rates the taste on a scale of 1 to 10. The data thus collected are presented in Fig. 15-1.

What should the market research data tell us? If there is truly no difference in flavor, we would expect, in a large survey, the number of consumers who rate the new-batter taste as better than the original-batter taste to be equal to the number of consumers who think the new taste is worse than the old taste. In other words, if there is truly no difference between the taste of the original batter and the taste

---

[2] A number of sources are available if you should want to dig deeper into nonparametric methods. See, for example, W. J. Conover, *Practical Nonparametric Statistics,* John Wiley & Sons, Inc., New York, 1971.

## FIGURE 15-1 Data for the sign test

Taste ratings by 10 consumers of chicken coated with original batter and chicken coated with new batter (10 indicates a "very good taste," and 1 indicates a "very poor taste")

| Consumer | Taste ratings | | Sign of difference between new coating and original coating ($y - x$) |
| | Original batter ($x$) | New batter ($y$) | |
| --- | --- | --- | --- |
| R. MacDonald | 3 | 9 | + |
| G. Price | 5 | 5 | 0 |
| B. King | 3 | 6 | + |
| L. J. Silver | 1 | 3 | + |
| P. P. Gino | 5 | 10 | + |
| E. J. McGee | 8 | 4 | − |
| S. White | 2 | 2 | 0 |
| E. Fudd | 8 | 5 | − |
| Y. Sam | 4 | 6 | + |
| M. Muffett | 6 | 7 | + |

$n$ = number of relevant observations
  = number of plus signs + number of minus signs
  = 6 + 2
  = 8

$r$ = the number of fewer signs
  = 2

of the new batter, we would expect the median difference between the two taste ratings to be zero. That is, the probability of getting consumers who would report better taste would be equal to the probability of selecting consumers who would report a worse taste.

**Stating the null and alternative hypotheses** As in the case of any hypothesis test, the first step in a sign test is to state the null and alternative hypotheses. Two-tailed or one-tailed sign tests may be conducted, and this fact, of course, determines the form of the alternative hypothesis.

The *null hypothesis* to be tested in our example is that the new ingredients have no effect on the taste of the chicken: a positive sign indicating a taste improvement is just as likely as a negative sign indicating a loss of flavor when the difference between the two taste ratings for each subject is determined. The *alternative hypothesis* in our example is that the new batter improves taste. We thus have a right-tailed test, and the alternative hypothesis is to the effect that there is more than a 50 percent chance that a person will report that the new batter has a better flavor than the original batter. Thus, the statistical hypotheses are:

$$H_0: p = 0.5$$
$$H_1: p > 0.5$$

where $p$ is the probability of getting a taste improvement

**Selecting the level of significance** After stating the null and alternative hypotheses, the second step is to establish a criterion for rejecting or accepting the null hypothesis. Let's assume that for our example, the risk of erroneously rejecting the null hypothesis when it is actually true is to be limited to no more than 5 percent. Thus, the level of significance is $\alpha = .05$.

**Determining the sign of differences between paired observations** After the null and alternative hypotheses have been determined, and after the level of significance has been selected, the next step is to systematically subtract one observation from the other observation and then record whether the difference is positive (an improvement in taste) or negative (a loss in flavor). The last column of Fig. 15-1 shows the sign of the difference for each subject when the taste rating for the original batter is subtracted *from* the taste rating for the new batter. In the case of the first subject, R. MacDonald, the taste rating for the new batter is greater or *better* than the taste rating for the original batter; thus, there is a *positive* sign. In situations where there is no change in taste ratings, a zero representing a tie is recorded.

**Counting the frequency of signs** The next step is to tally up the pluses, the minuses, and the zeros. Figure 15-1 shows 6 pluses, 2 minuses, and 2 zeros—which means that 6 consumers thought there was a taste improvement, 2 thought flavor had declined, and 2 perceived no change. After the tally, *we designate the lesser sum of the two signs as r*. In the case of Fig. 15-1, we have $r = 2$ because there are only 2 negatives compared with the 6 positive signs.

**Determining the likelihood of observed sample results** The only subjects or paired observations relevant for analysis are those where taste differences (positive or negative) have been recorded. In our case, only 8 of the 10 pairs of data are relevant for analysis, and thus we have $n = 8$. (The responses of Price and White are not included in the analysis because they provide no indication of a difference one way or another.) From the 8 relevant subjects or paired observations, one would expect four of the differences to be positive and four of the differences to be negative *if* the null hypothesis is true. On the basis of the two negative responses in Fig. 15-1 and the nature of a right-tailed test, we must ask ourselves the following question: What is the chance of having at most only 2 out of 8 subjects perceiving a negative taste change when in fact the null hypothesis is true (where 50 percent of the sample should record a positive change and where 50 percent should record a negative change)? Formulation of the answer begins by referring to the Binomial Probability Distribution (if $n$ is small) in Appendix 1.[3] Since we have 8 relevant subjects, we look for that section of the table where $n = 8$ and $r = 2$. After locating that section, look in the column where $p = .50$—a value that stems from the null hypothesis. We see that the probability of getting *at most* only 2 out

---

[3] When the sample size is relatively large—i.e., $\geq 30$—the normal approximation to the binomial distribution may be used. Actually, there is little difference in results in using the normal approximation to the binomial in sign tests when the sample size is greater than 20.

of 8 subjects reporting a negative change is .1445, which is a summation of the probabilities of getting 0 out of 8 (.0039), 1 out of 8 (.0312), and 2 out of 8 (.1094). In other words, if there were truly no difference in taste between the original and new batters, the chances of getting at most only 2 out of 8 subjects reporting a loss of taste would be only 14.5 percent.

**Drawing a statistical conclusion concerning the null hypothesis** The question now is whether or not the sample probability result of .1445 is sufficient for us to accept the null hypothesis that there is no significant difference in consumer taste ratings. Although the probability of getting at most only 2 out of 8 consumers to report negatively on the new batter mix is not particularly high at .1445, it is higher than the .05 level of significance specified earlier in the problem. That is, the sample probability result would have had to be less than .05 for us to have rejected the null hypothesis.

In summary, then, the *decision rule* to follow in small-sample sign test situations in making a statistical decision is:

Accept the $H_0$ if $\alpha \lessgtr$ the probability of the sample results.

*or*

Reject $H_0$ and accept $H_1$ if $\alpha >$ the probability of the sample results.

Since, in our example, .05 < .1445, we would accept the null hypothesis; the new batter mix cannot be said to be a significant improvement over the original recipe.[4]

**Sign Test Procedure with Large Samples**

If the sample size is reasonably large, and if the normal approximation to the binomial distribution may be used, the same decision rules discussed in Chapter 8 apply, and the critical ratio ($Z$ value) is computed as follows:

$$CR = \frac{2R - n}{\sqrt{n}} \tag{15-1}$$

where $R$ = number of positive signs

$n$ = number of relevant paired observations

Suppose, for example, that in our Chicken Out problem there had been 33 consumers in the sample. Assume also that the following results had been obtained:

$$
\left.
\begin{aligned}
+ \text{ differences} &= 18 \\
- \text{ differences} &= 12
\end{aligned}
\right\} n = 30
$$
$$
\underline{\phantom{+} 0 \text{ differences} = \phantom{0}3}
$$
$$
\text{total} = 33
$$

---

[4] If we had been conducting a two-tailed test, we would have *doubled* the probabilities obtained from the binomial table prior to making the statistical decision. For example, if we were conducting a two-tailed test on the Chicken Out data, the sample results would be two times .1445, or .2890.

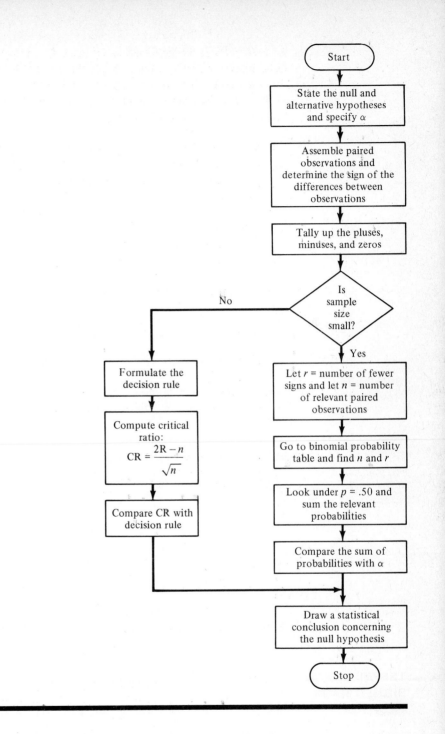

**FIGURE 15-2**

**Procedures for conducting sign tests**

The flowchart contains the following elements:

Start

State the null and alternative hypotheses and specify $\alpha$

Assemble paired observations and determine the sign of the differences between observations

Tally up the pluses, minuses, and zeros

Is sample size small?

No →

Formulate the decision rule

Compute critical ratio:
$$CR = \frac{2R - n}{\sqrt{n}}$$

Compare CR with decision rule

Yes →

Let $r$ = number of fewer signs and let $n$ = number of relevant paired observations

Go to binomial probability table and find $n$ and $r$

Look under $p = .50$ and sum the relevant probabilities

Compare the sum of probabilities with $\alpha$

Draw a statistical conclusion concerning the null hypothesis

Stop

If a right-tailed test is to be made, the hypotheses would remain unchanged. And if the .05 level of significance is used, the decision rule could be stated in the following familiar format:

Accept $H_0$ if CR $\leqslant 1.64$.

*or*

Reject $H_0$ and accept $H_1$ if CR $> 1.64$.

The critical ratio is computed as follows:

$$CR = \frac{2R - n}{\sqrt{n}}$$

$$= \frac{2(18) - 30}{\sqrt{30}}$$

$$= \frac{36 - 30}{5.477}$$

$$= 1.095$$

Since $1.095 < 1.64$, the null hypothesis would be accepted. In this case, the conclusion would be that there is no significant difference between the taste ratings of the two batters.

Figure 15-2 summarizes the sign test procedures outlined in this section.

<table>
<tr><td>Self-testing<br>Review 15-1</td><td>1   What is a sign test?</td></tr>
</table>

**Self-testing Review 15-1**

**1** What is a sign test?

**2** "Only a one-tailed hypothesis test may be conducted in a sign test." Comment on this statement.

**3** How many observations for each subject are required for a sign test?

**4** What is the null hypothesis in a sign test?

**5 a** What probability distribution is used in testing the hypotheses of sign tests when the sample size is small?
   **b** When the sample size exceeds 30?

**6 a** If the differences between paired data used in a sign test are 5 positives, 7 negatives, and 6 ties or zeros, we have $n = 18$ and $r = 7$. True or false?
   **b** In a right-tailed test at the .10 level, should the null hypothesis be accepted using the data given in **a**?

**7** If the differences between paired data used in a sign test are 16 positives, 26 negatives, and 4 zeros, what would be the statistical decision for a two-tailed test at the .05 level of significance?

**THE WILCOXON SIGNED RANK TEST**

While the sign test focuses solely upon the *direction* of the differences within pairs, the Wilcoxon signed rank test (named after Frank Wilcoxon, the statistician who first proposed it in the 1940s) is employed when the *magnitude* as well as the direction of the differences is relevant to determine whether there are true differences between pairs of data drawn from one sample or two related samples. When we wish to incorporate the *size* of differences in addition to the direction of the differences into our decision making, the Wilcoxon signed rank test should be used.

## Wilcoxon Signed Rank Test Procedure

Let's use the Chicken Out example again. Suppose that the management of the firm wants to make a decision concerning the new batter mix based not just on how many persons thought the new batter improved taste but also on the *amount* of taste improvement from the new batter. The Wilcoxon signed rank test is appropriate, and the data for analysis are drawn from Fig. 15-1 and are reproduced in Fig. 15-3.

**Stating the hypotheses and $\alpha$** As you may have anticipated, we must state the hypotheses and the desired level of significance. The null hypothesis in this case is that there is no difference between the tastes of the original and new batters. Therefore, in a large sample, the number of positive signs should equal the number of negative signs. Since this is a right-tailed test, the alternative hypothesis states that the taste of the new batter is better than the taste of the original mix. Thus, the hypotheses might be written as follows:

$H_0$: The two batters are equally tasty (or tasteless?).

$H_1$: The new batter mix is better tasting.

In addition, for this example, we will reject the null hypothesis at the .01 level of significance.

**Determining the size and sign of differences between paired data**
After stating the hypotheses and determining the significance level, the next step is to prepare the raw data for testing. The *size* and *sign* of the differences between the paired data are computed, and these are shown in the third column of Fig. 15-3. For

### FIGURE 15-3 Computations for Wilcoxon signed rank tests

| Consumers | (1) Original batter taste score | (2) New batter taste score | (3) Difference: new batter rating less original batter rating | (4) Rank irrespective of sign | (5) Signed rank Positive | (6) Signed rank Negative |
|---|---|---|---|---|---|---|
| R. MacDonald | 3 | 9 | +6 | 8 | +8 | |
| G. Price | 5 | 5 | 0 | (ignore) | | |
| B. King | 3 | 6 | +3 | 4.5 | +4.5 | |
| L. J. Silver | 1 | 3 | +2 | 2.5 | +2.5 | |
| P. P. Gino | 5 | 10 | +5 | 7 | +7 | |
| E. J. McGee | 8 | 4 | −4 | 6 | | −6 |
| S. White | 2 | 2 | 0 | (ignore) | | |
| E. Fudd | 8 | 5 | −3 | 4.5 | | −4.5 |
| Y. Sam | 4 | 6 | +2 | 2.5 | +2.5 | |
| M. Muffett | 6 | 7 | +1 | 1 | +1 | |
| | | | | | +25.5 | −10.5 |

$n$ = number of relevant observations
  = number of plus signs + number of minus signs
  = 6 + 2
  = 8

$T$ = the smaller of the two rank sums
  = 10.5

example, McGee initially gave the taste of the original batter an 8 score but felt that the taste of the new batter merited only a 4. Thus, the recorded difference for McGee is −4. Differences for the other consumers are recorded in a similar fashion.

**Ranking the differences irrespective of the signs** In the next step, we temporarily *ignore* the plus and minus signs in column 3 and rank the *absolute* values of the differences. A rank of 1 is assigned to the *smallest* difference; a rank of 2 is given to the next smallest value; and so on. (Differences of zero are ignored.) Since the two taste scores for Muffett had the least differences, that *difference, irrespective of the direction,* was assigned the rank of 1. In the cases of Silver and Sam, who are tied for second and third ranks with differences of 2, we assign to each a rank of 2.5, which is the *average* of the ranks 2 and 3. This procedure is continued until all differences have been ranked.

**Affixing signs to the assigned ranks** The next step is to *affix the sign of each difference* (as shown in column 3, Fig. 15-3) *to its rank* (as shown in column 4). This step results in the figures in the last two columns of Fig. 15-3. For example, the size of the difference between the data for Gino was assigned a rank of 7, and since the difference was positive, a corresponding +7 is recorded. Signed ranks are similarly produced for the other consumers.

**Summing the ranks** The last step before hypothesis testing is to add up all the positive ranks and then add up all the negative ranks. The *smaller of the two sums is designated as the computed $T$ value*. Since the sum of the negative ranks is 10.5 and the sum of the positive ranks is 25.5, the sum of 10.5 is designated as the computed value of $T$. (As a check on your accuracy, the sum of the positive and negative ranks—i.e., 25.5 + 10.5—must, of course, equal the sum of the ranks in column 4 of Fig. 15-3.)

**Drawing a statistical conclusion concerning the null hypothesis** We can now proceed to test the null hypothesis by comparing the computed $T$ value with the appropriate table value of $T$ for a given level of significance. Under the condition that the null hypothesis is true, the $T$ table in Appendix 8 provides the values of $T$ at corresponding $\alpha$ levels of .01 and .05 for both one-tailed and two-tailed tests. Since we tallied eight ranks (ties do not count), we have $n = 8$. For a one-tailed test where $n = 8$ and $\alpha = .01$, the table value of $T$ is 1. *If the computed $T$ value is equal to or less than the table $T$ value, the null hypothesis should be rejected.* Since our computed $T$ value equals 10.5, and since this statistic is greater than the table $T$ value of 1, the null hypothesis cannot be rejected. It must be concluded that the new batter mix produces no significant taste improvement over the original batter.

For your convenience, the entire procedure for the Wilcoxon signed rank test is summarized in Fig. 15-4.

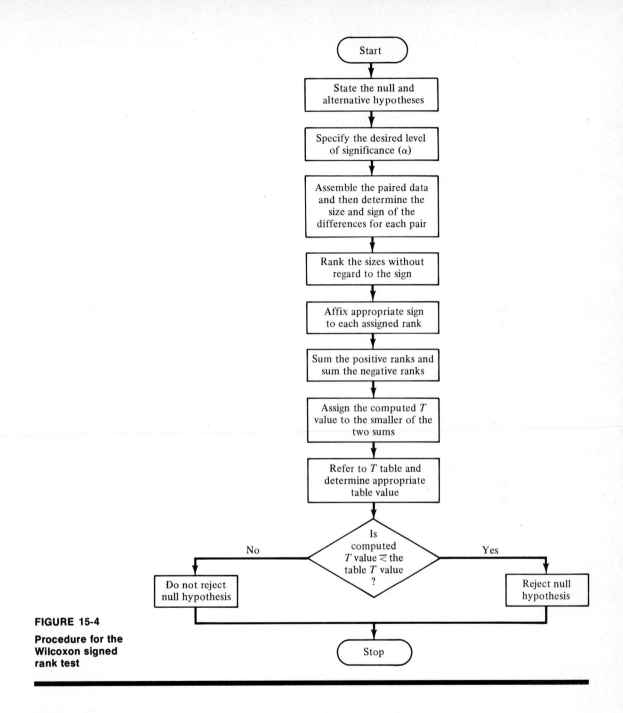

**FIGURE 15-4**

**Procedure for the Wilcoxon signed rank test**

The flowchart contains the following steps:

Start

→ State the null and alternative hypotheses

→ Specify the desired level of significance ($\alpha$)

→ Assemble the paired data and then determine the size and sign of the differences for each pair

→ Rank the sizes without regard to the sign

→ Affix appropriate sign to each assigned rank

→ Sum the positive ranks and sum the negative ranks

→ Assign the computed $T$ value to the smaller of the two sums

→ Refer to $T$ table and determine appropriate table value

→ Is computed $T$ value $\lessgtr$ the table $T$ value?

No → Do not reject null hypothesis

Yes → Reject null hypothesis

→ Stop

**Self-testing Review 15-2**

1  How does the Wilcoxon signed rank test differ from the sign test?

2  What is the null hypothesis in the Wilcoxon signed rank test?

3  In a Wilcoxon signed rank test, when you initially rank the differences between paired data, you ignore the sign of the differences. True or false?

**4** If the sum of the positive ranks and the sum of the negative ranks were 25 and 20 respectively, we would have a computed $T$ value of 25. True or false?

**5** What is the critical $T$-table value to use when $n = 32$, $\alpha = .05$, and a two-tailed test is being conducted?

**6** If the computed $T$ value is less than the table $T$ value, the null hypothesis is rejected. True or false? `

## THE MANN-WHITNEY TEST

With the sign test and the Wilcoxon signed rank test, paired data drawn from one sample or two closely related samples can be analyzed for significant differences. In the situation where one wishes to test the null hypothesis that there is no true difference between two sets of data and where the data are drawn from two *independent* samples, one may conduct the Mann-Whitney test. This test is often referred to as the $U$ test, since a statistic called $U$ is computed for testing the null hypothesis.

## The Mann-Whitney Test Procedure

Let us assume that the alumni director of a business school is compiling biographical data on alumni who graduated 10 years earlier. After receiving spotty returns from a mail survey, the director wants to know whether persons who concentrated in marketing are earning more than persons who concentrated in finance. Figure 15-5 shows that the alumni director has received salary data from 8 ($n_1 = 8$) marketing majors and 12 ($n_2 = 12$) finance majors.

**Stating the hypotheses and $\alpha$** As with other types of hypothesis tests, the first step in conducting the Mann-Whitney test is to state the null and alternative hypotheses and the specified level of significance. In this case, the null hypothesis

### FIGURE 15-5 Data for the Mann-Whitney test

Salary of 10-year graduates who majored in marketing and who majored in finance

| Marketing major | Annual income, $ (in thousands) | Income rank | Finance major | Annual income, $ (in thousands) | Income rank |
|---|---|---|---|---|---|
| G. Price | 22.4 | 15 | W. Lee | 21.9 | 14 |
| J. Jones | 17.8 | 3 | M. Galper | 16.8 | 1 |
| M. Doe | 26.5 | 16 | D. Lemons | 28.0 | 17 |
| K. Seller | 19.3 | 8 | T. Grady | 19.5 | 10 |
| S. Martin | 18.2 | 5.5 | P. Davis | 18.2 | 5.5 |
| J. Dreher | 21.1 | 13 | D. Henry | 17.9 | 4 |
| B. DeVito | 19.7 | 11 | B. Ruth | 35.8 | 19 |
| R. Coyne | 43.5 | 20 | J. P. Getty | 20.5 | 12 |
| | | | A. Carnegie | 18.7 | 7 |
| | | | J. Carter | 19.4 | 9 |
| | | | G. Ford | 17.3 | 2 |
| | | | R. Frank | 32.9 | 18 |
| $n_1 = 8$ | | $R_1 = 91.5$ | $n_2 = 12$ | | $R_2 = 118.5$ |

is that after 10 years, there is no difference between the salaries of the marketing majors and the salaries of the finance majors—i.e., $H_0$: The salaries of both majors are equal. Since a right-tailed test is to be made, the alternative hypothesis is that the salaries of the marketing majors are higher than the salaries of the finance majors 10 years after graduation—i.e., $H_1$: The salaries of the marketing majors are greater than those of the finance majors. Furthermore, the alumni director desires a significance level of $\alpha = .01$.

**Ranking data irrespective of sample category** After assembling the data, the next step is to *assign ranks to the entire set of income figures irrespective of major*. Since the annual salary of alumnus Galper is the lowest of the salaries of the 20 persons who responded, that salary is assigned a rank of 1. And since Coyne reported the highest income of either major, that income is assigned a rank of 20.

**Summing the ranks under each sample category and computing the U statistic** After the ranks have been assigned to all the data, the income ranks for each major should be summed. For all the marketing majors, the sum of the ranks, $R_1$, is 91.5, and the sum of the ranks of all the finance majors, $R_2$, is 118.5. We are now ready to compute the $U$ statistic. Either of the following formulas may be used in the computation of $U$:

$$U = n_1 n_2 + \frac{n_1(n_1 + 1)}{2} - R_1 \qquad (15\text{-}2)$$

*or*

$$U = n_1 n_2 + \frac{n_2(n_2 + 1)}{2} - R_2 \qquad (15\text{-}3)$$

where $R_1$ = the sum of ranks assigned to the sample with size $n_1$
$\qquad$ $R_2$ = the sum of ranks assigned to the sample with size $n_2$

These two formulas will most likely result in two different values for $U$. *The value which is selected for U in hypothesis testing is the lesser of the two values.* In using formula (15-2), we have

$$U = 8(12) + \frac{8(8 + 1)}{2} - 91.5 = 40.5$$

And with formula (15-3), we have

$$U = 8(12) + \frac{12(12 + 1)}{2} - 118.5 = 55.5$$

Consequently, the value assigned to $U$ for testing the null hypothesis is 40.5, which is the lesser of the two computed values. To check whether our computation of the $U$ value is correct, the following formula may be used:

$$\text{The lesser value of } U = n_1 n_2 - \text{the larger value of } U \qquad (15\text{-}4)$$

Note that in our example

$$U = 8(12) - 55.5$$
$$= 40.5$$

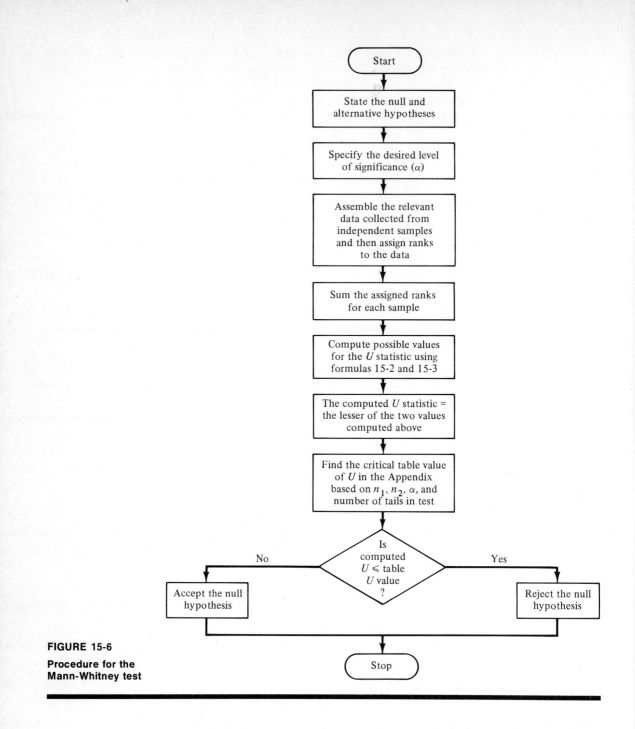

**FIGURE 15-6**

**Procedure for the Mann-Whitney test**

## Drawing a statistical conclusion concerning the null hypothesis

After computing the $U$ statistic, we are now ready to formally test the null hypothesis. Essentially, this test involves comparing the computed $U$ value with the expected table $U$ value that would be appropriate if the null hypothesis were true. The tables in Appendix 9 provide values of $U$ for the appropriate $n_1$, $n_2$, and $\alpha$ under the condition that the null hypothesis is valid. *The decision rule is:*

Reject the null hypothesis if the computed $U$ value is *equal to or less than* the appropriate value in the $U$ table.

In our example, we have $n_1 = 8$, $n_2 = 12$, and a desired level of confidence of .01 in a one-tailed test. The appropriate $U$ value from the second table in Appendix 9 is 17. Since the computed $U$ statistic equals 40.5 and is obviously greater than 17, the null hypothesis cannot be rejected. It may be concluded that there is no real salary difference between the alumni who majored in marketing and the alumni who majored in finance.

Figure 15-6 illustrates the procedure for conducting the Mann-Whitney test.

**Self-testing Review 15-3**

1 How does the Mann-Whitney test differ from the sign test?

2 How does the Mann-Whitney test differ from the $U$ test?

3 In the Mann-Whitney test, the sizes of the two independent samples must always be equal to each other. True or false?

4 When assigning ranks to relevant data in a Mann-Whitney test, we temporarily ignore the sample category of the specific pieces of data. True or false?

5 If the computed $U$ value is equal to or less than the table value of $U$, the null hypothesis is rejected. True or false?

6 What is the critical table value of $U$ when $n_1 = 12$, $n_2 = 13$, $\alpha = .05$, and a two-tailed test is being made?

**RUNS TEST FOR RANDOMNESS**

A financier is curious as to whether the recent increases and/or decreases in the daily Dow Jones Industrial Average (DJIA) are truly random or whether there is any order or *pattern* to the changes that might affect her portfolio. To satisfy her curiosity, the financier could conduct a *runs test for randomness*. The purpose of a runs test is to determine if randomness exists or if there is an underlying pattern in a sequence of sample data. The test is based upon the number of *runs* or *series* of identical types of results in sequential data. For example, if the financier noticed that for 15 consecutive business days the DJIA had a string or run of 15 consecutive losses, she might readily conclude that there is a pattern in the behavior of the stock market. Unfortunately, decision making isn't always as clear-cut as the preceding sentence might suggest. Therefore, the runs test is another hypothesis-testing procedure that is designed to assist decision makers.

**Procedure for Conducting a Runs Test**

Suppose the DJIA for the most recent 15 consecutive business days reflected the following changes:

Day: 1 2 3 4 5 6 7 8 9 10 11 12 13 14 15
Change: + + − − + + + + + − − + + − +

The plus signs indicate an increase over the preceding day, while the negative signs reflect a decrease over the preceding day.

**Stating the null and alternative hypotheses** The hypotheses for our runs test are:

$H_0$: There is randomness in the DJIA sequential data under analysis.

$H_1$: There is a pattern in the sequential DJIA data under analysis.

The runs test is designed to detect a pattern in the sequence of data, but it cannot tell us the *nature* of the pattern. Thus, in our example, the test might tell us that there is a pattern to stock market changes, but we cannot conclude from the runs test whether the pattern is in an upward or downward direction.

**Counting the number of runs** On the basis of the sequence of signs, can the financier conclude randomness, or is there *any* pattern? (She is not concerned here with the specific type of pattern.) The first step to answering this question is to count the number of runs. This is done in the following manner, using the preceding data:

Change: $\boxed{+\ +}\ \boxed{-\ -}\ \boxed{+\ +\ +\ +\ +}\ \boxed{-\ -}\ \boxed{+\ +}\ \boxed{-}\ \boxed{+}$
Run:  1      2          3          4      5    6  7

There are seven runs in the sequence of data. The first run is a series of two pluses; the second run is a series of two minuses; the third run is a group of five pluses; and so on. Thus, we may state that $r$ (the number of runs) $= 7$. Given our data, do the seven runs indicate a random movement in the stock market, or is there a possible pattern to the runs?

**Counting the frequency of occurrence** The next step in a runs test is to first identify the number of elements of one type of data (which is labeled $n_1$) and then identify the number of elements of the other type of data (which is labeled $n_2$). For our data, we have 10 pluses (and thus $n_1 = 10$) and 5 negatives (therefore $n_2 = 5$). If there had been a case of no change in the DJIA (i.e., a tie), that case would not have been counted.

**Drawing a statistical conclusion** If $n_1$ and $n_2$ are each equal to or less than 20,[5] we begin the test of the null hypothesis by referring to the tables in Appendix 10. These tables are based on the assumption that $H_0$ is true, and they provide critical values of $r$ based upon $n_1$, $n_2$, and a level of significance of .05. The following *decision rule* is used to compare the sample $r$ value with the table $r$ value:

The null hypothesis should be rejected if the sample $r$ value is *equal to or less than* the appropriate $r$ value from Appendix 10, table (*a*); the $H_0$ should also be

---

[5] Runs test procedures are available when $n_1$ or $n_2$ are $> 20$, but space constraints prevent us from considering these procedures in this text.

rejected if the sample $r$ value is *equal to or greater than* the table $r$ value found in Appendix 10, table (*b*).

Since we have $n_1 = 10$ and $n_2 = 5$, the corresponding $r$ value from Appendix 10, table (*a*), is 3, and the $r$ value from Appendix 10, table (*b*), is 12. Appendix 10 thus tells us that in a random sequence of 15 observations where 10 pluses and 5 minuses are noted, the chances of getting 3 or less or 12 or more runs is only 5 percent. Since the sample $r$ is 7 and thus falls between the table values, we cannot reject the null hypothesis. Seven runs are quite likely in a *random sequence* of 15 observations that are similar to our sample data. The financier may therefore conclude that there has been no detectable pattern of behavior in the stock market for the past 15 days.

Figure 15-7 summarizes the procedure for conducting a runs test for randomness.

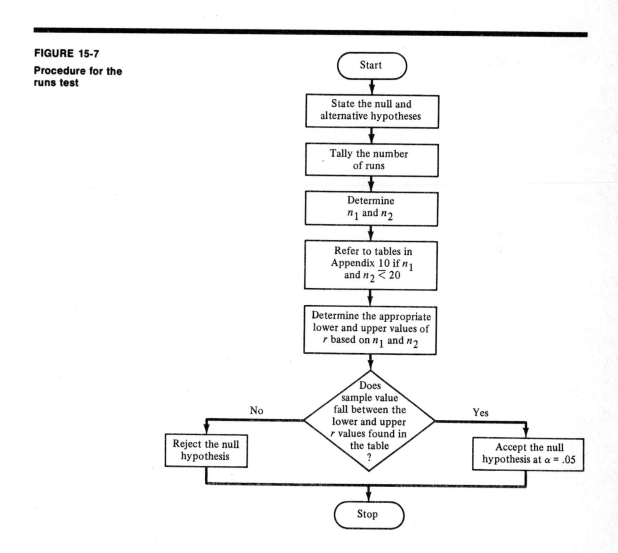

**FIGURE 15-7**

**Procedure for the runs test**

1  What is a runs test?

2  Does a runs test need two independent samples?

3  What is the alternative hypothesis in a runs test?

4  The runs test is only concerned with detecting a pattern; it is not concerned with the type or direction of the pattern, if any is detected. True or false?

5  If the sample $r$ value is less than the lower table $r$ value or greater than the upper table $r$ value, the alternative hypothesis is accepted. True or false?

6  The tables in Appendix 10 may be used when $n_1$ and $n_2$ are each greater than 20. True or false?

7  In a runs test for randomness, there are 10 runs in the data sequence. The value of $n_1$ is 19, and the value of $n_2$ is 14. At the .05 level, should the $H_0$ be accepted?

## SPEARMAN RANK CORRELATION COEFFICIENT

The Spearman rank correlation coefficient, $r_s$, is a measure of the closeness of association between two ordinal variables; that is, $r_s$ is a measure of the degree of relationship between *ranked* data. The coefficient of correlation ($r$) found in Chapter 14 was computed using the actual values of $X$ and $Y$; the Spearman measure we will now consider uses rank values for $X$ and $Y$ rather than actual values.

## Procedure for Computing the Spearman Rank Correlation Coefficient

The Ajax Insurance Corporation has been operating a sales refresher course designed to improve the performance of its sales representatives. A number of classes have completed the course. In an attempt to assess the value of the program, the sales training manager wants to determine if there is a relationship between performance in the program and subsequent performance in generating annual sales. Figure 15-8 shows the data collected by the sales training manager on 11 ($n = 11$) program graduates.

**Ranking the data**  As a first step, the manager ranked each of the 11 representatives according to his or her performance in the sales course. A rank of 1 was assigned to the person with the best performance; a rank of 2 was given to the next best graduate; etc. Then, each salesperson was ranked according to sales performance in the subsequent year. A rank of 1 was assigned to the person who had the most sales; a rank of 2 was given to the one with the next highest sales; etc. For example, sales representative Steele was considered the best among the persons who attended the sales course, and Steele had the fourth highest sales in the 12 months following completion of the program.

**Computing the differences between ranks**  The next step is the systematic computation of the differences between ranks. These differences, labeled $D$, are shown in the third column of Fig. 15-8. Since sales representative McCabe achieved a rank of 5 for course performance but had a *lesser* rank of 7 for sales performance, the difference assigned to McCabe is $-2$.

**Computing $r_s$**  After computing $D$ for each person, the manager is ready to compute the Spearman measure, which is defined as follows:

## FIGURE 15-8 Data for the computation of the Spearman rank correlation coefficient

| Salesperson | Course performance rank (1) | Annual sales rank (2) | Difference between ranks (1 − 2) D (3) | $D^2$ (4) |
|---|---|---|---|---|
| Steele | 1 | 4 | −3 | 9 |
| Spier | 2 | 6 | −4 | 16 |
| Devine | 3 | 1 | 2 | 4 |
| Hanlon | 4 | 2 | 2 | 4 |
| McCabe | 5 | 7 | −2 | 4 |
| Braman | 6 | 10 | −4 | 16 |
| Seville | 7 | 3 | 4 | 16 |
| McNally | 8 | 5 | 3 | 9 |
| Reid | 9 | 8 | 1 | 1 |
| Silva | 10 | 9 | 1 | 1 |
| Gould | 11 | 11 | 0 | 0 |
| | | | $\Sigma D = 0$ | $\Sigma D^2 = 80$ |

$$r_s = 1 - \left(\frac{6\Sigma D^2}{n(n^2 - 1)}\right)$$

$$= 1 - \left(\frac{6(80)}{11(121 - 1)}\right)$$

$$= 1 - .364$$

$$= .636$$

$$r_s = 1 - \left(\frac{6\Sigma D^2}{n(n^2 - 1)}\right) \tag{15-5}$$

To compute $r_s$, we must square the differences between ranks and then sum the squared differences—i.e., perform the operations represented by $\Sigma D^2$ in the numerator of formula (15-5). The last column in Fig. 15-8 shows the sum of the squared differences. The computations shown in Fig. 15-8 give us a value of $r_s$ of .636.

As a basis for interpreting $r_s$, you should keep in mind that when $r_s$ (like $r$ in the last chapter) is zero, there is no correlation. And, like $r$ in Chapter 14, when $r_s$ is +1.00 or −1.00, there is perfect correlation. In our example, therefore, the manager might conclude that there is a correlation between course performance and subsequent sales activity.

## Testing $r_s$ for Significance

A more formal test may be conducted to determine whether there is truly a statistical relationship as suggested by $r_s$. A null hypothesis may be established to the effect that there is no relationship between course performance and sales performance—i.e., $H_0$: $r_s = 0$. Since the training manager prefers to believe that the course improves selling ability, a right-tailed test is appropriate, and the alternative hypothesis is that there is a positive relationship between course performance and sales performance—i.e., $H_1$: $r_s > 0$. Let us assume that we wish to conduct the test at

$\alpha = .05$. The essential question in our hypothesis test is how likely is it that we could have obtained the sample value of $r_s = .636$ if there was truly no relationship between the two variables?

If the sample size is larger than 10, we may conduct a hypothesis test by computing the critical ratio (CR) as follows:

$$CR = r_s\sqrt{\frac{n-2}{1-r_s^2}} \tag{15-6}$$

On the basis of our data, we have

$$CR = .636\sqrt{\frac{11-2}{1-.636^2}}$$

$$= .636\sqrt{\frac{9}{1-.404}}$$

$$= +2.47$$

After computing the critical ratio, we are now ready to draw a conclusion based upon the following *decision rule* for a right-tailed test at the .05 level of significance:

Accept the $H_0$ if the CR $\lessgtr$ the appropriate $t$ value.

*or*

Reject the $H_0$ and accept the $H_1$ if the CR $>$ the appropriate $t$ value.

What is the appropriate $t$ value? The appropriate $t$ value, in case you have forgotten, is found in Appendix 4. The correct $df$ (degrees of freedom) row to select at this time is determined by using $n - 2$, since we have two variables (course performance and sales performance). The levels of significance given in the columns of Appendix 4 are for one-tailed tests. Thus, in our example the $t$ value for $n - 2$ $(11 - 2)$ or 9 degrees of freedom at the .05 level is 1.833. Since our CR $= 2.47$, and since the appropriate $t$ value $= 1.833$, the null hypothesis is rejected. We can conclude that there is a statistical relationship between participation in the sales course and subsequent sales performance.

Figure 15-9 summarizes the procedure for computing $r_s$ and testing its significance.

**Self-testing Review 15-5**

1  What is the Spearman rank correlation coefficient?

2  What may be concluded if $r_s = +1.36$?

3  When $n > 10$, the significance of $r_s$ may be tested through the use of formula (15-6). True or false?

4  If $\Sigma D^2 = 566$ and $n = 16$, what is $r_s$?

5  a  If $r_s = .67$ and $n = 13$, then CR $= 2.43$. True or false?
   b  At the .01 level of significance, would the $H_0$ be accepted with a one-tailed test? (Use the data given in part **a** of this question.)

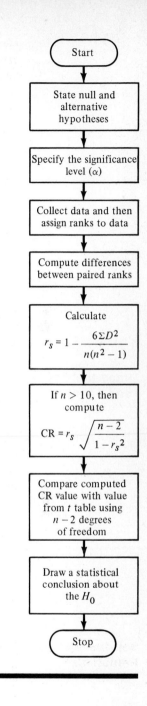

**FIGURE 15-9**

**Procedure for computing and testing the Spearman rank correlation coefficient**

The flowchart reads:

Start → State null and alternative hypotheses → Specify the significance level ($\alpha$) → Collect data and then assign ranks to data → Compute differences between paired ranks →

Calculate
$$r_s = 1 - \frac{6\Sigma D^2}{n(n^2 - 1)}$$

If $n > 10$, then compute
$$CR = r_s \sqrt{\frac{n - 2}{1 - r_s^2}}$$

→ Compare computed CR value with value from $t$ table using $n - 2$ degrees of freedom → Draw a statistical conclusion about the $H_0$ → Stop

## SUMMARY

A researcher is often limited by the imprecise quantitative properties of the data available for analysis purposes. For example, the data may come from only a few samples, and there may be little or no knowledge of the shape of the population distribution and its effect on the sampling distribution. When such problems arise, nonparametric techniques may be used. In this chapter, we have discussed only a few of the common nonparametric techniques, and we have limited our attention primarily to small-sample cases.

When one wishes to find out if there are significant differences between paired rank data which are drawn from one sample or from two related samples, the sign test or the Wilcoxon signed rank test may be appropriate. If the *magnitude* or size of the differences between paired data is to be considered in decision making, the Wilcoxon signed rank test should be employed; if only the direction of differences is needed for a decision, the sign test is sufficient.

The Mann-Whitney test, or *U* test, should be employed when differences between paired rank data are under study *and* when the data are drawn from two independent samples.

In the case of a single sample with sequential data, a runs test for randomness may be conducted. This test is designed to detect the presence or absence of pattern or order in the sequential data.

Finally, the Spearman rank correlation coefficient was discussed. This measure is a correlation coefficient for paired rank data. The computation of $r_s$ provides a measure of association between two variables.

## Important Terms and Concepts

1 Distribution-free statistics

2 Nonparametric statistics

3 Ordinal data

4 Nominal data

5 Sign test

6 Tied observations

7 Wilcoxon signed rank test

8 Direction of differences

9 Magnitude of differences

10 Wilcoxon $T$ statistic

11 Mann-Whitney test

12 Mann-Whitney $U$ statistic

13 Runs test for randomness

14 Spearman rank correlation coefficient

15 $CR = \dfrac{2R - n}{\sqrt{n}}$

16 $U = n_1 n_2 + \dfrac{n_1(n_1 + 1)}{2} - R_1$

17 $U = n_1 n_2 + \dfrac{n_2(n_2 + 1)}{2} - R_2$

18 $r_s = 1 - \left( \dfrac{6 \Sigma D^2}{n(n^2 - 1)} \right)$

19 $CR = r_s \sqrt{\dfrac{n - 2}{1 - r_s^2}}$

## Problems

1 Polly Esta, owner of Natural Textiles, Ltd., is seriously concerned about the chronically low daily output of her factory workers. Therefore, she has devised a bonus plan and wants to find out if this bonus plan will result in any improvement. (She is not overly concerned at this time with the degree of productivity improvement.) In an experiment, eight workers were offered the bonus plan. Their daily output before and after the bonus plan is shown below:

| Worker | Unit output before bonus plan | Unit output after bonus plan |
|--------|------------------------------|-----------------------------|
| Harris Tweed | 80 | 85 |
| Stitch N. Tyme | 75 | 75 |
| Les Hemm | 65 | 71 |
| Tom Taylor | 82 | 79 |
| Chuck Moore | 56 | 68 |
| Tex Tile | 70 | 86 |
| John Trim | 73 | 71 |
| Mat Wool | 62 | 59 |

**a** What is the alternative hypothesis?

**b** What is $n$? What is $r$?

**c** On the basis of the $H_0$, what is the probability of obtaining a value as high as $r$ given the size of $n$?

**d** If the null hypothesis is rejected, what is the level of significance?

**2** The marketing director of the National Shampoo Company is wondering if producing the existing green shampoo in a darker shade will improve consumers' perception of product effectiveness. At this point, the director simply wants to determine the feasibility of developing the idea further and is just looking for some degree of improvement in perception of product effectiveness. Data have been collected from seven persons; each has rated the original green shampoo and the darker-shade shampoo. A 1-to-10 scale was used where 1 stood for "very ineffective" and 10 signified "most effective." The data are shown below:

| Consumer | Effectiveness rating for original-color shampoo | Effectiveness rating for darker-color shampoo |
|----------|-----------------------------------------------|----------------------------------------------|
| Abe Bell | 4 | 2 |
| Will Ling | 6 | 6 |
| Peg Brown | 7 | 4 |
| Dan D. Ruff | 5 | 6 |
| Sue Weese | 9 | 8 |
| Jack Sprat | 1 | 3 |
| Jim Hawkins | 3 | 8 |

**a** What is the alternative hypothesis?

**b** What is the value of $n$? The value of $r$?

**c** If the $H_0$ is rejected, what is the probability of error in judgment?

**3** The Cal Q. Leighter Computer Company employs 500 salespeople. In an attempt to reduce the amount of time needed to close a sale, the company has produced a visual-aid package to be used in sales presentations. So far, only 10 salespeople have requested and used the visual-aid package. When each of these salespeople made the request to use the package, he or she was asked to estimate the amount of time usually needed in a presentation to make a sale.

After each one used the visual aids for 2 months, he or she was again asked to estimate how much time it took to make a sale. The data are shown below:

| Salesperson | Average time needed to make a sale | |
|---|---|---|
| | Before use of visual aids | After use of visual aids |
| A | 23 | 17 |
| B | 45 | 43 |
| C | 36 | 36 |
| D | 42 | 37 |
| E | 25 | 20 |
| F | 33 | 39 |
| G | 28 | 31 |
| H | 25 | 21 |
| I | 35 | 27 |
| J | 30 | 40 |

**a** What is the alternative hypothesis?
**b** If the .05 level is specified, would the $H_0$ be rejected?

4 Assume we are conducting a sign test, and in determining the differences between paired data, we have the following facts:

| Subject | Differences between paired observations |
|---|---|
| A | + |
| B | − |
| C | − |
| D | + |
| E | − |
| F | 0 |

The alternative hypothesis is that the probability of getting a negative sign is greater than .50. If the null hypothesis is rejected, what is the level of significance?

5 Assume that you have the following facts for a sign test: $n = 15$, $r = 3$, two-tailed test, and a required $\alpha = .05$. Would the null hypothesis be rejected?

6 Conduct a sign test based upon the following data:

| Subject | Differences between paired observations |
|---|---|
| a | + |
| b | − |
| c | − |
| d | − |
| e | − |
| f | − |

If we assume that the alternative hypothesis is that a negative sign is more likely than a positive sign, at what level of significance would the $H_0$ be rejected?

7 The Bovine Dairy Association sponsored a series of 30-second commercials promoting milk consumption. Eighteen stores were asked to record their unit sales of milk for the week prior to the campaign. After the campaign appeared on television, the same 18 stores were asked to report their sales for the week immediately following the airing of the commercials. The data were as follows:

| | Unit weekly sales | |
|---|---|---|
| Store name | Before campaign | After campaign |
| Jones | 124 | 136 |
| Ma & Pa | 107 | 105 |
| Granny's | 82 | 89 |
| Ralph's | 114 | 128 |
| J & A | 940 | 1,080 |
| Korner | 75 | 85 |
| Superette | 105 | 105 |
| Mike's | 94 | 95 |
| Buy More | 865 | 985 |
| Value | 620 | 820 |
| Pete's | 80 | 75 |
| Foodco | 750 | 725 |
| Koop | 330 | 350 |
| Speedy | 110 | 112 |
| Walt's | 125 | 120 |
| Big Bag | 400 | 425 |
| Pay Now | 400 | 450 |
| Plus | 175 | 215 |

a Conduct a sign test at the .10 level.
b Conduct a Wilcoxon signed rank test at the .05 level.
c Assume that 18 additional stores had been contacted and had recorded "before" and "after" sales data. Assume also that the following results had been obtained:

$$+ \text{ differences} = 24$$
$$- \text{ differences} = 10$$
$$0 \text{ differences} = \underline{2}$$
$$36$$

Conduct a sign test with $\alpha = .05$.

8 The True Grit Sand Company has two operating pits in the Boston area. It has always been the suspicion of the owner of True Grit that location B is more productive than location A simply because of the geography of the areas; that is, the difference in productivity between the two areas is not attributable

to differences in personnel or machines. In an effort to confirm this belief, the owner monitored the weekly output of 12 workers in location A and then transferred these workers to location B. The output of these 12 workers in location B was also monitored for a week. The results are shown below:

| Worker name | Weekly output | |
| | Location A | Location B |
| --- | --- | --- |
| Spade | 100 | 105 |
| Dozer | 150 | 145 |
| Trukk | 160 | 163 |
| Graider | 95 | 95 |
| Levell | 110 | 118 |
| Bobb | 87 | 90 |
| Pile | 135 | 143 |
| Rock | 125 | 129 |
| Pebble | 98 | 86 |
| Sands | 142 | 145 |
| Dunes | 110 | 85 |
| Gravell | 130 | 132 |

**a** Conduct a sign test with $\alpha = .05$.
**b** Conduct a Wilcoxon signed rank test with $\alpha = .01$.

9 A clinical pharmacist is wondering if a new pain-killer is effective for persons with chronic pain. She believes that the new drug should significantly reduce the amount of pain. She wishes to record not only the change in pain that occurred after the dosage of the drug but also the amount of the change. Using an accepted measurement instrument, she recorded the pain level of 8 patients before and after the drug had been administered. A high score on this scale corresponds to a high degree of pain. The data are:

| Patient | Pain level before taking the drug | Pain level after taking the drug |
| --- | --- | --- |
| A | 14 | 8 |
| B | 15 | 9 |
| C | 10 | 11 |
| D | 12 | 10 |
| E | 11 | 11 |
| F | 13 | 9 |
| G | 12 | 11 |
| H | 10 | 10 |

**a** What are the null and alternative hypotheses?
**b** Assuming that the pharmacist specified that $\alpha = .05$, what should be concluded about the effectiveness of the new drug?

**10** Assume that you are conducting a Wilcoxon signed rank test and that the differences between paired observations are as follows:

| Subject | Differences between paired observations |
|---------|-----------------------------------------|
| A | +3 |
| B | 0 |
| C | −1 |
| D | +8 |
| E | +4 |
| F | −2 |
| G | +1 |
| H | +6 |

**a** What is the sum of the positive ranks?
**b** What is the sum of the negative ranks?
**c** What is the computed $T$ value?
**d** With a two-tailed test and a specified $\alpha$ of .05, would you reject the null hypothesis?

**11** Assume that the alternative hypothesis in a test is as follows:

$H_1$: The probability of a decrease is greater than the probability of an increase.

Conduct a Wilcoxon signed rank test at $\alpha = .01$ for the following data:

| Subject | Differences between paired data |
|---------|---------------------------------|
| 1 | +6 |
| 2 | −9 |
| 3 | +2 |
| 4 | −4 |
| 5 | −3 |
| 6 | +1 |
| 7 | 0 |
| 8 | −1 |
| 9 | −5 |
| 10 | +3 |
| 11 | −2 |
| 12 | 0 |
| 13 | −7 |
| 14 | −10 |

**12** Use the data in problem **1** to conduct a two-tailed Wilcoxon signed rank test at the .05 level of significance.

**13** Conduct a two-tailed Wilcoxon signed rank test on the data in problem **3**. Use the .01 level of significance.

**14** A high school counselor is wondering if persons who scored high on the verbal section of a college entrance exam will perform just as well in a business school as those who scored high on the math section of the same entrance exam. The grade-point average (GPA) of a sample of students was selected (a 4.0 represents an A and 1.0 represents a D), and the following data were obtained:

| Students with high verbal scores | | Students with high math scores | |
|---|---|---|---|
| Name | College GPA | Name | College GPA |
| Chipps | 2.4 | Boole | 3.1 |
| Hawthorne | 3.2 | Pythags | 2.3 |
| Walden | 3.9 | Chebyshev | 1.9 |
| Canterbury | 1.6 | Bayes | 2.1 |
| Emerson | 2.2 | Sine | 2.7 |
| Jones | 2.5 | Cosine | 3.6 |
| Smith | 2.4 | | |

Conduct a two-tailed test at the .01 level of significance.

**15** The Flatt Tire Company has been testing a new emergency tire inflator which is supposed to be significantly faster than the leading competitor's inflator. Motorists were randomly selected for the product testing. Some of the subjects were assigned to use the new product, while the remainder were to use the leading competitor's product. The seconds required to inflate a tire are recorded below:

| Flatt tire inflator | Leading competitor's inflator |
|---|---|
| 17 | 23 |
| 16 | 21 |
| 21 | 32 |
| 19 | 21 |
| 15 | 19 |
| 14 | 20 |
| 16 | 21 |
| 16 | 22 |
| 23 | |

Assuming that you are conducting the research, what statistical decision would you make at the .05 level?

**16** A job counselor believes that college graduates tend to be more satisfied in their jobs than non-college graduates. A job-satisfaction test was administered to subjects classified under each category. (A high score indicates a high degree of job satisfaction.) The following results were obtained:

| Subject | Non-college graduate | Subject | College graduate |
|---------|---------------------|---------|------------------|
| a | 102 | aa | 78 |
| b | 87 | bb | 93 |
| c | 93 | cc | 101 |
| d | 98 | dd | 85 |
| e | 95 | ee | 84 |
| f | 101 | ff | 77 |
| g | 92 | gg | 92 |
| h | 85 | hh | 86 |
| i | 88 | | |
| j | 95 | | |
| k | 97 | | |
| l | 96 | | |

Make a statistical decision at the .05 level.

**17** A psychologist hypothesized that students from high school A tended to be more aggressive than students from high school B. A psychological test was administered to randomly selected students from each school. A high score on this test represented a high degree of aggression. The following results were obtained:

| High school A | | High school B | |
|---------------|------------|---------------|------------|
| Student | Test score | Student | Test score |
| Jim Jungle | 43 | Frank Mild | 47 |
| Mike Tuff | 56 | John Plain | 68 |
| Bill Bully | 31 | Bobby Blah | 39 |
| Sam Shove | 30 | Ken Kwiat | 29 |
| Tom Truant | 41 | Carl Calm | 36 |
| Steve Skipp | 38 | Dave Dull | 42 |
| | | Gary Good | 33 |
| | | Kurt Kind | 54 |

Make a statistical decision at the .05 level.

**18** Gary Gullible was asked by a friend to engage in flipping a coin for money. Gary's friend would win a dollar each time a head (H) appeared, while Gary would win a dollar for each tail (T) that appeared. After 20 flips, poor Gary was down by $6. Since the coin was provided by the friend, Gary began to wonder if the coin was "loaded." Here is the sequence of results:

T T T H H H T H H H H H T T T H H H H H

What can you tell Gary with $\alpha = .05$?

**19** The Economics Institute has developed a new forecasting model and is anxious to learn if errors between its estimates and actual results are truly random or if there is a pattern to the errors. A series of 25 estimates was generated and compared with actual results. The errors of overestimation (+) and underestimation (−) are shown below:

$$+ + - + - + - - - - - + + - - - - + + + + + - + +$$

What conclusions can you reach at the .05 level?

**20** Over a 22-day period, the supervisor of a trenching crew monitored productivity to determine how the crew was performing relative to its quota. The supervisor was interested in learning if the daily work above (+) or below (−) the quota was following a random pattern. The observations were as follows:

$$- - - - - - + - - - - + + - + - - - - - + + -$$

Make a statistical decision at the .05 level.

**21** Conduct a runs test at $\alpha = .05$ for the following data series:

$$+ + + + - - - - - - + + + + - - + + + + - + + + + + + + +$$

**22** Conduct a runs test at $\alpha = .05$ for the following data series:

H H H H H T H H H T T T T T T H H H H H

**23** Abner Doubleplay, pitching coach for the Plainville Cougars, has noticed that in recent years some of the successful pitchers in the league were overweight. This has led Abner to wonder if weight affects pitching performance. Weight figures and winning percentages have been gathered for 21 league pitchers. The heaviest pitcher was assigned a weight rank of 1, and the pitcher with the highest winning percentage was assigned a winning rank of 1. The results are shown below:

| Pitcher | Weight rank | Winning rank |
|---|---|---|
| a | 3 | 6 |
| b | 7 | 1 |
| c | 15 | 21 |
| d | 10 | 2 |
| e | 2 | 9 |
| f | 13 | 13 |
| g | 6 | 8 |
| h | 21 | 5 |
| i | 8.5 | 19 |
| j | 1 | 12 |
| k | 12 | 4 |
| l | 14 | 14 |
| m | 17 | 18 |
| n | 4 | 20 |
| o | 18 | 11 |
| p | 8.5 | 7 |
| q | 11 | 16 |
| r | 20 | 3 |
| s | 19 | 15 |
| t | 16 | 17 |
| u | 5 | 10 |

What conclusion can you draw at the .05 level of significance?

24 Mickey Bubbles, sales manager of the Cool Cola Bottling Corporation, wants to know how strong a relationship (if any) there is between daily temperature and sales on corresponding days. Because of poor record-keeping procedures, Mickey must make do with rank data (where the hottest day has been assigned a 1 and the highest sales figure has been assigned a 1). Fifteen days were selected randomly, and the paired data are as follows:

| Temperature rank | Sales rank |
|---|---|
| 6 | 5 |
| 11 | 12 |
| 4 | 2 |
| 7 | 7 |
| 1 | 4 |
| 12 | 14 |
| 8 | 10 |
| 2 | 1 |
| 15 | 15 |
| 14 | 13 |
| 5 | 3 |
| 10 | 9 |
| 13 | 11 |
| 9 | 8 |
| 3 | 6 |

What conclusion may be drawn at the .01 level?

25 A psychologist believes that those who score high on a need-achievement test will likely have a high salary to match. To test this theory, the psychologist has administered questionnaires to 17 persons and has ranked the data in such a way that the highest value in each category has been assigned a 1. The paired data are:

| Need-achievement rank | Salary rank |
|---|---|
| 1 | 3 |
| 8 | 7 |
| 4 | 2 |
| 10 | 12 |
| 12 | 9 |
| 2 | 1 |
| 13 | 11 |
| 6 | 6 |
| 16 | 17 |
| 11 | 13 |
| 14 | 15 |
| 3 | 5 |
| 9 | 10 |
| 7 | 8 |
| 15 | 14 |
| 17 | 16 |
| 5 | 4 |

What conclusion may be reached at the .01 level?

26 According to news reports, people in a mountain region of the country of Placebo claim that many of their neighbors live past 100 years of age. The Information Minister of Placebo claims that such longevity is related to the consumption of native cucumbers. Professor Pry doubts that there is any association (either positive or negative) between the age of a person and the annual consumption of cucumbers. The Placebo government has permitted Professor Pry to randomly select and interview 15 residents of the mountain region. Due to the lack of official records, the information provided by the subjects on age and cucumber consumption was not in a precise format. Therefore, the following data from the subjects must be converted to ordinal data for analysis:

| Subject | Reported age | Reported annual cucumber consumption |
|---------|--------------|--------------------------------------|
| Ben Dover | 102 | 156 |
| Stan Strait | 136 | 175 |
| Al Bowe | 98 | 134 |
| Rip V. Winkle | 110 | 143 |
| Nee Kapp | 106 | 129 |
| I. Claude Jawn | 156 | 164 |
| Howard Hertz | 92 | 124 |
| S. Keemo | 89 | 110 |
| Hugo First | 143 | 160 |
| Rip Mend | 124 | 109 |
| Red Hott | 94 | 95 |
| Hott N. Tott | 105 | 120 |
| Sy N. Nara | 117 | 133 |
| Ive Haddit | 108 | 119 |
| Hal Widdit | 97 | 101 |

If a rank of 1 is assigned to the lowest value in each category, and if $\alpha = .01$, what conclusion should be made?

**Topics for Discussion**

1 What are nonparametric statistics?

2 What are some examples of nominal data and ordinal data?

3 How does the sign test differ from the Wilcoxon signed rank test?

4 What do the Wilcoxon signed rank test and the $U$ test have in common?

5 "The results of a runs test do not permit us to make conclusions about the types of patterns in sequential data if patterns are detected." Discuss this statement.

6 What is the major difference between the parametric and the nonparametric correlation coefficients?

**Answers to Self-testing Review Questions**

15-1

1 A sign test is conducted to determine if there are real differences between paired ordinal data drawn from a single sample or two closely related samples; the test is based upon the signs of differences between pairs of data.

**2** The statement is incorrect. A two-tailed test may be conducted.

**3** In order to conduct a sign test, it is necessary to observe or measure each subject twice.

**4** The null hypothesis in a sign test is that the probability of a positive sign occurring is equal to the probability of a negative sign occurring. In other words, the median difference between the paired data should be zero.

**5 a** The binomial probability distribution should be used in a sign test when the sample size is small.

**b** The normal approximation to the binomial probability distribution may be used in this case.

**6 a** False. The number of relevant data is 12 ($n = 12$). Also, $r$ equals 5, which is the smaller sum of the two signs.

**b** $H_0$: $p = 0.5$
$H_1$: $p > 0.5$

$\alpha = .10$

With $n = 12$ and $r = 5$, the sum of the relevant probabilities is .3867 (.0002 + .0029 + .0161 + .0537 + .1204 + .1934). Since .10 < .3867, the $H_0$ is accepted.

**7** $H_0$: $p = 0.5$
$H_1$: $p \neq 0.5$

$\alpha = .05$

Decision rule: Accept $H_0$ if CR falls between $\pm 1.96$.

$$\text{CR} = \frac{2R - n}{\sqrt{n}} = \frac{2(16) - 42}{\sqrt{42}} = \frac{-10}{\sqrt{6.481}} = -1.543$$

Decision: Accept $H_0$, since the CR falls between $\pm 1.96$.

## 15-2

**1** The Wilcoxon signed rank test incorporates the magnitude as well as the direction of the differences between paired ordinal data.

**2** The null hypothesis in a Wilcoxon signed rank test is that there is no real difference between the paired data.

**3** True.

**4** False. The statistic $T$ is the lesser of the two sums of the ranks, and therefore we have $T = 20$.

**5** The table $T$ value is 159.

**6** True.

### 15-3

1 The data for a Mann-Whitney test are collected from independent samples, whereas the data for a sign test are collected from one sample or two related samples.

2 The Mann-Whitney test and the $U$ test are the same.

3 False. The samples in a Mann-Whitney test do not have to be equal.

4 True. In the initial ranking of the data, the data from the two groups are aggregated and then ranked irrespective of the sample category.

5 True.

6 The critical table value of $U$ is 41.

### 15-4

1 A runs test is designed to determine the existence or nonexistence of a pattern in sequential data.

2 The data for a runs test are drawn from one sample.

3 The alternative hypothesis in a runs test is that there is a pattern in the sequence of data.

4 True.

5 True.

6 False. Use of the tables is permitted when the size of each sample is equal to or less than 20.

7 The $H_0$ should be rejected. The lower table $r$ value is 11, and since the sample $r$ of 10 falls below this table value, the $H_0$ cannot be accepted.

### 15-5

1 The $r_s$ is a measure of association between ordinal data.

2 It must be concluded that the coefficient was computed incorrectly, since $r_s$ may have values between $-1.00$ and $+1.00$ only.

3 True.

4 The answer is computed as follows:

$$r_s = 1 - \left( \frac{6\Sigma D^2}{n(n^2 - 1)} \right) = 1 - \left( \frac{6(566)}{16(256 - 1)} \right) = .1677$$

5    a    False. CR $= 2.99$.

$$CR = .67\sqrt{\frac{13 - 2}{1 - .67^2}} = 2.99$$

    b    The $H_0$ would be rejected. The $t$ value at 11 $df$ and $\alpha = .01$ is 2.718, which is $<$ the CR of 2.99.

# CHAPTER 16

# WHERE DO WE GO FROM HERE?

**There comes a time when one asks even of Shakespeare, even of Beethoven, "Is this all?"**

**— Aldous Huxley**

After you have just spent one or two terms wrestling with statistical methodology, you may not be too thrilled to learn that you have only scratched the surface. Yet, alas, such is the case. Unfortunately, books that attempt to cover the field of statistics comprehensively tend to resemble encyclopedias and, like encyclopedias, are used by some people as reference works and by others as doorstops. Even the more comprehensive books make no claims of completeness. After all, such topics as time-series analysis, correlation, and index numbers have each been the subject of entire books.

It is obviously impossible, then, to cover in detail in one chapter all the topics not covered previously in this book. That is not our intent. Rather, we shall describe several other important methods of quantitative analysis, give examples of their use, and discuss their limitations.

More specifically, we shall briefly consider (1) *sample design*, (2) *multiple regression*, (3) *decision theory*, (4) *linear programming*, (5) *inventory models*, (6) *waiting line theory, and* (7) *simulation*.

**SAMPLE DESIGN**

In Chapter 6 mention was made of stratified samples and cluster samples; all the discussion of estimation and statistical inference was for simple random sampling. But in many cases it is either impossible or impractical to take a simple random sample.

If there is a large amount of variation in a population, a very large random sample might be required to yield the desired sampling error. The cost of such a large sample could be completely prohibitive. In a case like this it might be possible to divide the population into groups of like elements so that the variation within each group is relatively small. Samples could then be taken from each of these groups. Since the variation in each group is less than the variation for the entire population, the sample size required for this *stratified sample* would be smaller than that required for a random sample.

Suppose, for example, that you wish to take a sample of stores in a particular city to estimate total retail sales in the city for the previous month. The amount of variation in this population would be tremendous, ranging all the way from the very large sales figures for large department stores and supermarkets to the very small sales figures for small neighborhood stores. However, if the stores were *stratified* into four or five groups according to size, the amount of variation in each group would be relatively small. *Stratified sampling is used* by public accountants to estimate the value of an individual item. It is also used in personnel studies with employees being stratified according to job classification. In short, this type of design is useful *when a very heterogeneous population can be stratified into fairly homogeneous groups*.

In order to take either a random sample or a stratified sample, it is necessary to have some kind of complete listing of the population being sampled. *This is often an impossibility*. For example, we might wish to take a sample of employees of retail stores in a city. It is often impossible to draw up a complete listing of all these people, but we could draw up a list of all retail stores in the city. We could then take a sample of the stores, and the employees of the sampled stores could be interviewed. *This type of sample design is known as cluster sampling*, since we are dividing the population into clusters of elements and then drawing samples of the clusters. *This type of design is often used in market surveys and election polls*. Clusters of people are sampled by drawing a sample of city blocks and then interviewing the people who live on these blocks.

*Multistage sampling* is simply a variation of cluster sampling in which only a sample of the second-stage units is selected. With regard to the sample of store employees discussed above, this would mean that for each store taken into the sample, a *sample* of the employees of that store would be selected. This is often an advantageous design, since there is likely to be more variation among clusters than within clusters, and multistage sampling allows more clusters to be taken into the sample.

It should be understood that *choosing the best sample design does not assure the validity of the sample results.* Errors can come from many sources—the way a question is worded, the way an interviewer asks the question, or the manner in which the population is defined. Some years ago a large oil company spent a considerable sum of money on a survey to determine brand preferences and buying motives for motor oil. The data that were collected turned out to be quite useless, simply because the company forgot to ask the people who were interviewed if they owned an automobile. The expense incurred in designing the sample could not remedy this defect.

## MULTIPLE REGRESSION

**Most economists think of God as working great multiple regressions in the sky.**
**— E. R. Fiedler**

In Chapter 14, the concepts of regression and correlation were discussed, but this discussion was limited to a consideration of simple regression and correlation. Many students may wonder about this terminology, since they see nothing simple about it, but in this case the word "simple" is part of the statistical jargon meaning that only two variables were used in the analysis. Sometimes the amount of correlation in a two-variable analysis will not be sufficient for reliable estimates to be made of the dependent variable. In such cases it may be that the *addition* of one or more independent variables will explain enough *additional variation* to make reliable estimates possible. This is a multiple regression. Thus, *in a multiple regression there is still only one dependent variable, but there are two or more independent variables.*

*Multiple regressions have been used extensively to build forecasting models.* Such models have been built to forecast gross national product, automobile sales, gasoline sales, and demand for park and recreational facilities. Some other examples of multiple regression and correlation applications are correlating crop yield with rainfall and temperature, correlating production levels with the number of workers and number of hours of overtime, and correlating the yield of a chemical process with the time and temperature of the process.

There are *two problems that might arise to limit the use of multiple regressions.* The *first* is the fact that two of the independent variables might themselves be correlated. For example, in trying to forecast gasoline sales one might correlate gasoline sales with the number of automobiles in use and consumer income, but it is obvious that as consumer income increases, the number of automobiles in use may also increase. The *second* problem arises when we correlate time series. In this case our basic assumption of independence of the observations is violated. For example, if our dependent variable is new-automobile sales, it is obvious that sales in any one year are not independent of sales in other years. New-car sales in 1982 are affected by sales in 1981 and affect sales in 1983. The presence of either of these problems makes it impossible to give an accurate interpretation to the correlation measures computed in the analysis.

## DECISION THEORY

In Chapter 8 procedures were explained for testing statistical hypotheses. As a result of these procedures, decisions were reached to either accept or reject the stated hypothesis. Modern decision theory also takes into consideration the *monetary values of the actions which might be taken* and formulates a procedure for making the best decision under conditions of uncertainty.

To illustrate this procedure, let's consider the following simple decision problem: The manager of a small grocery store must decide how many loaves of bread to stock each day. On the basis of past experience, she knows that the store has never sold less than 11 nor more than 14 loaves per day. *The procedure for analyzing this decision problem is as follows:*

1 *Construct a payoff table.* Since sales always range from 11 to 14, the alternative courses of action available to the store manager are to stock 11, 12, 13, or 14 loaves. A payoff, or profit, table can be constructed which will show the amount of profit that would be made for each alternative course of action at each possible level of demand.

2 *Compute the expected monetary value of each action.* On the basis of past sales records, a probability distribution for demand can be developed showing the probabilities of selling 11, 12, 13, and 14 loaves in a day. The expected monetary value of each action can be computed by applying these probabilities to the payoff table. The expected monetary value of an action is the amount of average profit one would expect to make if this action were followed day in and day out over a long period of time.

3 *Choose the action that has the greatest expected monetary value.* An important aspect of modern decision theory is the use of additional information to *revise* the original probability distribution. For example, a manager trying to decide whether or not to market a new product can assess the probabilities of success or failure of the product on the basis of experience with similar products in the past, but he would probably like to have some more information before making a decision. Such information could be obtained by taking a survey of department store buyers or consumers. This new information could then be incorporated into the analysis through the use of a particular probability theorem formulated by Thomas Bayes to revise the original probabilities. Because of the extensive use of Bayes' theorem, modern decision theory is commonly referred to as *Bayesian decision theory.*

It is obvious that *a decision analysis is no better than the probability distribution used for computing the expected monetary values.* The decision maker should remember that these probabilities are based on what has happened in the past and may not reflect the current situation. There are many external factors that can cause the relevant probability to change. The invention of nylon changed the probability distribution for the demand for silk. Improvements in the quality of a raw material will change the probability distribution for the percentage of defective products that will be produced. The use of out-of-date probabilities would obviously lead to a bad decision.

# LINEAR PROGRAMMING

*Linear programming may be defined as a technique for maximizing or minimizing a linear function subject to certain linear constraints.* While this definition might sound like a lot of jargonese double-talk, it is really not as formidable as it sounds. We are generally trying to maximize profits or minimize costs, and when we speak of constraints, we simply mean that there are limits on the values the variables can take. For example, in a profit maximization problem, there is a limit to the amount of each product we can produce. *To illustrate the uses of linear programming, let's consider the following four examples:*

1  A firm produces three products which have different profits per unit. Given information on the production time required by each product in each department and the total productive capacity of each department, linear programming can be used to determine the product mix that will maximize profit.

2  A company must blend together three ingredients to make a cattle feed which must meet certain specifications as to protein and vitamin content. Given the cost of each of the ingredients and information on the protein and vitamin content of each ingredient, linear programming can be used to determine the minimum-cost mixture which will meet the required nutritional specifications.

3  A company has three factories and six warehouses in different parts of the country. Linear programming can be used to determine which factories should ship to which warehouses in order to minimize the total transportation cost.

4  A factory has five orders to fill. The profitability of each order will vary depending on which of the available machines is used to do the work. Linear programming can be used to assign the jobs to machines in a way that will maximize the total profit.

The above examples are meant to be simply illustrative of the uses of linear programming; they do not constitute a complete and exhaustive listing of all possible applications. All these illustrations deal with business problems, and even so, we have not come close to listing all the more common business applications. Outside the business area, linear programming has been used in dealing with health problems, pollution problems, and problems of welfare economics. In short, *linear programming is an allocation model, and as such, it should be considered any time there is a problem concerning the allocation of scarce resources.*

Since linear programming allows us to determine the combination of factors that will give us the maximum profit or the minimum cost, it is obviously a technique that greatly aids the decision-making process. But like all quantitative techniques, it is useful *only* if it is used in the proper circumstances. The *important limitation on the use of linear programming is* implied in the definition stated at the outset of our discussion; that is, *the functions must be linear.* If the cost (or profit) function is nonlinear and we try to use linear programming to minimize (or maximize) it, the results will be meaningless.

# INVENTORY MODELS

Businesspeople must make many critical decisions, but perhaps no decisions are more critical or more difficult to make than those concerning inventories. In this regard, managers can commit two costly mistakes—they can have too much inventory on hand, or they can have too little inventory on hand. If they are *overstocked,* they incur the extra cost of storing the unneeded inventory, and they lose money on left-over items they cannot sell. If they are *understocked,* they lose the profit they could have made on sales that were lost.

Mathematical models can be used to aid in solving these and other problems relating to inventory control. These inventory models are simply equations, sometimes coupled with relevant probability distributions, which describe the particular inventory system being studied.

*One important use* of inventory models is the determination of the economic order quantity. This is the quantity that should be ordered each time inventory is replenished in order to minimize the total of ordering costs and inventory carrying costs. Each time an order is placed, certain administrative costs are incurred — e.g., the clerical cost of preparing the purchase order, the bookkeeping costs, the cost of checking in the order and verifying the invoice, and the cost of writing the check to pay for the order. Inventory carrying costs include the cost of storage and ware-housing, insurance on the inventory, and interest on the investment in the inventory. Obviously, the ordering cost could be minimized by ordering the entire quantity needed at one time, but this would mean carrying a very large inventory at a very high carrying cost. Carrying costs could be kept to a minimum by ordering small quantities each time, but this would require a large number of orders and, therefore, a very high ordering cost. A model that describes the total cost of the inventory system allows us to determine the order quantity that will minimize this total cost.

Inventory models are also used to determine the inventory level at which new orders should be placed. If this reorder point is placed too high, the average level of inventory and the resulting carrying costs will be unnecessarily large. On the other hand, if the reorder point is set too low, the inventory might be depleted before the new order arrives, and the firm would lose sales and customer goodwill. An inventory model that incorporates a probability distribution for inventory usage makes it possible to determine the reorder point that minimizes the total of expected shortage costs and carrying costs.

If an inventory model is to yield useful results, it must accurately describe the inventory system *under study*. Before using a model developed for another system, it is essential that the two systems be compared to make sure they are identical. Many textbooks on management quote what they refer to as the economic lot size formula. In reality, there is no such thing as *the* economic lot size formula. This is really *an* economic lot size formula for an inventory system that meets the following conditions: (1) The demand for the product is known, (2) the demand is constant, and (3) the cost of running short is so great that shortages can *never* be allowed to occur. If this formula is used in a situation in which one or more of these assumptions are not valid, the result will most certainly not be a quantity that minimizes costs.

## WAITING LINE THEORY

If you have tried to fight your way out of a crowded supermarket or tried to make a bank deposit on Friday afternoon, you do not have to be told what a waiting line is. You might be surprised, though, to know that there is a body of theory available for dealing with such problems. This body of theory makes it possible to answer such questions as (1) What is the average length of the waiting line? (2) What is the average time a person will have to wait for service? (3) What is the probability that there will be more than a certain number of people waiting in line? (4) What is the probability that a person will have to wait in line more than a certain amount of time?

*Waiting line models have been used to solve a wide variety of practical problems.* They have been used to make determinations as to the number of check-out counters to keep open in a supermarket, the number of teller's windows to keep open in a bank, the number of maintenance personnel to employ in a factory, the number of tollgates to keep open on a turnpike, or the number of nurses required at a hospital station. As in the case of all other quantitative techniques, *these methods are useful only when applied in situations in which they fit the problem under consideration.* A manager of engineering in an aerospace firm related the following story, which illustrates this point.

It seems that this manager always had a long line of engineers waiting at the blueprint room to check out materials. Since these men were in a fairly high salary bracket, the time they spent waiting in line cost the company a considerable sum of money. A research team was called in to do a waiting line analysis and determine the best number of clerks to employ in the blueprint room. This team spent several days collecting data on the number of engineers arriving at the blueprint room and the average time required for a clerk to fill a request. They then went back to their offices to feed all these numbers to their computers. Late one afternoon the personnel manager happened to walk by the blueprint room and saw the situation. The next morning he replaced the two women who had been working there with two men, and there has not been a waiting line since. So, it seems that in spite of all the numbers generated by the research team, two important figures were left out of the model.

## SIMULATION

In the physical sciences, experiments may be performed in a laboratory using small models of a process or an operation. Many complex variations may be possible in these tests, and the results show the scientist what happens under certain controlled conditions. Simulation is similar to scientific experimentation. Perhaps Fig. 16-1 will clarify the meaning of simulation. At its base, Fig. 16-1 rests on reality or fact. In complex situations, few people (if any) fully understand all aspects of the situation; therefore, theories are developed which may focus attention on only part of the complex whole. In some situations models may be built or conceived in order to test or represent a theory. Finally, simulation is the use of a model in the attempt to identify and/or reflect the behavior of a real person, process, or system.

FIGURE 16-1

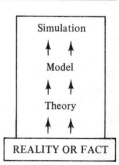

In organizations, administrators may evaluate proposed projects or strategies by constructing theoretical models. They can then determine what happens to these models under certain conditions or when certain assumptions are tested. Simulation is thus a trial-and-error problem-solving approach; it is also a planning aid that may be of considerable value to organizations.

Simulation models have helped top business executives decide, for example, whether or not to expand operations by acquiring a new plant. Among the dozens of complicating variables that would have to be incorporated into such models are facts and assumptions about (1) the present and potential size of the total market, (2) the present and potential company share of this total market, (3) product selling prices, and (4) the investment required to achieve various production levels. Thus, simulation has helped top executives in their strategic planning and decision-making activities.

Simulation has also been useful in (1) establishing design parameters for aircraft and space vehicles, (2) refining medical treatment techniques, (3) teaching students, (4) planning urban improvements, and (5) planning transportation systems where variables such as expected road usage, effects of highway construction on traffic load, and use of one-way streets are considered.

## SUMMARY

In this chapter we have briefly surveyed several quantitative methods of analysis. No attempt has been made to explain the methodology for using these techniques, but illustrations of their use have been given to show the types of problems that can be solved. Much attention has also been given to the limitations of these analyses. It is important to understand that a quantitative analysis will not make a decision for you. It simply provides more and better information on which to base a decision.

## Important Terms and Concepts

1 Stratified sample
2 Cluster sample
3 Multistage sampling
4 Multiple regression
5 Bayesian decision theory

6 Linear programming
7 Inventory models
8 Waiting line models
9 Simulation

## Topics for Discussion

1 **a** How may a stratified sample be used?
  **b** A cluster sample?

2 **a** How may multiple regression be used?
  **b** What factors limit the use of multiple regression?

3 What are the steps in decision theory?

4 How may linear programming be used?

5 **a** What are inventory models?
  **b** How may inventory models be used?

6 **a** What is simulation?
  **b** How may simulation models be used?

APPENDIXES

# APPENDIX 1
## SELECTED VALUES OF THE BINOMIAL PROBABILITY DISTRIBUTION

$$P(r) = {_n}C_r\,(p)^r(q)^{n-r}$$

*Example:* If $p = .30$, $n = 5$, and $r = 2$, then $P(r) = .3087$. (When $p$ is greater than .50, the value of $P(r)$ is found by locating the table for the specified $n$ and using $n - r$ in place of the given $r$ and $1 - p$ in place of the specified $p$.)

| n | r | .01 | .05 | .10 | .15 | .20 | .25 | .30 | .35 | .40 | .45 | .50 |
|---|---|-----|-----|-----|-----|-----|-----|-----|-----|-----|-----|-----|
| 1 | 0 | .9900 | .9500 | .9000 | .8500 | .8000 | .7500 | .7000 | .6500 | .6000 | .5500 | .5000 |
|   | 1 | .0100 | .0500 | .1000 | .1500 | .2000 | .2500 | .3000 | .3500 | .4000 | .4500 | .5000 |
| 2 | 0 | .9801 | .9025 | .8100 | .7225 | .6400 | .5625 | .4900 | .4225 | .3600 | .3025 | .2500 |
|   | 1 | .0198 | .0950 | .1800 | .2550 | .3200 | .3750 | .4200 | .4550 | .4800 | .4950 | .5000 |
|   | 2 | .0001 | .0025 | .0100 | .0225 | .0400 | .0625 | .0900 | .1225 | .1600 | .2025 | .2500 |
| 3 | 0 | .9703 | .8574 | .7290 | .6141 | .5120 | .4219 | .3430 | .2746 | .2160 | .1664 | .1250 |
|   | 1 | .0294 | .1354 | .2430 | .3251 | .3840 | .4219 | .4410 | .4436 | .4320 | .4084 | .3750 |
|   | 2 | .0003 | .0071 | .0270 | .0574 | .0960 | .1406 | .1890 | .2389 | .2880 | .3341 | .3750 |
|   | 3 | .0000 | .0001 | .0010 | .0034 | .0080 | .0156 | .0270 | .0429 | .0640 | .0911 | .1250 |
| 4 | 0 | .9606 | .8145 | .6561 | .5220 | .4096 | .3164 | .2401 | .1785 | .1296 | .0915 | .0625 |
|   | 1 | .0388 | .1715 | .2916 | .3685 | .4096 | .4219 | .4116 | .3845 | .3456 | .2995 | .2500 |
|   | 2 | .0006 | .0135 | .0486 | .0975 | .1536 | .2109 | .2646 | .3105 | .3456 | .3675 | .3750 |
|   | 3 | .0000 | .0005 | .0036 | .0115 | .0258 | .0469 | .0756 | .1115 | .1536 | .2005 | .2500 |
|   | 4 | .0000 | .0000 | .0001 | .0005 | .0016 | .0039 | .0081 | .0150 | .0256 | .0410 | .0625 |
| 5 | 0 | .9510 | .7738 | .5905 | .4437 | .3277 | .2373 | .1681 | .1160 | .0778 | .0503 | .0312 |
|   | 1 | .0480 | .2036 | .3280 | .3915 | .4096 | .3955 | .3602 | .3124 | .2592 | .2059 | .1562 |
|   | 2 | .0010 | .0214 | .0729 | .1382 | .2048 | .2637 | .3087 | .3364 | .3456 | .3369 | .3125 |
|   | 3 | .0000 | .0011 | .0081 | .0244 | .0512 | .0879 | .1323 | .1811 | .2304 | .2757 | .3125 |
|   | 4 | .0000 | .0000 | .0004 | .0022 | .0064 | .0146 | .0284 | .0488 | .0768 | .1128 | .1562 |
|   | 5 | .0000 | .0000 | .0000 | .0001 | .0003 | .0010 | .0024 | .0053 | .0102 | .0185 | .0312 |
| 6 | 0 | .9415 | .7351 | .5314 | .3771 | .2621 | .1780 | .1176 | .0754 | .0467 | .0277 | .0156 |
|   | 1 | .0571 | .2321 | .3543 | .3993 | .3932 | .3560 | .3025 | .2437 | .1866 | .1359 | .0938 |
|   | 2 | .0014 | .0305 | .0984 | .1762 | .2458 | .2966 | .3241 | .3280 | .3110 | .2780 | .2344 |
|   | 3 | .0000 | .0021 | .0146 | .0415 | .0819 | .1318 | .1852 | .2355 | .2765 | .3032 | .3125 |
|   | 4 | .0000 | .0001 | .0012 | .0055 | .0154 | .0330 | .0595 | .0951 | .1382 | .1861 | .2344 |
|   | 5 | .0000 | .0000 | .0001 | .0004 | .0015 | .0044 | .0102 | .0205 | .0369 | .0609 | .0938 |
|   | 6 | .0000 | .0000 | .0000 | .0000 | .0001 | .0002 | .0007 | .0018 | .0041 | .0083 | .0156 |
| 7 | 0 | .9321 | .6983 | .4783 | .3206 | .2097 | .1335 | .0824 | .0490 | .0280 | .0152 | .0078 |
|   | 1 | .0659 | .2573 | .3720 | .3960 | .3670 | .3115 | .2471 | .1848 | .1306 | .0872 | .0547 |
|   | 2 | .0020 | .0406 | .1240 | .2097 | .2753 | .3115 | .3177 | .2985 | .2613 | .2140 | .1641 |
|   | 3 | .0000 | .0036 | .0230 | .0617 | .1147 | .1730 | .2269 | .2679 | .2903 | .2918 | .2734 |
|   | 4 | .0000 | .0002 | .0026 | .0109 | .0287 | .0577 | .0972 | .1442 | .1935 | .2388 | .2734 |
|   | 5 | .0000 | .0000 | .0002 | .0012 | .0043 | .0115 | .0250 | .0466 | .0774 | .1172 | .1641 |
|   | 6 | .0000 | .0000 | .0000 | .0001 | .0004 | .0013 | .0036 | .0084 | .0172 | .0320 | .0547 |
|   | 7 | .0000 | .0000 | .0000 | .0000 | .0000 | .0001 | .0002 | .0006 | .0016 | .0037 | .0078 |

SOURCE: Adapted from Leonard J. Kazmier, *Statistical Analysis for Business and Economics*, 2d ed., copyright © 1973 by McGraw-Hill, Inc. Used by permission of McGraw-Hill Book Company.

| $n$ | $r$ | .01 | .05 | .10 | .15 | .20 | .25 | $p$ .30 | .35 | .40 | .45 | .50 |
|---|---|---|---|---|---|---|---|---|---|---|---|---|
| 8 | 0 | .9227 | .6634 | .4305 | .2725 | .1678 | .1002 | .0576 | .0319 | .0168 | .0084 | .0039 |
| | 1 | .0746 | .2793 | .3826 | .3847 | .3355 | .2670 | .1977 | .1373 | .0896 | .0548 | .0312 |
| | 2 | .0026 | .0515 | .1488 | .2376 | .2936 | .3115 | .2065 | .2587 | .2090 | .1569 | .1094 |
| | 3 | .0001 | .0054 | .0331 | .0839 | .1468 | .2076 | .2541 | .2786 | .2787 | .2568 | .2188 |
| | 4 | .0000 | .0004 | .0046 | .0185 | .0459 | .0865 | .1361 | .1875 | .2322 | .2627 | .2734 |
| | 5 | .0000 | .0000 | .0004 | .0026 | .0092 | .0231 | .0467 | .0808 | .1239 | .1719 | .2188 |
| | 6 | .0000 | .0000 | .0000 | .0002 | .0011 | .0038 | .0100 | .0217 | .0413 | .0403 | .1094 |
| | 7 | .0000 | .0000 | .0000 | .0000 | .0001 | .0004 | .0012 | .0033 | .0079 | .0164 | .0312 |
| | 8 | .0000 | .0000 | .0000 | .0000 | .0000 | .0000 | .0001 | .0002 | .0007 | .0017 | .0039 |
| 9 | 0 | .9135 | .6302 | .3874 | .2316 | .1342 | .0751 | .0404 | .0207 | .0101 | .0046 | .0020 |
| | 1 | .0830 | .2985 | .3874 | .3679 | .3020 | .2253 | .1556 | .1004 | .0605 | .0339 | .0176 |
| | 2 | .0034 | .0629 | .1722 | .2597 | .3020 | .3003 | .2668 | .2162 | .1612 | .1110 | .0703 |
| | 3 | .0001 | .0077 | .0446 | .1069 | .1762 | .2336 | .2668 | .2716 | .2508 | .2119 | .1641 |
| | 4 | .0000 | .0006 | .0074 | .0283 | .0661 | .1168 | .1715 | .2194 | .2508 | .2600 | .2461 |
| | 5 | .0000 | .0000 | .0008 | .0050 | .0165 | .0389 | .0735 | .1181 | .1672 | .2128 | .2461 |
| | 6 | .0000 | .0000 | .0001 | .0006 | .0028 | .0087 | .0210 | .0424 | .0743 | .1160 | .1641 |
| | 7 | .0000 | .0000 | .0000 | .0000 | .0003 | .0012 | .0039 | .0098 | .0212 | .0407 | .0703 |
| | 8 | .0000 | .0000 | .0000 | .0000 | .0000 | .0001 | .0004 | .0013 | .0035 | .0083 | .0176 |
| | 9 | .0000 | .0000 | .0000 | .0000 | .0000 | .0000 | .0000 | .0001 | .0003 | .0008 | .0020 |
| 10 | 0 | .9044 | .5987 | .3487 | .1969 | .1074 | .0563 | .0282 | .0135 | .0060 | .0025 | .0010 |
| | 1 | .0914 | .3151 | .3874 | .3474 | .2684 | .1877 | .1211 | .0725 | .0403 | .0207 | .0098 |
| | 2 | .0042 | .0746 | .1937 | .2759 | .3020 | .2816 | .2335 | .1757 | .1209 | .0763 | .0439 |
| | 3 | .0001 | .0105 | .0574 | .1298 | .2013 | .2503 | .2668 | .2522 | .2150 | .1665 | .1172 |
| | 4 | .0000 | .0010 | .0112 | .0401 | .0881 | .1460 | .2001 | .2377 | .2508 | .2384 | .2051 |
| | 5 | .0000 | .0001 | .0015 | .0085 | .0264 | .0584 | .1029 | .1536 | .2007 | .2340 | .2461 |
| | 6 | .0000 | .0000 | .0001 | .0012 | .0055 | .0162 | .0368 | .0689 | .1115 | .1596 | .2051 |
| | 7 | .0000 | .0000 | .0000 | .0001 | .0008 | .0031 | .0090 | .0212 | .0425 | .0746 | .1172 |
| | 8 | .0000 | .0000 | .0000 | .0000 | .0001 | .0004 | .0014 | .0043 | .0106 | .0229 | .0439 |
| | 9 | .0000 | .0000 | .0000 | .0000 | .0000 | .0000 | .0001 | .0005 | .0016 | .0042 | .0098 |
| | 10 | .0000 | .0000 | .0000 | .0000 | .0000 | .0000 | .0000 | .0000 | .0001 | .0003 | .0010 |
| 11 | 0 | .8953 | .5688 | .3138 | .1673 | .0859 | .0422 | .0198 | .0088 | .0036 | .0014 | .0005 |
| | 1 | .0995 | .3293 | .3835 | .3248 | .2362 | .1549 | .0932 | .0518 | .0266 | .0125 | .0054 |
| | 2 | .0050 | .0867 | .2131 | .2866 | .2953 | .2581 | .1998 | .1395 | .0887 | .0513 | .0269 |
| | 3 | .0002 | .0137 | .0710 | .1517 | .2215 | .2581 | .2568 | .2254 | .1774 | .1259 | .0806 |
| | 4 | .0000 | .0010 | .0112 | .0401 | .0881 | .1460 | .2001 | .2377 | .2508 | .2384 | .2051 |
| | 5 | .0000 | .0001 | .0025 | .0132 | .0388 | .0803 | .1321 | .1830 | .2207 | .2360 | .2256 |
| | 6 | .0000 | .0000 | .0003 | .0023 | .0097 | .0268 | .0566 | .0985 | .1471 | .1931 | .2256 |
| | 7 | .0000 | .0000 | .0000 | .0003 | .0017 | .0064 | .0173 | .0379 | .0701 | .1128 | .1611 |
| | 8 | .0000 | .0000 | .0000 | .0000 | .0002 | .0011 | .0037 | .0102 | .0234 | .0462 | .0806 |
| | 9 | .0000 | .0000 | .0000 | .0000 | .0000 | .0001 | .0005 | .0018 | .0052 | .0126 | .0269 |
| | 10 | .0000 | .0000 | .0000 | .0000 | .0000 | .0000 | .0000 | .0002 | .0007 | .0021 | .0054 |
| | 11 | .0000 | .0000 | .0000 | .0000 | .0000 | .0000 | .0000 | .0000 | .0000 | .0002 | .0005 |

| n | r | .01 | .05 | .10 | .15 | .20 | .25 | p .30 | .35 | .40 | .45 | .50 |
|---|---|-----|-----|-----|-----|-----|-----|-----|-----|-----|-----|-----|
| 12 | 0 | .8864 | .5404 | .2824 | .1422 | .0687 | .0317 | .0138 | .0057 | .0022 | .0008 | .0002 |
| | 1 | .1074 | .3413 | .3766 | .3012 | .2062 | .1267 | .0712 | .0368 | .0174 | .0075 | .0029 |
| | 2 | .0060 | .0988 | .2301 | .2924 | .2835 | .2323 | .1678 | .1088 | .0639 | .0339 | .0161 |
| | 3 | .0002 | .0173 | .0852 | .1720 | .2362 | .2581 | .2397 | .1954 | .1419 | .0923 | .0537 |
| | 4 | 0000 | .0021 | .0213 | .0683 | .1329 | .1936 | .2311 | .2367 | .2128 | .1700 | .1204 |
| | 5 | .0000 | .0002 | .0038 | .0193 | .0532 | .1032 | .1585 | .2039 | .2270 | .2225 | .1934 |
| | 6 | .0000 | .0000 | .0005 | .0040 | .0155 | .0401 | .0792 | .1281 | .1766 | .2124 | .2256 |
| | 7 | .0000 | .0000 | .0000 | .0006 | .0033 | .0115 | .0291 | .0591 | .1009 | .1489 | .1934 |
| | 8 | .0000 | .0000 | .0000 | .0001 | .0005 | .0024 | .0078 | .0199 | .0420 | .0762 | .1208 |
| | 9 | .0000 | .0000 | .0000 | .0000 | .0001 | .0004 | .0015 | .0048 | .0125 | .0277 | .0537 |
| | 10 | .0000 | .0000 | .0000 | .0000 | .0000 | .0000 | .0002 | .0008 | .0025 | .0068 | .0161 |
| | 11 | .0000 | .0000 | .0000 | .0000 | .0000 | .0000 | .0000 | .0001 | .0003 | .0010 | .0029 |
| | 12 | .0000 | .0000 | .0000 | .0000 | .0000 | .0000 | .0000 | .0000 | .0000 | .0001 | .0002 |
| 13 | 0 | .8775 | .5133 | .2542 | .1209 | .0550 | .0238 | .0097 | .0037 | .0013 | .0004 | .0001 |
| | 1 | .1152 | .3512 | .3672 | .2774 | .1787 | .1029 | .0540 | .0259 | .0113 | .0045 | .0016 |
| | 2 | .0070 | .1109 | .2448 | .2937 | .2680 | .2059 | .1388 | .0836 | .0453 | .0220 | .0095 |
| | 3 | .0003 | .0214 | .0997 | .1900 | .2457 | .2517 | .2181 | .1651 | .1107 | .0660 | .0349 |
| | 4 | .0000 | .0028 | .0277 | .0838 | .1535 | .2097 | .2337 | .2222 | .1845 | .1350 | .0873 |
| | 5 | .0000 | .0003 | .0055 | .0266 | .0691 | .1258 | .1803 | .2154 | .2214 | .1989 | .1571 |
| | 6 | .0000 | .0000 | .0008 | .0063 | .0230 | .0559 | .1030 | .1546 | .1968 | .2169 | .2095 |
| | 7 | .0000 | .0000 | .0001 | .0011 | .0058 | .0186 | .0442 | .0833 | .1312 | .1775 | .2095 |
| | 8 | .0000 | .0000 | .0001 | .0001 | .0011 | .0047 | .0142 | .0336 | .0656 | .1089 | .1571 |
| | 9 | .0000 | .0000 | .0000 | .0000 | .0001 | .0009 | .0034 | .0101 | .0243 | .0495 | .0873 |
| | 10 | .0000 | .0000 | .0000 | .0000 | .0000 | .0001 | .0006 | .0022 | .0065 | .0162 | .0349 |
| | 11 | .0000 | .0000 | .0000 | .0000 | .0000 | .0000 | .0001 | .0003 | .0012 | .0036 | .0095 |
| | 12 | .0000 | .0000 | .0000 | .0000 | .0000 | .0000 | .0000 | .0000 | .0001 | .0005 | .0016 |
| | 13 | .0000 | .0000 | .0000 | .0000 | .0000 | .0000 | .0000 | .0000 | .0000 | .0000 | .0001 |
| 14 | 0 | .8687 | .4877 | .2288 | .1028 | .0440 | .0178 | .0068 | .0024 | .0008 | .0002 | .0001 |
| | 1 | .1229 | .3593 | .3559 | .2539 | .1539 | .0832 | .0407 | .0181 | .0073 | .0027 | .0009 |
| | 2 | .0081 | .1229 | .2570 | .2912 | .2501 | .1802 | .1134 | .0634 | .0317 | .0141 | .0056 |
| | 3 | .0003 | .0259 | .1142 | .2056 | .2501 | .2402 | .1943 | .1366 | .0845 | .0462 | .0222 |
| | 4 | .0000 | .0037 | .0349 | .0998 | .1720 | .2202 | .2290 | .2022 | .1549 | .1040 | .0611 |
| | 5 | .0000 | .0004 | .0078 | .0352 | .0860 | .1468 | .1963 | .2178 | .2066 | .1701 | .1222 |
| | 6 | .0000 | .0000 | .0013 | .0093 | .0322 | .0734 | .1262 | .1759 | .2066 | .2088 | .1833 |
| | 7 | .0000 | .0000 | .0002 | .0019 | .0092 | .0280 | .0618 | .1082 | .1574 | .1952 | .2095 |
| | 8 | .0000 | .0000 | .0000 | .0003 | .0020 | .0082 | .0232 | .0510 | .0918 | .1398 | .1833 |
| | 9 | .0000 | .0000 | .0000 | .0000 | .0003 | .0018 | .0066 | .0183 | .0408 | .0762 | .1222 |
| | 10 | .0000 | .0000 | .0000 | .0000 | .0000 | .0003 | .0014 | .0049 | .0136 | .0312 | .0611 |
| | 11 | .0000 | .0000 | .0000 | .0000 | .0000 | .0000 | .0002 | .0010 | .0033 | .0093 | .0222 |
| | 12 | .0000 | .0000 | .0000 | .0000 | .0000 | .0000 | .0000 | .0001 | .0005 | .0019 | .0056 |
| | 13 | .0000 | .0000 | .0000 | .0000 | .0000 | .0000 | .0000 | .0000 | .0001 | .0002 | .0009 |
| | 14 | .0000 | .0000 | .0000 | .0000 | .0000 | .0000 | .0000 | .0000 | .0000 | .0000 | .0001 |

APPENDIX 1/SELECTED VALUES OF THE BINOMIAL PROBABILITY DISTRIBUTION

| n | r | .01 | .05 | .10 | .15 | .20 | .25 | .30 | .35 | .40 | .45 | .50 |
|---|---|-----|-----|-----|-----|-----|-----|-----|-----|-----|-----|-----|
| 15 | 0 | .8601 | .4633 | .2059 | .0874 | .0352 | .0134 | .0047 | .0016 | .0005 | .0001 | .0000 |
|    | 1 | .1303 | .3658 | .3432 | .2312 | .1319 | .0668 | .0305 | .0126 | .0047 | .0016 | .0005 |
|    | 2 | .0092 | .1348 | .2669 | .2856 | .2309 | .1559 | .0916 | .0476 | .0219 | .0090 | .0032 |
|    | 3 | .0004 | .0307 | .1285 | .2184 | .2501 | .2252 | .1700 | .1110 | .0634 | .0318 | .0139 |
|    | 4 | .0000 | .0049 | .0428 | .1156 | .1876 | .2252 | .2186 | .1792 | .1268 | .0780 | .0417 |
|    | 5 | .0000 | .0006 | .0105 | .0499 | .1032 | .1651 | .2061 | .2123 | .1859 | .1404 | .0916 |
|    | 6 | .0000 | .0000 | .0019 | .0132 | .0430 | .0917 | .1472 | .1906 | .2066 | .1914 | .1527 |
|    | 7 | .0000 | .0000 | .0003 | .0030 | .0138 | .0393 | .0811 | .1319 | .1771 | .2013 | .1964 |
|    | 8 | .0000 | .0000 | .0000 | .0005 | .0035 | .0131 | .0348 | .0710 | .1181 | .1647 | .1964 |
|    | 9 | .0000 | .0000 | .0000 | .0001 | .0007 | .0034 | .0116 | .0298 | .0612 | .1048 | .1527 |
|    | 10 | .0000 | .0000 | .0000 | .0000 | .0001 | .0007 | .0030 | .0096 | .0245 | .0515 | .0916 |
|    | 11 | .0000 | .0000 | .0000 | .0000 | .0000 | .0001 | .0006 | .0024 | .0074 | .0191 | .0417 |
|    | 12 | .0000 | .0000 | .0000 | .0000 | .0000 | .0000 | .0001 | .0004 | .0016 | .0052 | .0139 |
|    | 13 | .0000 | .0000 | .0000 | .0000 | .0000 | .0000 | .0000 | .0001 | .0003 | .0010 | .0032 |
|    | 14 | .0000 | .0000 | .0000 | .0000 | .0000 | .0000 | .0000 | .0000 | .0000 | .0001 | .0005 |
|    | 15 | .0000 | .0000 | .0000 | .0000 | .0000 | .0000 | .0000 | .0000 | .0000 | .0000 | .0000 |
| 16 | 0 | .8515 | .4401 | .1853 | .0743 | .0281 | .0100 | .0033 | .0010 | .0003 | .0001 | .0000 |
|    | 1 | .1376 | .3706 | .3294 | .2097 | .1126 | .0535 | .0228 | .0087 | .0030 | .0009 | .0002 |
|    | 2 | .0104 | .1463 | .2745 | .2775 | .2111 | .1336 | .0732 | .0353 | .0150 | .0056 | .0018 |
|    | 3 | .0005 | .0359 | .1423 | .2285 | .2463 | .2079 | .1465 | .0888 | .0468 | .0215 | .0085 |
|    | 4 | .0000 | .0061 | .0514 | .1311 | .2001 | .2252 | .2040 | .1553 | .1014 | .0572 | .0278 |
|    | 5 | .0000 | .0008 | .0137 | .0555 | .1201 | .1802 | .2099 | .2008 | .1623 | .1123 | .0667 |
|    | 6 | .0000 | .0001 | .0028 | .0180 | .0550 | .1101 | .1649 | .1982 | .1983 | .1684 | .1222 |
|    | 7 | .0000 | .0000 | .0004 | .0045 | .0197 | .0524 | .1010 | .1524 | .1889 | .1969 | .1746 |
|    | 8 | .0000 | .0000 | .0001 | .0009 | .0055 | .0197 | .0487 | .0923 | .1417 | .1812 | .1964 |
|    | 9 | .0000 | .0000 | .0000 | .0001 | .0012 | .0058 | .0185 | .0442 | .0840 | .1318 | .1746 |
|    | 10 | .0000 | .0000 | .0000 | .0000 | .0002 | .0014 | .0056 | .0167 | .0392 | .0755 | .1222 |
|    | 11 | .0000 | .0000 | .0000 | .0000 | .0000 | .0002 | .0013 | .0049 | .0142 | .0337 | .0667 |
|    | 12 | .0000 | .0000 | .0000 | .0000 | .0000 | .0000 | .0002 | .0011 | .0040 | .0115 | .0278 |
|    | 13 | .0000 | .0000 | .0000 | .0000 | .0000 | .0000 | .0000 | .0002 | .0008 | .0029 | .0085 |
|    | 14 | .0000 | .0000 | .0000 | .0000 | .0000 | .0000 | .0000 | .0000 | .0001 | .0005 | .0018 |
|    | 15 | .0000 | .0000 | .0000 | .0000 | .0000 | .0000 | .0000 | .0000 | .0000 | .0001 | .0002 |
|    | 16 | .0000 | .0000 | .0000 | .0000 | .0000 | .0000 | .0000 | .0000 | .0000 | .0000 | .0000 |
| 17 | 0 | .8429 | .4181 | .1668 | .0631 | .0225 | .0075 | .0023 | .0007 | .0002 | .0000 | .0000 |
|    | 1 | .1447 | .3741 | .3150 | .1893 | .0957 | .0426 | .0169 | .0060 | .0019 | .0005 | .0001 |
|    | 2 | .0117 | .1575 | .2800 | .2673 | .1914 | .1136 | .0581 | .0260 | .0102 | .0035 | .0010 |
|    | 3 | .0006 | .0415 | .1556 | .2359 | .2393 | .1893 | .1245 | .0701 | .0341 | .0144 | .0052 |
|    | 4 | .0000 | .0076 | .0605 | .1457 | .2093 | .2209 | .1868 | .1320 | .0796 | .0411 | .0182 |
|    | 5 | .0000 | .0010 | .0175 | .0668 | .1361 | .1914 | .2081 | .1849 | .1379 | .0875 | .0472 |
|    | 6 | .0000 | .0001 | .0039 | .0236 | .0680 | .1276 | .1784 | .1991 | .1839 | .1432 | .0944 |
|    | 7 | .0000 | .0000 | .0007 | .0065 | .0267 | .0668 | .1201 | .1685 | .1927 | .1841 | .1484 |
|    | 8 | .0000 | .0000 | .0001 | .0014 | .0084 | .0279 | .0644 | .1134 | .1606 | .1883 | .1855 |
|    | 9 | .0000 | .0000 | .0000 | .0003 | .0021 | .0093 | .0276 | .0611 | .1070 | .1540 | .1855 |

| n | r | .01 | .05 | .10 | .15 | .20 | .25 | .30 | .35 | .40 | .45 | .50 |
|---|---|------|------|------|------|------|------|------|------|------|------|------|
| 17 | 10 | .0000 | .0000 | .0000 | .0000 | .0004 | .0025 | .0095 | .0263 | .0571 | .1008 | .1484 |
|  | 11 | .0000 | .0000 | .0000 | .0000 | .0001 | .0005 | .0026 | .0090 | .0242 | .0525 | .0944 |
|  | 12 | .0000 | .0000 | .0000 | .0000 | .0000 | .0001 | .0006 | .0024 | .0081 | .0215 | .0472 |
|  | 13 | .0000 | .0000 | .0000 | .0000 | .0000 | .0000 | .0001 | .0005 | .0021 | .0068 | .0182 |
|  | 14 | .0000 | .0000 | .0000 | .0000 | .0000 | .0000 | .0000 | .0001 | .0004 | .0016 | .0052 |
|  | 15 | .0000 | .0000 | .0000 | .0000 | .0000 | .0000 | .0000 | .0000 | .0001 | .0003 | .0010 |
|  | 16 | .0000 | .0000 | .0000 | .0000 | .0000 | .0000 | .0000 | .0000 | .0000 | .0000 | .0001 |
|  | 17 | .0000 | .0000 | .0000 | .0000 | .0000 | .0000 | .0000 | .0000 | .0000 | .0000 | .0000 |
| 18 | 0 | .8345 | .3972 | .1501 | .0536 | .0180 | .0056 | .0016 | .0004 | .0001 | .0003 | .0010 |
|  | 1 | .1517 | .3763 | .3002 | .1704 | .0811 | .0338 | .0126 | .0042 | .0012 | .0003 | .0001 |
|  | 2 | .0130 | .1683 | .2835 | .2556 | .1723 | .0958 | .0458 | .0190 | .0069 | .0022 | .0006 |
|  | 3 | .0007 | .0473 | .1680 | .2406 | .2297 | .1704 | .1046 | .0547 | .0246 | .0095 | .0001 |
|  | 4 | .0000 | .0093 | .0700 | .1592 | .2153 | .2130 | .1681 | .1104 | .0614 | .0291 | .0117 |
|  | 5 | .0000 | .0014 | .0218 | .0787 | .1507 | .1988 | .2017 | .1664 | .1146 | .0666 | .0327 |
|  | 6 | .0000 | .0002 | .0052 | .0301 | .0816 | .1436 | .1873 | .1941 | .1655 | .1181 | .0708 |
|  | 7 | .0000 | .0000 | .0010 | .0091 | .0350 | .0820 | .1376 | .1792 | .1892 | .1657 | .1214 |
|  | 8 | .0000 | .0000 | .0002 | .0022 | .0120 | .0376 | .0811 | .1327 | .1734 | .1864 | .1669 |
|  | 9 | .0000 | .0000 | .0000 | .0004 | .0033 | .0139 | .0386 | .0794 | .1284 | .1694 | .1855 |
|  | 10 | .0000 | .0000 | .0000 | .0001 | .0008 | .0042 | .0149 | .0385 | .0771 | .1248 | .1669 |
|  | 11 | .0000 | .0000 | .0000 | .0000 | .0001 | .0010 | .0046 | .0151 | .0374 | .0742 | .1214 |
|  | 12 | .0000 | .0000 | .0000 | .0000 | .0000 | .0002 | .0012 | .0047 | .0145 | .0354 | .0708 |
|  | 13 | .0000 | .0000 | .0000 | .0000 | .0000 | .0000 | .0002 | .0012 | .0045 | .0134 | .0327 |
|  | 14 | .0000 | .0000 | .0000 | .0000 | .0000 | .0000 | .0000 | .0002 | .0011 | .0039 | .0117 |
|  | 15 | .0000 | .0000 | .0000 | .0000 | .0000 | .0000 | .0000 | .0000 | .0002 | .0009 | .0031 |
|  | 16 | .0000 | .0000 | .0000 | .0000 | .0000 | .0000 | .0000 | .0000 | .0000 | .0001 | .0006 |
|  | 17 | .0000 | .0000 | .0000 | .0000 | .0000 | .0000 | .0000 | .0000 | .0000 | .0000 | .0001 |
|  | 18 | .0000 | .0000 | .0000 | .0000 | .0000 | .0000 | .0000 | .0000 | .0000 | .0000 | .0000 |
| 19 | 0 | .8262 | .3774 | .1351 | .0456 | .0144 | .0042 | .0011 | .0003 | .0001 | .0000 | .0000 |
|  | 1 | .1586 | .3774 | .2852 | .1529 | .0685 | .0268 | .0093 | .0029 | .0008 | .0002 | .0000 |
|  | 2 | .0144 | .1787 | .2852 | .2428 | .1540 | .0803 | .0358 | .0138 | .0046 | .0013 | .0003 |
|  | 3 | .0008 | .0533 | .1796 | .2428 | .2182 | .1517 | .0869 | .0422 | .0175 | .0062 | .0018 |
|  | 4 | .0000 | .0112 | .0798 | .1714 | .2182 | .2023 | .1491 | .0909 | .0467 | .0203 | .0074 |
|  | 5 | .0000 | .0018 | .0266 | .0907 | .1636 | .2023 | .1916 | .1468 | .0933 | .0497 | .0222 |
|  | 6 | .0000 | .0002 | .0069 | .0374 | .0955 | .1574 | .1916 | .1844 | .1451 | .0949 | .0518 |
|  | 7 | .0000 | .0000 | .0014 | .0122 | .0443 | .0974 | .1525 | .1844 | .1797 | .1443 | .0961 |
|  | 8 | .0000 | .0000 | .0002 | .0032 | .0166 | .0487 | .0981 | .1489 | .1797 | .1771 | .1442 |
|  | 9 | .0000 | .0000 | .0000 | .0007 | .0051 | .0198 | .0514 | .0980 | .1464 | .1771 | .1762 |
|  | 10 | .0000 | .0000 | .0000 | .0001 | .0013 | .0066 | .0220 | .0528 | .0976 | .1449 | .1762 |
|  | 11 | .0000 | .0000 | .0000 | .0000 | .0003 | .0018 | .0077 | .0233 | .0532 | .0970 | .1442 |
|  | 12 | .0000 | .0000 | .0000 | .0000 | .0000 | .0004 | .0022 | .0083 | .0237 | .0529 | .0961 |
|  | 13 | .0000 | .0000 | .0000 | .0000 | .0000 | .0001 | .0005 | .0024 | .0085 | .0233 | .0518 |
|  | 14 | .0000 | .0000 | .0000 | .0000 | .0000 | .0000 | .0001 | .0006 | .0024 | .0082 | .0222 |
|  | 15 | .0000 | .0000 | .0000 | .0000 | .0000 | .0000 | .0000 | .0001 | .0005 | .0022 | .0074 |
|  | 16 | .0000 | .0000 | .0000 | .0000 | .0000 | .0000 | .0000 | .0000 | .0001 | .0005 | .0018 |

APPENDIX 1/SELECTED VALUES OF THE BINOMIAL PROBABILITY DISTRIBUTION

| n | r | .01 | .05 | .10 | .15 | .20 | .25 | p .30 | .35 | .40 | .45 | .50 |
|---|---|-----|-----|-----|-----|-----|-----|-----|-----|-----|-----|-----|
| 19 | 17 | .0000 | .0000 | .0000 | .0000 | .0000 | .0000 | .0000 | .0000 | .0000 | .0001 | .0003 |
|  | 18 | .0000 | .0000 | .0000 | .0000 | .0000 | .0000 | .0000 | .0000 | .0000 | .0000 | .0000 |
|  | 19 | .0000 | .0000 | .0000 | .0000 | .0000 | .0000 | .0000 | .0000 | .0000 | .0000 | .0000 |
| 20 | 0 | .8179 | .3585 | .1216 | .0388 | .0115 | .0032 | .0008 | .0002 | .0000 | .0000 | .0000 |
|  | 1 | .1652 | .3774 | .2702 | .1368 | .0576 | .0211 | .0068 | .0020 | .0005 | .0001 | .0000 |
|  | 2 | .0159 | .1887 | .2852 | .2293 | .1369 | .0669 | .0278 | .0100 | .0031 | .0008 | .0002 |
|  | 3 | .0010 | .0596 | .1901 | .2428 | .2054 | .1339 | .0718 | .0323 | .0123 | .0040 | .0011 |
|  | 4 | .0000 | .0133 | .0898 | .1821 | .2182 | .1897 | .1304 | .0738 | .0350 | .0139 | .0046 |
|  | 5 | .0000 | .0022 | .0319 | .1028 | .1746 | .2023 | .1789 | .1272 | .0746 | .0365 | .0148 |
|  | 6 | .0000 | .0003 | .0089 | .0454 | .1091 | .1686 | .1916 | .1712 | .1244 | .0746 | .0370 |
|  | 7 | .0000 | .0000 | .0020 | .0160 | .0545 | .1124 | .1643 | .1844 | .1659 | .1221 | .0739 |
|  | 8 | .0000 | .0000 | .0004 | .0046 | .0222 | .0609 | .1144 | .1614 | .1797 | .1623 | .1201 |
|  | 9 | .0000 | .0000 | .0001 | .0011 | .0074 | .0271 | .0654 | .1158 | .1597 | .1771 | .1602 |
|  | 10 | .0000 | .0000 | .0000 | .0002 | .0020 | .0099 | .0308 | .0686 | .1171 | .1593 | .1762 |
|  | 11 | .0000 | .0000 | .0000 | .0000 | .0005 | .0030 | .0120 | .0336 | .0710 | .1185 | .1602 |
|  | 12 | .0000 | .0000 | .0000 | .0000 | .0001 | .0008 | .0039 | .0136 | .0355 | .0727 | .1201 |
|  | 13 | .0000 | .0000 | .0000 | .0000 | .0000 | .0002 | .0010 | .0045 | .0146 | .0366 | .0739 |
|  | 14 | .0000 | .0000 | .0000 | .0000 | .0000 | .0000 | .0002 | .0012 | .0049 | .0150 | .0370 |
|  | 15 | .0000 | .0000 | .0000 | .0000 | .0000 | .0000 | .0000 | .0003 | .0013 | .0049 | .0148 |
|  | 16 | .0000 | .0000 | .0000 | .0000 | .0000 | .0000 | .0000 | .0000 | .0003 | .0013 | .0046 |
|  | 17 | .0000 | .0000 | .0000 | .0000 | .0000 | .0000 | .0000 | .0000 | .0000 | .0002 | .0011 |
|  | 18 | .0000 | .0000 | .0000 | .0000 | .0000 | .0000 | .0000 | .0000 | .0000 | .0000 | .0002 |
|  | 19 | .0000 | .0000 | .0000 | .0000 | .0000 | .0000 | .0000 | .0000 | .0000 | .0000 | .0000 |
|  | 20 | .0000 | .0000 | .0000 | .0000 | .0000 | .0000 | .0000 | .0000 | .0000 | .0000 | .0000 |
| 25 | 0 | .7778 | .2774 | .0718 | .0172 | .0038 | .0008 | .0001 | .0000 | .0000 | .0000 | .0000 |
|  | 1 | .1964 | .3650 | .1994 | .0759 | .0236 | .0063 | .0014 | .0003 | .0000 | .0000 | .0000 |
|  | 2 | .0238 | .2305 | .2659 | .1607 | .0708 | .0251 | .0074 | .0018 | .0004 | .0001 | .0000 |
|  | 3 | .0018 | .0930 | .2265 | .2174 | .1358 | .0641 | .0243 | .0076 | .0019 | .0004 | .0001 |
|  | 4 | .0001 | .0269 | .1384 | .2110 | .1867 | .1175 | .0572 | .0224 | .0071 | .0018 | .0004 |
|  | 5 | .0000 | .0060 | .0646 | .1564 | .1960 | .1645 | .1030 | .0506 | .0199 | .0063 | .0016 |
|  | 6 | .0000 | .0010 | .0239 | .0920 | .1633 | .1828 | .1472 | .0908 | .0442 | .0172 | .0053 |
|  | 7 | .0000 | .0001 | .0072 | .0441 | .1108 | .1654 | .1712 | .1327 | .0800 | .0381 | .0143 |
|  | 8 | .0000 | .0000 | .0018 | .0175 | .0623 | .1241 | .1651 | .1607 | .1200 | .0701 | .0322 |
|  | 9 | .0000 | .0000 | .0004 | .0058 | .0294 | .0781 | .1336 | .1635 | .1511 | .1084 | .0609 |
|  | 10 | .0000 | .0000 | .0000 | .0016 | .0118 | .0417 | .0916 | .1409 | .1612 | .1419 | .0974 |
|  | 11 | .0000 | .0000 | .0000 | .0004 | .0040 | .0189 | .0536 | .1034 | .1465 | .1583 | .1328 |
|  | 12 | .0000 | .0000 | .0000 | .0000 | .0012 | .0074 | .0268 | .0650 | .1140 | .1511 | .1550 |
|  | 13 | .0000 | .0000 | .0000 | .0000 | .0003 | .0025 | .0115 | .0350 | .0760 | .1236 | .1550 |
|  | 14 | .0000 | .0000 | .0000 | .0000 | .0000 | .0007 | .0042 | .0161 | .0434 | .0867 | .1328 |

APPENDIX 1/SELECTED VALUES OF THE BINOMIAL PROBABILITY DISTRIBUTION

| n | r | .01 | .05 | .10 | .15 | .20 | .25 | .30 | .35 | .40 | .45 | .50 |
|---|---|-----|-----|-----|-----|-----|-----|-----|-----|-----|-----|-----|
| 25 | 15 | .0000 | .0000 | .0000 | .0000 | .0000 | .0002 | .0013 | .0064 | .0212 | .0520 | .0974 |
| | 16 | .0000 | .0000 | .0000 | .0000 | .0000 | .0000 | .0004 | .0021 | .0088 | .0266 | .0609 |
| | 17 | .0000 | .0000 | .0000 | .0000 | .0000 | .0000 | .0001 | .0006 | .0031 | .0115 | .0322 |
| | 18 | .0000 | .0000 | .0000 | .0000 | .0000 | .0000 | .0000 | .0001 | .0009 | .0042 | .0143 |
| | 19 | .0000 | .0000 | .0000 | .0000 | .0000 | .0000 | .0000 | .0000 | .0002 | .0013 | .0053 |
| | 20 | .0000 | .0000 | .0000 | .0000 | .0000 | .0000 | .0000 | .0000 | .0000 | .0001 | .0016 |
| | 21 | .0000 | .0000 | .0000 | .0000 | .0000 | .0000 | .0000 | .0000 | .0000 | .0000 | .0004 |
| | 22 | .0000 | .0000 | .0000 | .0000 | .0000 | .0000 | .0000 | .0000 | .0000 | .0000 | .0001 |
| 30 | 0 | .7397 | .2146 | .0424 | .0076 | .0012 | .0002 | .0000 | .0000 | .0000 | .0000 | .0000 |
| | 1 | .2242 | .3389 | .1413 | .0404 | .0093 | .0018 | .0003 | .0000 | .0000 | .0000 | .0000 |
| | 2 | .0328 | .2586 | .2277 | .1034 | .0337 | .0086 | .0018 | .0003 | .0000 | .0000 | .0000 |
| | 3 | .0031 | .1270 | .2361 | .1703 | .0785 | .0269 | .0072 | .0015 | .0003 | .0000 | .0000 |
| | 4 | .0002 | .0451 | .1771 | .2028 | .1325 | .0604 | .0208 | .0056 | .0012 | .0002 | .0000 |
| | 5 | .0000 | .0124 | .1023 | .1861 | .1723 | .1047 | .0464 | .0157 | .0041 | .0008 | .0001 |
| | 6 | .0000 | .0027 | .0474 | .1368 | .1795 | .1455 | .0829 | .0353 | .0115 | .0029 | .0006 |
| | 7 | .0000 | .0005 | .0180 | .0828 | .1538 | .1662 | .1219 | .0652 | .0263 | .0081 | .0019 |
| | 8 | .0000 | .0001 | .0058 | .0420 | .1106 | .1593 | .1501 | .1009 | .0505 | .0191 | .0055 |
| | 9 | .0000 | .0000 | .0016 | .0181 | .0676 | .1298 | .1573 | .1328 | .0823 | .0382 | .0133 |
| | 10 | .0000 | .0000 | .0004 | .0067 | .0355 | .0909 | .1416 | .1502 | .1152 | .0656 | .0280 |
| | 11 | .0000 | .0000 | .0001 | .0022 | .0161 | .0551 | .1103 | .1471 | .1396 | .0976 | .0509 |
| | 12 | .0000 | .0000 | .0000 | .0006 | .0064 | .0291 | .0749 | .1254 | .1474 | .1265 | .0806 |
| | 13 | .0000 | .0000 | .0000 | .0001 | .0022 | .0134 | .0444 | .0935 | .1360 | .1433 | .1115 |
| | 14 | .0000 | .0000 | .0000 | .0000 | .0007 | .0054 | .0231 | .0611 | .1101 | .1424 | .1354 |
| | 15 | .0000 | .0000 | .0000 | .0000 | .0002 | .0019 | .0106 | .0351 | .0783 | .1242 | .1445 |
| | 16 | .0000 | .0000 | .0000 | .0000 | .0000 | .0006 | .0042 | .0177 | .0489 | .0953 | .1354 |
| | 17 | .0000 | .0000 | .0000 | .0000 | .0000 | .0002 | .0015 | .0079 | .0269 | .0642 | .1115 |
| | 18 | .0000 | .0000 | .0000 | .0000 | .0000 | .0000 | .0005 | .0031 | .0129 | .0379 | .0806 |
| | 19 | .0000 | .0000 | .0000 | .0000 | .0000 | .0000 | .0001 | .0010 | .0054 | .0196 | .0509 |
| | 20 | .0000 | .0000 | .0000 | .0000 | .0000 | .0000 | .0000 | .0003 | .0020 | .0088 | .0280 |
| | 21 | .0000 | .0000 | .0000 | .0000 | .0000 | .0000 | .0000 | .0001 | .0006 | .0034 | .0133 |
| | 22 | .0000 | .0000 | .0000 | .0000 | .0000 | .0000 | .0000 | .0000 | .0002 | .0012 | .0055 |
| | 23 | .0000 | .0000 | .0000 | .0000 | .0000 | .0000 | .0000 | .0000 | .0000 | .0003 | .0019 |
| | 24 | .0000 | .0000 | .0000 | .0000 | .0000 | .0000 | .0000 | .0000 | .0000 | .0001 | .0006 |
| | 25 | .0000 | .0000 | .0000 | .0000 | .0000 | .0000 | .0000 | .0000 | .0000 | .0000 | .0001 |

# APPENDIX 2
## AREAS UNDER THE STANDARD NORMAL PROBABILITY DISTRIBUTION

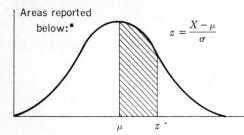

$$z = \frac{X - \mu}{\sigma}$$

Areas reported below:*

*damn it Janet!*

| z | .00 | .01 | .02 | .03 | .04 | .05 | .06 | .07 | .08 | .09 |
|---|-----|-----|-----|-----|-----|-----|-----|-----|-----|-----|
| 0.0 | .0000 | .0040 | .0080 | .0120 | .0160 | .0199 | .0239 | .0279 | .0319 | .0359 |
| 0.1 | .0398 | .0438 | .0478 | .0517 | .0557 | .0596 | .0636 | .0675 | .0714 | .0753 |
| 0.2 | .0793 | .0832 | .0871 | .0910 | .0948 | .0987 | .1026 | .1064 | .1103 | .1141 |
| 0.3 | .1179 | .1217 | .1255 | .1293 | .1331 | .1368 | .1406 | .1443 | .1480 | .1517 |
| 0.4 | .1554 | .1591 | .1628 | .1664 | .1700 | .1736 | .1772 | .1808 | .1844 | .1879 |
| 0.5 | .1915 | .1950 | .1985 | .2019 | .2054 | .2088 | .2123 | .2157 | .2190 | .2224 |
| 0.6 | .2257 | .2291 | .2324 | .2357 | .2389 | .2422 | .2454 | .2486 | .2518 | .2549 |
| 0.7 | .2580 | 2.612 | .2642 | .2673 | .2704 | .2734 | .2764 | .2794 | .2823 | .2852 |
| 0.8 | .2881 | .2910 | .2939 | .2967 | .2995 | .3023 | .3051 | .3078 | .3106 | .3133 |
| 0.9 | .3159 | .3186 | .3212 | .3238 | .3264 | .3289 | .3315 | .3340 | .3365 | .3389 |
| 1.0 | .3413 | .3438 | .3461 | .3485 | .3508 | .3531 | .3554 | .3577 | .3599 | .3621 |
| 1.1 | .3643 | .3665 | .3686 | .3708 | .3729 | .3749 | .3770 | .3790 | .3810 | .3830 |
| 1.2 | .3849 | .3869 | .3888 | .3907 | .3925 | .3944 | .3962 | .3980 | .3997 | .4014 |
| 1.3 | .4032 | .4049 | .4066 | .4082 | .4099 | .4115 | .4131 | .4147 | .4162 | .4177 |
| 1.4 | .4192 | .4207 | .4222 | .4236 | .4251 | .4265 | .4279 | .4292 | .4306 | .4319 |
| 1.5 | .4332 | .4345 | .4357 | .4370 | .4382 | .4394 | .4406 | .4418 | .4429 | .4441 |
| 1.6 | .4452 | .4463 | .4474 | .4484 | .4495 | .4505 | .4515 | .4525 | .4535 | .4545 |
| 1.7 | .4554 | .4564 | .4573 | .4582 | .4591 | .4599 | .4608 | .4616 | .4625 | .4633 |
| 1.8 | .4641 | .4649 | .4656 | .4664 | .4671 | .4678 | .4686 | .4693 | .4699 | .4706 |
| 1.9 | .4713 | .4719 | .4726 | .4732 | .4738 | .4744 | .4750 | .4756 | .4761 | .4767 |
| 2.0 | .4772 | .4778 | .4783 | .4788 | .4793 | .4798 | .4803 | .4808 | .4812 | .4817 |
| 2.1 | .4821 | .4826 | .4830 | .4834 | .4838 | .4842 | .4846 | .4850 | .4854 | .4857 |
| 2.2 | .4861 | .4864 | .4868 | .4871 | .4875 | .4878 | .4881 | .4884 | .4887 | .4890 |
| 2.3 | .4893 | .4896 | .4898 | .4901 | .4904 | .4906 | .4909 | .4911 | .4913 | .4916 |
| 2.4 | .4918 | .4920 | .4922 | .4925 | .4927 | .4929 | .4931 | .4932 | .4934 | .4936 |
| 2.5 | .4938 | .4940 | .4941 | .4943 | .4945 | .4946 | .4948 | .4949 | .4951 | .4952 |
| 2.6 | .4953 | .4955 | .4956 | .4957 | .4959 | .4960 | .4961 | .4962 | .4963 | .4964 |
| 2.7 | .4965 | .4966 | .4967 | .4968 | .4969 | .4970 | .4971 | .4972 | .4973 | .4974 |
| 2.8 | .4974 | .4975 | .4976 | .4977 | .4977 | .4978 | .4979 | .4979 | .4980 | .4981 |
| 2.9 | .4981 | .4982 | .4983 | .4983 | .4984 | .4984 | .4985 | .4985 | .4986 | .4986 |
| 3.0 | .4987 | .4987 | .4987 | .4988 | .4989 | .4989 | .4989 | .4989 | .4990 | .4990 |
| 3.5 | .4997 | | | | | | | | | |
| 4.0 | .4999683 | | | | | | | | | |

*4⁹⁵*

*.4990*

* Example: For $z = 1.96$, the shaded area is 0.4750 out of the total area of 1.0000.

# APPENDIX 3
## A BRIEF TABLE OF RANDOM NUMBERS

| | | | | | | | |
|---|---|---|---|---|---|---|---|
| 10097 | 85017 | 84532 | 13618 | 23157 | 86952 | 02438 | 76520 |
| 37542 | 16719 | 82789 | 69041 | 05545 | 44109 | 05403 | 64894 |
| 08422 | 65842 | 27672 | 82186 | 14871 | 22115 | 86529 | 19645 |
| 99019 | 76875 | 20684 | 39187 | 38976 | 94324 | 43204 | 09376 |
| 12807 | 93640 | 39160 | 41453 | 97312 | 41548 | 93137 | 80157 |
| | | | | | | | |
| 66065 | 99478 | 70086 | 71265 | 11742 | 18226 | 29004 | 34072 |
| 31060 | 65119 | 26486 | 47353 | 43361 | 99436 | 42753 | 45571 |
| 85269 | 70322 | 21592 | 48233 | 93806 | 32584 | 21828 | 02051 |
| 63573 | 58133 | 41278 | 11697 | 49540 | 61777 | 67954 | 05325 |
| 73796 | 44655 | 81255 | 31133 | 36768 | 60452 | 38537 | 03529 |
| | | | | | | | |
| 98520 | 02295 | 13487 | 98662 | 07092 | 44673 | 61303 | 14905 |
| 11805 | 85035 | 54881 | 35587 | 43310 | 48897 | 48493 | 39808 |
| 83452 | 01197 | 86935 | 28021 | 61570 | 23350 | 65710 | 06288 |
| 88685 | 97907 | 19078 | 40646 | 31352 | 48625 | 44369 | 86507 |
| 99594 | 63268 | 96905 | 28797 | 57048 | 46359 | 74294 | 87517 |
| | | | | | | | |
| 65481 | 52841 | 59684 | 67411 | 09243 | 56092 | 84369 | 17468 |
| 80124 | 53722 | 71399 | 10916 | 07959 | 21225 | 13018 | 17727 |
| 74350 | 11434 | 51908 | 62171 | 93732 | 26958 | 02400 | 77402 |
| 69916 | 62375 | 99292 | 21177 | 72721 | 66995 | 07289 | 66252 |
| 09893 | 28337 | 20923 | 87929 | 61020 | 62841 | 31374 | 14225 |
| | | | | | | | |
| 91499 | 38631 | 79430 | 62421 | 97959 | 67422 | 69992 | 68479 |
| 80336 | 49172 | 16332 | 44670 | 35089 | 17691 | 89246 | 26940 |
| 44104 | 89232 | 57327 | 34679 | 62235 | 79655 | 81336 | 85157 |
| 12550 | 02844 | 15026 | 32439 | 58537 | 48274 | 81330 | 11100 |
| 63606 | 40387 | 65406 | 37920 | 08709 | 60623 | 2237 | 16505 |
| | | | | | | | |
| 61196 | 80240 | 44177 | 51171 | 08723 | 39323 | 05798 | 26457 |
| 15474 | 44910 | 99321 | 72173 | 56239 | 04595 | 10836 | 95270 |
| 94557 | 33663 | 86347 | 00926 | 44915 | 34823 | 51770 | 67897 |
| 42481 | 86430 | 19102 | 37420 | 41976 | 76559 | 24358 | 97344 |
| 23523 | 31379 | 68588 | 81675 | 15694 | 43438 | 36879 | 73208 |
| | | | | | | | |
| 04493 | 98086 | 32533 | 17767 | 14523 | 52494 | 24826 | 75246 |
| 00549 | 33185 | 04805 | 05431 | 94598 | 97654 | 16232 | 64051 |
| 35963 | 80951 | 68953 | 99634 | 81949 | 15307 | 00406 | 26898 |
| 59808 | 79752 | 02529 | 40200 | 73742 | 08391 | 49140 | 45427 |
| 46058 | 18633 | 99970 | 67348 | 49329 | 95236 | 32537 | 01390 |
| | | | | | | | |
| 32179 | 74029 | 74717 | 17674 | 90446 | 00597 | 45240 | 87379 |
| 69234 | 54178 | 10805 | 35635 | 45266 | 61406 | 41941 | 20117 |
| 19565 | 11664 | 77602 | 99817 | 28573 | 41430 | 96382 | 01758 |
| 45155 | 48324 | 32135 | 26803 | 16213 | 14938 | 71961 | 19476 |
| 94864 | 69074 | 45753 | 20505 | 78317 | 31994 | 98145 | 36168 |

SOURCE: Leonard K. Kazmier, *Statistical Analysis for Business and Economics*, 2d ed., copyright © 1973 by McGraw-Hill, Inc. Used with permission of McGraw-Hill Book Company.

# APPENDIX 4
# AREAS FOR $t$ DISTRIBUTIONS

Areas reported below:*

$$t = \frac{\overline{X} - \mu}{s_{\overline{X}}}$$

Proportion of area (one tail)

(sample size) degrees of freedom →

| df | 0.10 | 0.05 | 0.025 | 0.01 | 0.005 |
|----|------|------|-------|------|-------|
| 1 | 3.078 | 6.314 | 12.706 | 31.821 | 63.657 |
| 2 | 1.886 | 2.920 | 4.303 | 6.965 | 9.925 |
| 3 | 1.638 | 2.353 | 3.182 | 4.541 | 5.841 |
| 4 | 1.533 | 2.132 | 2.776 | 3.747 | 4.604 |
| 5 | 1.476 | 2.015 | 2.571 | 3.365 | 4.032 |
| 6 | 1.440 | 1.943 | 2.447 | 3.143 | 3.707 |
| 7 | 1.415 | 1.895 | 2.365 | 2.998 | 3.499 |
| 8 | 1.397 | 1.860 | 2.306 | 2.896 | 3.355 |
| 9 | 1.383 | 1.833 | 2.262 | 2.821 | 3.250 |
| 10 | 1.372 | 1.812 | 2.228 | 2.764 | 3.169 |
| 11 | 1.363 | 1.796 | 2.201 | 2.718 | 3.106 |
| 12 | 1.356 | 1.782 | 2.179 | 2.681 | 3.055 |
| 13 | 1.350 | 1.771 | 2.160 | 2.650 | 3.012 |
| 14 | 1.345 | 1.761 | 2.145 | 2.624 | 2.977 |
| 15 | 1.341 | 1.753 | 2.131 | 2.602 | 2.947 |
| 16 | 1.337 | 1.746 | 2.120 | 2.583 | 2.921 |
| 17 | 1.333 | 1.740 | 2.110 | 2.567 | 2.898 |
| 18 | 1.330 | 1.734 | 2.101 | 2.552 | 2.878 |
| 19 | 1.328 | 1.729 | 2.093 | 2.539 | 2.861 |
| 20 | 1.325 | 1.725 | 2.086 | 2.528 | 2.845 |
| 21 | 1.323 | 1.721 | 2.080 | 2.518 | 2.831 |
| 22 | 1.321 | 1.717 | 2.074 | 2.508 | 2.819 |
| 23 | 1.319 | 1.714 | 2.069 | 2.500 | 2.807 |
| 24 | 1.318 | 1.711 | 2.064 | 2.492 | 2.797 |
| 25 | 1.316 | 1.708 | 2.060 | 2.485 | 2.787 |
| 26 | 1.315 | 1.706 | 2.056 | 2.479 | 2.779 |
| 27 | 1.314 | 1.703 | 2.052 | 2.473 | 2.771 |
| 28 | 1.313 | 1.701 | 2.048 | 2.467 | 2.763 |
| 29 | 1.311 | 1.699 | 2.045 | 2.462 | 2.756 |
| 30 | 1.310 | 1.697 | 2.042 | 2.457 | 2.750 |
| 40 | 1.303 | 1.684 | 2.021 | 2.423 | 2.704 |
| 60 | 1.296 | 1.671 | 2.000 | 2.390 | 2.660 |
| 120 | 1.289 | 1.658 | 1.980 | 2.358 | 2.617 |
| ∞ | 1.282 | 1.645 | 1.960 | 2.326 | 2.576 |

* Example: For the shaded area to represent 0.05 of the total area of 1.0, the value of $t$ with 10 degrees of freedom is 1.812.

SOURCE: Abridged from Table IV of R. A. Fisher, *Statistical Methods for Research Workers*, 14th ed., copyright © 1972 by Hafner Press. Used with permission of Hafner Press.

## TABLES OF SQUARES AND SQUARE ROOTS

| N | $N^2$ | $\sqrt{N}$ | $\sqrt{10N}$ | N | $N^2$ | $\sqrt{N}$ | $\sqrt{10N}$ |
|---|---|---|---|---|---|---|---|
|   |   |   |   | 50 | 2 500 | 7.071 068 | 22.36068 |
| 1 | 1 | 1.000 000 | 3.162 278 | 51 | 2 601 | 7.141 428 | 22.58318 |
| 2 | 4 | 1.414 214 | 4.472 136 | 52 | 2 704 | 7.211 103 | 22.80351 |
| 3 | 9 | 1.732 051 | 5.477 226 | 53 | 2 809 | 7.280 110 | 23.02173 |
| 4 | 16 | 2.000 000 | 6.324 555 | 54 | 2 916 | 7.348 469 | 23.23790 |
| 5 | 25 | 2.236 068 | 7.071 068 | 55 | 3 025 | 7.416 198 | 23.45208 |
| 6 | 36 | 2.449 490 | 7.745 967 | 56 | 3 136 | 7.483 315 | 23.66432 |
| 7 | 49 | 2.645 751 | 8.366 600 | 57 | 3 249 | 7.549 834 | 23.87467 |
| 8 | 64 | 2.828 427 | 8.944 272 | 58 | 3 364 | 7.615 773 | 24.08319 |
| 9 | 81 | 3.000 000 | 9.486 833 | 59 | 3 481 | 7.681 146 | 24.28992 |
| 10 | 100 | 3.162 278 | 10.00000 | 60 | 3 600 | 7.745 967 | 24.49490 |
| 11 | 121 | 3.316 625 | 10.48809 | 61 | 3 721 | 7.810 250 | 24.69818 |
| 12 | 144 | 3.464 102 | 10.95445 | 62 | 3 844 | 7.874 008 | 24.89980 |
| 13 | 169 | 3.605 551 | 11.40175 | 63 | 3 969 | 7.937 254 | 25.09980 |
| 14 | 196 | 3.741 657 | 11.83216 | 64 | 4 096 | 8.000 000 | 25.29822 |
| 15 | 225 | 3.872 983 | 12.24745 | 65 | 4 225 | 8.062 258 | 25.49510 |
| 16 | 256 | 4.000 000 | 12.64911 | 66 | 4 356 | 8.124 038 | 25.69047 |
| 17 | 289 | 4.123 106 | 13.03840 | 67 | 4 489 | 8.185 353 | 25.88436 |
| 18 | 324 | 4.242 641 | 13.41641 | 68 | 4 624 | 8.246 211 | 26.07681 |
| 19 | 361 | 4.358 899 | 13.78405 | 69 | 4 761 | 8.306 824 | 26.26785 |
| 20 | 400 | 4.472 136 | 14.14214 | 70 | 4 900 | 8.366 600 | 26.45751 |
| 21 | 441 | 4.582 576 | 14.49138 | 71 | 5 041 | 8.426 150 | 26.64583 |
| 22 | 484 | 4.690 416 | 14.83240 | 72 | 5 184 | 8.485 281 | 26.83282 |
| 23 | 529 | 4.795 832 | 15.16575 | 73 | 5 329 | 8.544 004 | 27.01851 |
| 24 | 576 | 4.898 979 | 15.49193 | 74 | 5 476 | 8.602 325 | 27.20294 |
| 25 | 625 | 5.000 000 | 15.81139 | 75 | 5 625 | 8.660 254 | 27.38613 |
| 26 | 676 | 5.099 020 | 16.12452 | 76 | 5 776 | 8.717 798 | 27.56810 |
| 27 | 729 | 5.196 152 | 16.43168 | 77 | 5 929 | 8.774 964 | 27.74887 |
| 28 | 784 | 5.291 503 | 16.73320 | 78 | 6 084 | 8.831 761 | 27.92848 |
| 29 | 841 | 5.385 165 | 17.02939 | 79 | 6 241 | 8.888 194 | 28.10694 |
| 30 | 900 | 5.477 226 | 17.32051 | 80 | 6 400 | 8.944 272 | 28.28427 |
| 31 | 961 | 5.567 764 | 17.60682 | 81 | 6 561 | 9.000 000 | 28.46050 |
| 32 | 1 024 | 5.656 854 | 17.88854 | 82 | 6 724 | 9.055 385 | 28.63564 |
| 33 | 1 089 | 5.744 563 | 18.16590 | 83 | 6 889 | 9.110 434 | 28.80972 |
| 34 | 1 156 | 5.830 952 | 18.43909 | 84 | 7 056 | 9.165 151 | 28.98275 |
| 35 | 1 225 | 5.916 080 | 18.70829 | 85 | 7 225 | 9.219 544 | 29.15476 |
| 36 | 1 296 | 6.000 000 | 18.97367 | 86 | 7 396 | 9.273 618 | 29.32576 |
| 37 | 1 369 | 6.082 763 | 19.23538 | 87 | 7 569 | 9.327 379 | 29.49576 |
| 38 | 1 444 | 6.164 414 | 19.49359 | 88 | 7 744 | 9.380 832 | 29.66479 |
| 39 | 1 521 | 6.244 998 | 19.74842 | 89 | 7 921 | 9.433 981 | 29.83287 |
| 40 | 1 600 | 6.324 555 | 20.00000 | 90 | 8 100 | 9.486 833 | 30.00000 |
| 41 | 1 681 | 6.403 124 | 20.24846 | 91 | 8 281 | 9.539 392 | 30.16621 |
| 42 | 1 764 | 6.480 741 | 20.49390 | 92 | 8 464 | 9.591 663 | 30.33150 |
| 43 | 1 849 | 6.557 439 | 20.73644 | 93 | 8 649 | 9.643 651 | 30.49590 |
| 44 | 1 936 | 6.633 250 | 20.97618 | 94 | 8 836 | 9.695 360 | 30.65942 |
| 45 | 2 025 | 6.708 204 | 21.21320 | 95 | 9 025 | 9.746 794 | 30.82207 |
| 46 | 2 116 | 6.782 330 | 21.44761 | 96 | 9 216 | 9.797 959 | 30.98387 |
| 47 | 2 209 | 6.855 655 | 21.67948 | 97 | 9 409 | 9.848 858 | 31.14482 |
| 48 | 2 304 | 6.928 203 | 21.90890 | 98 | 9 604 | 9.899 495 | 31.30495 |
| 49 | 2 401 | 7.000 000 | 22.13594 | 99 | 9 801 | 9.949 874 | 31.46427 |
| 50 | 2 500 | 7.071 068 | 22.36068 | 100 | 10 000 | 10.00000 | 31.62278 |

| N | N² | √N | √10N | N | N² | √N | √10N |
|---|---|---|---|---|---|---|---|
| 100 | 10 000 | 10.00000 | 31.62278 | 150 | 22 500 | 12.24745 | 38.72983 |
| 101 | 10 201 | 10.04988 | 31.78050 | 151 | 22 801 | 12.28821 | 38.85872 |
| 102 | 10 404 | 10.09950 | 31.93744 | 152 | 23 104 | 12.32883 | 39.98718 |
| 103 | 10 609 | 10.14889 | 32.09361 | 153 | 23 409 | 12.36932 | 39.11521 |
| 104 | 10 816 | 10.19804 | 32.24903 | 154 | 23 716 | 12.40967 | 39.24283 |
| 105 | 11 025 | 10.24695 | 32.40370 | 155 | 24 025 | 12.44990 | 39.37004 |
| 106 | 11 236 | 10.29563 | 32.55764 | 156 | 24 336 | 12.40000 | 39.49684 |
| 107 | 11 449 | 10.34408 | 32.71085 | 157 | 24 649 | 12.52996 | 39.62323 |
| 108 | 11 664 | 10.39230 | 32.86335 | 158 | 24 964 | 12.56981 | 39.74921 |
| 109 | 11 881 | 10.44031 | 33.01515 | 159 | 25 281 | 12.60952 | 39.87480 |
| 110 | 12 100 | 10.48809 | 33.16625 | 160 | 25 600 | 12.64911 | 40.00000 |
| 111 | 12 321 | 10.53565 | 33.31666 | 161 | 25 921 | 12.68858 | 40.12481 |
| 112 | 12 544 | 10.58301 | 33.46640 | 162 | 26 244 | 12.72792 | 40.24922 |
| 113 | 12 769 | 10.63015 | 33.61547 | 163 | 26 569 | 12.76715 | 40.37326 |
| 114 | 12 996 | 10.67708 | 33.76389 | 164 | 26 896 | 12.80625 | 40.49691 |
| 115 | 13 225 | 10.72381 | 33.91165 | 165 | 27 225 | 12.84523 | 40.62019 |
| 116 | 13 456 | 10.77033 | 34.05877 | 166 | 27 556 | 12.88410 | 40.74310 |
| 117 | 13 689 | 10.81665 | 34.20526 | 167 | 27 889 | 12.92285 | 40.86563 |
| 118 | 13 924 | 10.86278 | 34.35113 | 168 | 28 224 | 12.96148 | 40.98780 |
| 119 | 14 161 | 10.90871 | 34.49638 | 169 | 28 561 | 13.00000 | 41.10961 |
| 120 | 14 400 | 10.95445 | 34.64102 | 170 | 28 900 | 13.03840 | 41.23106 |
| 121 | 14 641 | 11.00000 | 34.78505 | 171 | 29 241 | 13.07670 | 41.35215 |
| 122 | 14 884 | 11.04536 | 34.92850 | 172 | 29 584 | 13.11488 | 41.47288 |
| 123 | 15 129 | 11.09054 | 35.07136 | 173 | 29 929 | 13.15295 | 41.59327 |
| 124 | 15 376 | 11.13553 | 35.21363 | 174 | 30 276 | 13.19091 | 41.71331 |
| 125 | 15 625 | 11.18034 | 35.35534 | 175 | 30 625 | 13.22876 | 41.83300 |
| 126 | 15 876 | 11.22497 | 35.49648 | 176 | 30 976 | 13.26650 | 41.95235 |
| 127 | 16 129 | 11.26943 | 35.63706 | 177 | 31 329 | 13.30413 | 42.07137 |
| 128 | 16 384 | 11.31371 | 35.77709 | 178 | 31 684 | 13.34166 | 42.19005 |
| 129 | 16 641 | 11.35782 | 35.91657 | 179 | 32 041 | 13.37909 | 42.30839 |
| 130 | 16 900 | 11.40175 | 36.05551 | 180 | 32 400 | 13.41641 | 42.42641 |
| 131 | 17 161 | 11.44552 | 36.19392 | 181 | 32 761 | 13.45362 | 42.54409 |
| 132 | 17 424 | 11.48913 | 36.33180 | 182 | 33 124 | 13.49074 | 42.66146 |
| 133 | 17 689 | 11.53256 | 36.46917 | 183 | 33 489 | 13.52775 | 42.77850 |
| 134 | 17 956 | 11.57584 | 36.60601 | 184 | 33 856 | 13.56466 | 42.89522 |
| 135 | 18 225 | 11.61895 | 36.74235 | 185 | 34 225 | 13.60147 | 43.01163 |
| 136 | 18 496 | 11.66190 | 36.87818 | 186 | 34 596 | 13.63818 | 43.12772 |
| 137 | 18 769 | 11.70470 | 37.01351 | 187 | 34 969 | 13.67479 | 43.24350 |
| 138 | 19 044 | 11.74734 | 37.14835 | 188 | 35 344 | 13.71131 | 43.35897 |
| 139 | 19 321 | 11.78983 | 37.28270 | 189 | 35 721 | 13.74773 | 43.47413 |
| 140 | 19 600 | 11.83216 | 37.41657 | 190 | 36 100 | 13.78405 | 43.58899 |
| 141 | 19 881 | 11.87434 | 37.54997 | 191 | 36 481 | 13.82027 | 43.70355 |
| 142 | 20 164 | 11.91638 | 37.68289 | 192 | 36 864 | 13.85641 | 43.81780 |
| 143 | 20 449 | 11.95826 | 37.81534 | 193 | 37 249 | 13.89244 | 43.93177 |
| 144 | 20 736 | 12.00000 | 37.94733 | 194 | 37 636 | 13.92839 | 44.04543 |
| 145 | 21 025 | 12.04159 | 38.07887 | 195 | 38 025 | 13.96424 | 44.15880 |
| 146 | 21 316 | 12.08305 | 38.20995 | 196 | 38 416 | 14.00000 | 44.27189 |
| 147 | 21 609 | 12.12436 | 38.34058 | 197 | 38 809 | 14.03567 | 44.38468 |
| 148 | 21 904 | 12.16553 | 38.47077 | 198 | 39 204 | 14.07125 | 44.49719 |
| 149 | 22 201 | 12.20656 | 38.60052 | 199 | 39 601 | 14.10674 | 44.60942 |
| 150 | 22 500 | 12.24745 | 38.72983 | 200 | 40 000 | 14.14214 | 44.72136 |

| N | N² | $\sqrt{N}$ | $\sqrt{10N}$ | N | N² | $\sqrt{N}$ | $\sqrt{10N}$ |
|---|---|---|---|---|---|---|---|
| 200 | 40 000 | 14.14214 | 44.72136 | 250 | 62 500 | 15.81139 | 50.00000 |
| 201 | 40 401 | 14.17745 | 44.83302 | 251 | 63 001 | 15.84298 | 50.09990 |
| 202 | 40 804 | 14.21267 | 44.94441 | 252 | 63 504 | 15.87451 | 50.19960 |
| 203 | 41 209 | 14.24781 | 45.05552 | 253 | 64 009 | 15.90597 | 50.29911 |
| 204 | 41 616 | 14.28296 | 45.16636 | 254 | 64 516 | 15.93738 | 50.39841 |
| 205 | 42 025 | 14.31782 | 45.27693 | 255 | 65 025 | 15.96872 | 50.49752 |
| 206 | 42 436 | 14.35270 | 45.38722 | 256 | 65 536 | 16.00000 | 50.59644 |
| 207 | 42 849 | 14.38749 | 45.49725 | 257 | 66 049 | 16.03122 | 50.69517 |
| 208 | 43 264 | 14.42221 | 45.60702 | 258 | 66 564 | 16.06238 | 50.79370 |
| 209 | 43 681 | 14.45683 | 45.71652 | 259 | 67 081 | 16.09348 | 50.89204 |
| 210 | 44 100 | 14.49138 | 45.82576 | 260 | 67 600 | 16.12452 | 50.99020 |
| 211 | 44 521 | 14.52584 | 45.93474 | 261 | 68 121 | 16.15549 | 51.08816 |
| 212 | 44 944 | 14.56022 | 46.04346 | 262 | 68 644 | 16.18641 | 51.18594 |
| 213 | 45 369 | 14.59452 | 46.15192 | 263 | 69 169 | 16.21727 | 51.28353 |
| 214 | 45 796 | 14.62874 | 46.26013 | 264 | 69 696 | 16.24808 | 51.38093 |
| 215 | 46 225 | 14.66288 | 46.36809 | 265 | 70 225 | 16.27882 | 51.47815 |
| 216 | 46 656 | 14.69694 | 46.47580 | 266 | 70 756 | 16.30951 | 51.57519 |
| 217 | 47 089 | 14.73092 | 46.58326 | 267 | 71 289 | 16.34013 | 51.67204 |
| 218 | 47 524 | 14.76482 | 46.69047 | 268 | 71 824 | 16.37071 | 51.76872 |
| 219 | 47 961 | 14.79865 | 46.79744 | 269 | 72 361 | 16.40122 | 51.86521 |
| 220 | 48 400 | 14.83240 | 46.90415 | 270 | 72 900 | 16.43168 | 51.96152 |
| 221 | 48 841 | 14.86607 | 47.01064 | 271 | 73 441 | 16.46208 | 52.05766 |
| 222 | 49 284 | 14.89966 | 47.11688 | 272 | 73 984 | 16.49242 | 52.15362 |
| 223 | 49 729 | 14.93318 | 47.22288 | 273 | 74 529 | 16.52271 | 52.24940 |
| 224 | 50 176 | 14.96663 | 47.32864 | 274 | 75 076 | 16.55295 | 52.34501 |
| 225 | 50 625 | 15.00000 | 47.43416 | 275 | 75 625 | 16.58312 | 52.44044 |
| 226 | 51 076 | 15.03330 | 47.53946 | 276 | 76 176 | 16.61325 | 52.53570 |
| 227 | 51 529 | 15.06652 | 47.64452 | 277 | 76 729 | 16.64332 | 52.63079 |
| 228 | 51 984 | 15.09967 | 47.74935 | 278 | 77 284 | 16.67333 | 52.72571 |
| 229 | 52 441 | 15.13275 | 47.85394 | 279 | 77 841 | 16.70329 | 52.82045 |
| 230 | 52 900 | 15.16575 | 47.95832 | 280 | 78 400 | 16.73320 | 52.91503 |
| 231 | 53 361 | 15.19868 | 48.06246 | 281 | 78 961 | 16.76305 | 53.00943 |
| 232 | 53 824 | 15.23155 | 48.16638 | 282 | 79 524 | 16.79286 | 53.10367 |
| 233 | 54 289 | 15.26434 | 48.27007 | 283 | 80 089 | 16.82260 | 53.19774 |
| 234 | 54 756 | 15.29706 | 48.37355 | 284 | 80 656 | 16.85230 | 53.29165 |
| 235 | 55 225 | 15.32971 | 48.47680 | 285 | 81 225 | 16.88194 | 53.38539 |
| 236 | 55 696 | 15.36229 | 48.57983 | 286 | 81 796 | 16.91153 | 53.47897 |
| 237 | 56 169 | 15.39480 | 46.68265 | 287 | 82 369 | 16.94107 | 53.57238 |
| 238 | 56 644 | 15.42725 | 48.78524 | 288 | 82 944 | 16.97056 | 53.66563 |
| 239 | 57 121 | 15.45962 | 48.88763 | 289 | 83 521 | 17.00000 | 53.75872 |
| 240 | 57 600 | 15.49193 | 48.98979 | 290 | 84 100 | 17.02939 | 53.85165 |
| 241 | 58 081 | 15.52417 | 49.09175 | 291 | 84 681 | 17.05872 | 53.94442 |
| 242 | 58 564 | 15.55635 | 49.19350 | 292 | 85 264 | 17.08801 | 54.03702 |
| 243 | 59 049 | 15.58846 | 49.29503 | 293 | 85 849 | 17.11724 | 54.12947 |
| 244 | 59 536 | 15.52050 | 49.39636 | 294 | 86 436 | 17.14643 | 54.22177 |
| 245 | 60 025 | 15.65248 | 49.49747 | 295 | 87 025 | 17.17556 | 54.31390 |
| 246 | 60 516 | 15.68439 | 49.59839 | 296 | 87 616 | 17.20465 | 54.40588 |
| 247 | 61 009 | 15.71623 | 49.69909 | 297 | 88 209 | 17.23369 | 54.49771 |
| 248 | 61 504 | 15.74802 | 49.79960 | 298 | 88 804 | 17.26268 | 54.58938 |
| 249 | 62 001 | 15.77973 | 49.89990 | 299 | 89 401 | 17.29162 | 54.68089 |
| 250 | 62 500 | 15.81139 | 50.00000 | 300 | 90 000 | 17.32051 | 54.77226 |

| N | N² | $\sqrt{N}$ | $\sqrt{10N}$ | N | N² | $\sqrt{N}$ | $\sqrt{10N}$ |
|---|---|---|---|---|---|---|---|
| 300 | 90 000 | 17.32051 | 54.77226 | 350 | 122 500 | 18.70829 | 59.16080 |
| 301 | 90 601 | 17.34935 | 54.86347 | 351 | 123 201 | 18.73499 | 59.24525 |
| 302 | 91 204 | 17.37815 | 54.95453 | 352 | 123 904 | 18.76166 | 59.32959 |
| 303 | 91 809 | 17.40690 | 55.04544 | 353 | 124 609 | 18.78829 | 59.41380 |
| 304 | 92 416 | 17.43560 | 55.13620 | 354 | 125 316 | 18.81489 | 59.49790 |
| 305 | 93 025 | 17.46425 | 55.22681 | 355 | 126 025 | 18.84144 | 59.58188 |
| 306 | 93 636 | 17.49288 | 55.31727 | 356 | 126 736 | 18.86796 | 59.66574 |
| 307 | 94 249 | 17.52142 | 55.40758 | 357 | 127 449 | 18.89444 | 59.74948 |
| 308 | 94 864 | 17.54993 | 55.49775 | 358 | 128 164 | 18.92089 | 59.83310 |
| 309 | 95 481 | 17.57840 | 55.58777 | 359 | 128 881 | 18.94730 | 59.91661 |
| 310 | 96 100 | 17.60682 | 55.67764 | 360 | 129 600 | 18.97367 | 60.00000 |
| 311 | 96 721 | 17.63519 | 55.76737 | 361 | 130 321 | 19.00000 | 60.08328 |
| 312 | 97 344 | 17.66352 | 55.85696 | 362 | 131 044 | 19.02630 | 60.16644 |
| 313 | 97 969 | 17.69181 | 55.94640 | 363 | 131 769 | 19.05256 | 60.24948 |
| 314 | 98 596 | 17.72005 | 56.03670 | 364 | 132 496 | 19.07878 | 60.33241 |
| 315 | 99 225 | 17.74824 | 56.12486 | 365 | 133 225 | 19.10497 | 60.41523 |
| 316 | 99 856 | 17.77639 | 56.21388 | 366 | 133 956 | 19.13113 | 60.49793 |
| 317 | 100 489 | 17.80449 | 56.30275 | 367 | 134 689 | 19.15724 | 60.58052 |
| 318 | 101 124 | 17.83255 | 56.39149 | 368 | 135 424 | 19.18333 | 60.66300 |
| 319 | 101 761 | 17.86057 | 56.48008 | 369 | 136 161 | 19.20937 | 60.74537 |
| 320 | 102 400 | 17.88854 | 56.56854 | 370 | 136 900 | 19.23538 | 60.82763 |
| 321 | 103 041 | 17.91647 | 56.65686 | 371 | 137 641 | 19.26136 | 60.90977 |
| 322 | 103 684 | 17.94436 | 56.74504 | 372 | 138 384 | 19.28730 | 60.99180 |
| 323 | 104 329 | 17.97220 | 56.83309 | 373 | 139 129 | 19.31321 | 61.07373 |
| 324 | 104 976 | 18.00000 | 56.92100 | 374 | 139 876 | 19.33908 | 61.15554 |
| 325 | 105 625 | 18.02776 | 57.00877 | 375 | 140 625 | 19.36492 | 61.23724 |
| 326 | 106 276 | 18.05547 | 57.09641 | 376 | 141 376 | 19.39072 | 61.31884 |
| 327 | 106 929 | 18.08314 | 57.18391 | 377 | 142 129 | 19.41649 | 61.40033 |
| 328 | 107 584 | 18.11077 | 57.27128 | 378 | 142 884 | 19.44222 | 61.48170 |
| 329 | 108 241 | 18.13836 | 57.35852 | 379 | 143 641 | 19.46792 | 61.56298 |
| 330 | 108 900 | 18.16590 | 57.44563 | 380 | 144 000 | 19.49359 | 61.64414 |
| 331 | 109 561 | 18.19341 | 57.53260 | 381 | 145 161 | 19.51922 | 61.72520 |
| 332 | 110 224 | 18.22087 | 57.61944 | 382 | 145 924 | 19.54483 | 61.80615 |
| 333 | 110 889 | 18.24829 | 57.70615 | 383 | 146 689 | 19.57039 | 61.88699 |
| 334 | 111 556 | 18.27567 | 57.79273 | 384 | 147 456 | 19.59592 | 61.96773 |
| 335 | 112 225 | 18.30301 | 57.87918 | 385 | 148 225 | 19.62142 | 62.04837 |
| 336 | 112 896 | 18.33030 | 57.96551 | 386 | 148 996 | 19.64688 | 62.12890 |
| 337 | 113 569 | 18.35756 | 58.05170 | 387 | 149 769 | 19.67232 | 62.20932 |
| 338 | 114 224 | 18.38478 | 57.13777 | 388 | 150 544 | 19.69772 | 62.28965 |
| 339 | 114 921 | 18.41195 | 58.22371 | 389 | 151 321 | 19.72308 | 62.36986 |
| 340 | 115 600 | 18.43909 | 58.30952 | 390 | 152 100 | 19.74842 | 62.44998 |
| 341 | 116 281 | 18.46619 | 58.39521 | 391 | 152 881 | 19.77372 | 62.52999 |
| 342 | 116 694 | 18.49324 | 58.48077 | 392 | 153 664 | 19.79899 | 62.60990 |
| 343 | 117 649 | 18.52026 | 58.56620 | 393 | 154 449 | 19.82423 | 62.68971 |
| 344 | 118 336 | 18.54724 | 58.65151 | 394 | 155 236 | 19.84943 | 62.76942 |
| 345 | | | | | | | |
| 345 | 119 025 | 18.57418 | 58.73670 | 395 | 156 025 | 19.87461 | 62.84903 |
| 346 | 119 716 | 18.60108 | 58.82176 | 396 | 156 816 | 19.89975 | 62.92853 |
| 347 | 120 409 | 18.62794 | 58.90671 | 397 | 157 609 | 19.92486 | 63.00794 |
| 348 | 121 104 | 18.65476 | 58.99152 | 398 | 158 404 | 19.94994 | 63.08724 |
| 349 | 121 801 | 18.68154 | 59.07622 | 399 | 159 201 | 19.97498 | 63.16645 |
| 350 | 122 500 | 18.70829 | 59.16080 | 400 | 160 000 | 20.00000 | 63.24555 |

| N | N² | √N | √10N | N | N² | √N | √10N |
|---|-----|-----|------|---|-----|-----|------|
| 400 | 160 000 | 20.00000 | 63.24555 | 450 | 202 500 | 21.21320 | 67.08204 |
| 401 | 160 801 | 20.02498 | 63.32456 | 451 | 203 401 | 21.23676 | 67.15653 |
| 402 | 161 604 | 20.04994 | 63.40347 | 452 | 204 304 | 21.26029 | 67.23095 |
| 403 | 162 409 | 20.07486 | 63.48228 | 453 | 205 209 | 21.28380 | 67.30527 |
| 404 | 163 216 | 20.09975 | 63.56099 | 454 | 206 116 | 21.30728 | 67.37952 |
| 405 | 164 025 | 20.12461 | 63.63961 | 455 | 207 025 | 21.33073 | 67.45369 |
| 406 | 164 836 | 20.14944 | 63.71813 | 456 | 207 936 | 21.35416 | 67.52777 |
| 407 | 165 649 | 20.17424 | 63.79655 | 457 | 208 849 | 21.37756 | 67.60178 |
| 408 | 166 464 | 20.19901 | 63.87488 | 458 | 209 764 | 21.40093 | 67.67570 |
| 409 | 167 281 | 20.22375 | 63.95311 | 459 | 210 681 | 21.42429 | 67.74954 |
| 410 | 168 100 | 20.24846 | 64.03124 | 460 | 211 600 | 21.44761 | 67.82330 |
| 411 | 168 921 | 20.27313 | 64.10928 | 461 | 212 521 | 21.47091 | 67.89698 |
| 412 | 169 744 | 20.29778 | 64.18723 | 462 | 213 444 | 21.49419 | 67.97058 |
| 413 | 170 569 | 20.32240 | 64.26508 | 463 | 214 369 | 21.51743 | 68.04410 |
| 414 | 171 396 | 20.34699 | 64.34283 | 464 | 215 296 | 21.54066 | 68.11755 |
| 415 | 172 225 | 20.37155 | 64.42049 | 465 | 216 225 | 21.56386 | 68.19091 |
| 416 | 173 056 | 20.39608 | 64.49806 | 466 | 217 156 | 21.58703 | 68.26419 |
| 417 | 173 889 | 20.42058 | 64.57554 | 467 | 218 089 | 21.61018 | 68.33740 |
| 418 | 174 724 | 20.44505 | 64.65292 | 468 | 219 024 | 21.63331 | 68.41053 |
| 419 | 175 561 | 20.46949 | 64.73021 | 469 | 219 961 | 21.65641 | 68.48357 |
| 420 | 176 400 | 20.49390 | 64.80741 | 470 | 220 900 | 21.67948 | 68.55655 |
| 421 | 177 241 | 20.51828 | 64.88451 | 471 | 221 841 | 21.70253 | 68.62944 |
| 422 | 178 084 | 20.54264 | 64.96153 | 472 | 222 784 | 21.72556 | 68.70226 |
| 423 | 178 929 | 20.56696 | 65.03845 | 473 | 223 729 | 21.74856 | 68.77500 |
| 424 | 179 776 | 20.59126 | 65.11528 | 474 | 224 676 | 21.77154 | 68.84706 |
| 425 | 180 625 | 20.61553 | 65.19202 | 475 | 225 625 | 21.79449 | 68.92024 |
| 426 | 181 476 | 20.63977 | 65.26808 | 476 | 226 576 | 21.81742 | 68.99275 |
| 427 | 182 329 | 20.66398 | 65.34524 | 477 | 227 529 | 21.84033 | 69.06519 |
| 428 | 183 184 | 20.68816 | 65.42171 | 478 | 228 484 | 21.86321 | 69.13754 |
| 429 | 184 041 | 20.71232 | 65.49809 | 479 | 229 441 | 21.88607 | 69.20983 |
| 430 | 184 900 | 20.73644 | 65.57439 | 480 | 230 400 | 21.90800 | 69.28203 |
| 431 | 185 761 | 20.76054 | 65.65059 | 481 | 231 361 | 21.93171 | 69.35416 |
| 432 | 186 624 | 20.78461 | 65.72671 | 482 | 232 324 | 21.95450 | 69.42622 |
| 433 | 187 489 | 20.80865 | 65.80274 | 483 | 233 280 | 21.97726 | 69.40820 |
| 434 | 188 356 | 20.83267 | 65.87868 | 484 | 234 256 | 22.00000 | 69.57011 |
| 435 | 189 225 | 20.85665 | 65.95453 | 485 | 235 225 | 22.02272 | 69.64194 |
| 436 | 190 096 | 20.88061 | 66.03030 | 486 | 236 196 | 22.04541 | 69.71370 |
| 437 | 190 969 | 20.90454 | 66.10598 | 487 | 237 169 | 22.06808 | 69.78530 |
| 438 | 191 844 | 20.92845 | 66.18157 | 488 | 238 144 | 22.09072 | 69.85700 |
| 439 | 192 721 | 20.95233 | 66.25708 | 489 | 239 121 | 22.11334 | 69.92853 |
| 440 | 193 600 | 20.97618 | 66.33250 | 490 | 240 100 | 22.13594 | 70.00000 |
| 441 | 194 481 | 21.00000 | 66.40783 | 491 | 241 081 | 22.15852 | 70.07139 |
| 442 | 195 364 | 21.02380 | 66.48308 | 492 | 242 064 | 22.18107 | 70.14271 |
| 443 | 196 249 | 21.04757 | 66.55825 | 493 | 243 049 | 22.20360 | 70.21396 |
| 444 | 197 136 | 21.07131 | 66.63332 | 494 | 244 036 | 22.22611 | 70.28513 |
| 445 | 198 025 | 21.09502 | 66.70832 | 495 | 245 025 | 22.24860 | 70.35624 |
| 446 | 198 916 | 21.11871 | 66.78323 | 496 | 246 016 | 22.27106 | 70.42727 |
| 447 | 199 809 | 21.14237 | 66.85806 | 497 | 247 009 | 22.29350 | 70.49823 |
| 448 | 200 704 | 21.16601 | 66.93280 | 498 | 248 004 | 22.31519 | 70.56912 |
| 449 | 201 601 | 21.18962 | 67.00746 | 499 | 249 001 | 22.33831 | 70.63993 |
| 450 | 202 500 | 21.21320 | 67.08204 | 500 | 250 000 | 22.36068 | 70.71068 |

| N | $N^2$ | $\sqrt{N}$ | $\sqrt{10N}$ | N | $N^2$ | $\sqrt{N}$ | $\sqrt{10N}$ |
|---|---|---|---|---|---|---|---|
| 500 | 250 000 | 22.36068 | 70.71068 | 550 | 302 500 | 23.45208 | 74.16198 |
| 501 | 251 001 | 22.38303 | 70.78135 | 551 | 303 601 | 23.47339 | 74.22937 |
| 502 | 252 004 | 22.40536 | 70.85196 | 552 | 304 704 | 23.49468 | 74.29670 |
| 503 | 253 009 | 22.42766 | 70.92249 | 553 | 305 809 | 23.51595 | 74.36397 |
| 504 | 254 016 | 22.44994 | 70.99296 | 554 | 306 916 | 23.53720 | 74.43118 |
| 505 | 255 025 | 22.47221 | 71.06335 | 555 | 308 025 | 23.55844 | 74.49832 |
| 506 | 256 036 | 22.49444 | 71.13368 | 556 | 309 136 | 23.57965 | 74.56541 |
| 507 | 257 049 | 22.51666 | 71.20393 | 557 | 310 249 | 23.60085 | 74.63243 |
| 508 | 258 064 | 22.53886 | 71.27412 | 558 | 311 364 | 23.62202 | 74.69940 |
| 509 | 259 081 | 22.56103 | 71.34424 | 559 | 312 481 | 23.64318 | 74.76630 |
| 510 | 260 100 | 22.58318 | 71.41428 | 560 | 313 600 | 23.66432 | 74.83315 |
| 511 | 261 121 | 22.60531 | 71.48426 | 561 | 314 721 | 23.68544 | 74.89993 |
| 512 | 262 144 | 22.62742 | 71.55418 | 562 | 315 844 | 23.70654 | 74.96666 |
| 513 | 263 169 | 22.64950 | 71.62402 | 563 | 316 969 | 23.72762 | 75.03333 |
| 514 | 264 196 | 22.67157 | 71.69379 | 564 | 318 096 | 23.74868 | 75.09993 |
| 515 | 265 225 | 22.69361 | 71.76350 | 565 | 319 225 | 23.76973 | 75.16648 |
| 516 | 266 256 | 22.71563 | 71.83314 | 566 | 320 356 | 23.79075 | 75.23297 |
| 517 | 267 289 | 22.73763 | 71.90271 | 567 | 321 489 | 23.81176 | 75.29940 |
| 518 | 268 324 | 22.75961 | 71.97222 | 568 | 322 624 | 23.83275 | 75.36577 |
| 519 | 269 361 | 22.78157 | 72.04165 | 569 | 323 761 | 23.85372 | 75.43209 |
| 520 | 270 400 | 22.80351 | 72.11103 | 570 | 324 900 | 23.87467 | 75.49834 |
| 521 | 271 441 | 22.82542 | 72.18033 | 571 | 326 041 | 23.89561 | 75.56454 |
| 522 | 272 484 | 22.84732 | 72.24957 | 572 | 327 184 | 23.91652 | 75.63068 |
| 523 | 273 529 | 22.86919 | 72.31874 | 573 | 328 329 | 23.93742 | 75.69676 |
| 524 | 274 576 | 22.89105 | 72.38784 | 574 | 329 476 | 23.95830 | 75.76279 |
| 525 | 275 625 | 22.91288 | 72.45688 | 575 | 330 625 | 23.97916 | 75.82875 |
| 526 | 276 676 | 22.93469 | 72.52586 | 576 | 331 776 | 24.00000 | 75.89466 |
| 527 | 277 729 | 22.95648 | 72.59477 | 577 | 332 929 | 24.02082 | 75.96052 |
| 528 | 278 784 | 22.97825 | 72.66361 | 578 | 334 084 | 24.04163 | 76.02631 |
| 529 | 279 841 | 23.00000 | 72.73239 | 579 | 335 241 | 24.06242 | 76.09205 |
| 530 | 280 900 | 23.02173 | 72.80110 | 580 | 336 400 | 24.08319 | 76.15773 |
| 531 | 281 961 | 23.04344 | 72.86975 | 581 | 337 561 | 24.10394 | 76.22336 |
| 532 | 283 024 | 23.06513 | 72.93833 | 582 | 338 724 | 24.12468 | 76.28892 |
| 533 | 284 089 | 23.08679 | 73.00685 | 583 | 339 889 | 24.14539 | 76.35444 |
| 534 | 285 156 | 23.10844 | 73.07530 | 584 | 341 056 | 24.16609 | 76.41989 |
| 535 | 286 225 | 23.13007 | 73.14369 | 585 | 342 225 | 24.18677 | 76.48529 |
| 536 | 287 296 | 23.15167 | 73.21202 | 586 | 343 396 | 24.20744 | 76.55064 |
| 537 | 288 369 | 23.17326 | 73.28028 | 587 | 344 569 | 24.22808 | 76.61593 |
| 538 | 289 444 | 23.19483 | 73.34848 | 588 | 345 744 | 24.24871 | 76.68116 |
| 539 | 290 521 | 23.21637 | 73.41662 | 589 | 346 921 | 24.26932 | 76.74634 |
| 540 | 291 600 | 23.23790 | 73.48469 | 590 | 348 100 | 24.28992 | 76.81146 |
| 541 | 292 681 | 23.25941 | 73.55270 | 591 | 349 281 | 24.31049 | 76.87652 |
| 542 | 293 764 | 23.28089 | 73.62065 | 592 | 350 464 | 24.33105 | 76.94154 |
| 543 | 294 849 | 23.30236 | 73.68853 | 593 | 351 649 | 24.35159 | 77.00649 |
| 544 | 295 936 | 23.32381 | 73.75636 | 594 | 352 836 | 24.37212 | 77.07140 |
| 545 | 297 025 | 23.34524 | 73.82412 | 595 | 354 025 | 24.39262 | 77.13624 |
| 546 | 298 116 | 23.36664 | 73.89181 | 596 | 355 216 | 24.41311 | 77.20104 |
| 547 | 299 209 | 23.38803 | 73.95945 | 597 | 356 409 | 24.43358 | 77.26578 |
| 548 | 300 304 | 23.40940 | 74.02702 | 598 | 357 604 | 24.45404 | 77.33046 |
| 549 | 301 401 | 23.43075 | 74.09453 | 599 | 358 801 | 24.47448 | 77.39509 |
| 550 | 302 500 | 23.45208 | 74.16198 | 600 | 360 000 | 24.49490 | 77.45967 |

| N | N² | √N | √10N | N | N² | √N | √10N |
|---|---|---|---|---|---|---|---|
| 600 | 360 000 | 24.49490 | 77.45967 | 650 | 422 500 | 25.49510 | 80.62258 |
| 601 | 361 201 | 24.51530 | 77.52419 | 651 | 423 801 | 25.51470 | 80.68457 |
| 602 | 362 404 | 24.53569 | 77.58868 | 652 | 425 409 | 25.55386 | 80.80842 |
| 603 | 363 609 | 24.55606 | 77.65307 | 653 | 426 409 | 25.55386 | 80.80842 |
| 604 | 364 816 | 24.57641 | 77.71744 | 654 | 427 716 | 25.57342 | 80.87027 |
| 605 | 366 025 | 24.59675 | 77.78175 | 655 | 429 025 | 25.59297 | 80.93207 |
| 606 | 367 236 | 24.61707 | 77.84600 | 656 | 430 336 | 25.61250 | 80.99383 |
| 607 | 368 449 | 24.63737 | 77.91020 | 657 | 431 649 | 25.63201 | 81.05554 |
| 608 | 369 664 | 24.65766 | 77.97435 | 658 | 432 964 | 25.65151 | 81.11720 |
| 609 | 370 881 | 24.67793 | 78.03845 | 659 | 434 281 | 25.67100 | 81.17881 |
| 610 | 372 100 | 24.69818 | 78.10250 | 660 | 435 600 | 25.69047 | 81.24038 |
| 611 | 373 321 | 24.71841 | 78.16649 | 661 | 436 921 | 25.70992 | 81.30191 |
| 612 | 374 544 | 24.73863 | 78.23043 | 662 | 438 244 | 25.72936 | 81.36338 |
| 613 | 375 769 | 24.75884 | 78.29432 | 663 | 439 569 | 25.74879 | 81.42481 |
| 614 | 376 996 | 24.77902 | 78.35815 | 664 | 440 896 | 25.76820 | 81.48620 |
| 615 | 378 225 | 24.79919 | 78.42194 | 665 | 442 225 | 25.78759 | 81.54753 |
| 616 | 379 456 | 24.81935 | 78.48567 | 666 | 443 556 | 25.80698 | 81.60882 |
| 617 | 380 689 | 24.83948 | 78.54935 | 667 | 444 889 | 25.82634 | 81.67007 |
| 618 | 381 924 | 24.85961 | 78.61298 | 668 | 446 224 | 25.84570 | 81.73127 |
| 619 | 383 161 | 24.87971 | 78.67655 | 669 | 447 561 | 25.86503 | 81.79242 |
| 620 | 384 400 | 24.89980 | 78.74008 | 670 | 448 900 | 25.88436 | 81.85353 |
| 621 | 385 641 | 24.91987 | 78.80355 | 671 | 450 241 | 25.90367 | 81.91459 |
| 622 | 386 884 | 24.93993 | 78.86698 | 672 | 451 584 | 25.92296 | 81.97561 |
| 623 | 288 129 | 24.95997 | 78.93035 | 673 | 452 929 | 25.94224 | 82.03658 |
| 624 | 389 376 | 24.97999 | 78.99367 | 674 | 454 276 | 25.96151 | 82.09750 |
| 625 | 390 625 | 25.00000 | 79.05694 | 675 | 455 625 | 25.98076 | 82.15838 |
| 626 | 391 876 | 25.01999 | 79.12016 | 676 | 456 976 | 26.00000 | 82.21922 |
| 627 | 393 129 | 25.03997 | 79.18333 | 677 | 458 329 | 26.01922 | 82.28001 |
| 628 | 394 384 | 25.05993 | 79.24645 | 678 | 459 684 | 26.03843 | 82.34076 |
| 629 | 395 641 | 25.07987 | 79.30952 | 679 | 461 041 | 26.05763 | 82.40146 |
| 630 | 396 900 | 25.09980 | 79.37254 | 680 | 462 400 | 26.07681 | 82.46211 |
| 631 | 398 161 | 25.11971 | 79.43551 | 681 | 463 761 | 26.09598 | 82.42272 |
| 632 | 399 424 | 25.13961 | 79.49843 | 682 | 465 124 | 26.11513 | 82.58329 |
| 633 | 400 689 | 25.15949 | 79.56130 | 683 | 466 489 | 26.13427 | 82.64381 |
| 634 | 401 956 | 25.17936 | 79.62412 | 684 | 467 856 | 26.15339 | 82.70429 |
| 635 | 403 225 | 25.19921 | 79.68689 | 685 | 469 225 | 26.17250 | 82.76473 |
| 636 | 404 496 | 25.21904 | 79.74961 | 686 | 470 596 | 26.19160 | 82.82512 |
| 637 | 405 769 | 25.23886 | 79.81228 | 687 | 471 969 | 26.21068 | 82.88546 |
| 638 | 407 044 | 25.25866 | 79.87490 | 688 | 473 344 | 26.22975 | 82.94577 |
| 639 | 408 321 | 25.27845 | 79.93748 | 689 | 474 721 | 26.24881 | 83.00602 |
| 640 | 409 600 | 25.29822 | 80.00000 | 690 | 476 100 | 26.26785 | 83.06624 |
| 641 | 410 881 | 25.31798 | 80.06248 | 691 | 477 481 | 26.28688 | 83.12641 |
| 642 | 412 164 | 25.33772 | 80.12490 | 692 | 478 864 | 26.30589 | 83.18654 |
| 643 | 413 449 | 25.35744 | 80.18728 | 693 | 480 249 | 26.32489 | 83.24662 |
| 644 | 414 736 | 25.37716 | 80.24961 | 694 | 481 636 | 26.34388 | 83.30666 |
| 645 | 416 025 | 25.39685 | 80.31189 | 695 | 483 025 | 26.36285 | 83.36666 |
| 646 | 417 316 | 25.41653 | 80.37413 | 696 | 484 416 | 26.38181 | 83.42661 |
| 647 | 418 609 | 25.43619 | 80.43631 | 697 | 485 809 | 26.40076 | 83.48653 |
| 648 | 419 904 | 25.45584 | 80.49845 | 698 | 487 204 | 26.41969 | 83.54639 |
| 649 | 421 201 | 25.47548 | 80.56054 | 699 | 488 601 | 26.43861 | 83.60622 |
| 650 | 422 500 | 25.49510 | 80.62258 | 700 | 490 000 | 26.45751 | 83.66600 |

| N | N² | √N | √10N | N | N² | √N | √10N |
|-----|---------|----------|----------|-----|---------|----------|----------|
| 700 | 490 000 | 26.45751 | 83.66600 | 750 | 562 500 | 27.38613 | 86.60254 |
| 701 | 491 401 | 26.47640 | 83.72574 | 751 | 564 001 | 27.40438 | 86.66026 |
| 702 | 492 804 | 26.49528 | 83.78544 | 752 | 565 504 | 27.42262 | 86.71793 |
| 703 | 494 209 | 26.51415 | 83.84510 | 753 | 567 009 | 27.44085 | 86.77557 |
| 704 | 495 616 | 26.53300 | 83.90471 | 754 | 568 516 | 27.45906 | 86.83317 |
| 705 | 497 025 | 26.55184 | 83.96428 | 755 | 570 025 | 27.47726 | 86.89074 |
| 706 | 498 436 | 26.57066 | 84.02381 | 756 | 571 536 | 27.49545 | 86.94826 |
| 707 | 499 849 | 26.58947 | 84.08329 | 757 | 573 049 | 27.51363 | 87.00575 |
| 708 | 501 264 | 26.60827 | 84.14274 | 758 | 574 564 | 27.53180 | 87.06320 |
| 709 | 502 681 | 26.62705 | 84.20214 | 759 | 576 081 | 27.54995 | 87.12061 |
| 710 | 504 100 | 26.64583 | 84.26150 | 760 | 577 600 | 27.56810 | 87.17798 |
| 711 | 505 521 | 26.66458 | 84.32082 | 761 | 579 121 | 27.58623 | 87.23531 |
| 712 | 506 944 | 26.68333 | 84.38009 | 762 | 580 644 | 27.60435 | 87.29261 |
| 713 | 508 369 | 26.70206 | 84.43933 | 763 | 582 169 | 27.62245 | 87.34987 |
| 714 | 509 796 | 26.72078 | 84.49852 | 764 | 583 696 | 27.64055 | 87.40709 |
| 715 | 511 225 | 26.73948 | 84.55767 | 765 | 585 225 | 27.65863 | 87.46428 |
| 716 | 512 656 | 26.75818 | 84.61578 | 766 | 586 756 | 27.67671 | 87.52143 |
| 717 | 514 089 | 26.77686 | 84.67585 | 767 | 588 289 | 27.69476 | 87.57854 |
| 718 | 515 524 | 26.79552 | 84.73488 | 768 | 589 824 | 27.71281 | 87.63561 |
| 719 | 516 961 | 26.81418 | 84.79387 | 769 | 591 361 | 27.73085 | 87.69265 |
| 720 | 518 400 | 26.83282 | 84.85281 | 770 | 592 900 | 27.74887 | 87.74964 |
| 721 | 519 841 | 26.85144 | 84.91172 | 771 | 594 441 | 27.76689 | 87.80661 |
| 722 | 521 284 | 26.87006 | 84.97058 | 772 | 595 984 | 27.78489 | 87.86353 |
| 723 | 522 729 | 26.88866 | 85.02941 | 773 | 597 529 | 27.80288 | 87.92042 |
| 724 | 524 176 | 26.90725 | 85.08819 | 774 | 599 076 | 27.82086 | 87.97727 |
| 725 | 525 625 | 26.92582 | 85.14693 | 775 | 600 625 | 27.83882 | 88.03408 |
| 726 | 527 076 | 26.94439 | 85.20563 | 776 | 602 176 | 27.85678 | 88.09086 |
| 727 | 528 529 | 26.96294 | 85.26429 | 777 | 603 729 | 27.87472 | 88.14760 |
| 728 | 529 984 | 26.98148 | 85.32292 | 778 | 605 284 | 27.89265 | 88.20431 |
| 729 | 531 411 | 27.00000 | 85.38150 | 779 | 606 841 | 27.91057 | 88.26098 |
| 730 | 532 900 | 27.01851 | 85.44004 | 780 | 608 400 | 27.92848 | 88.31761 |
| 731 | 534 361 | 27.03701 | 85.49854 | 781 | 609 961 | 27.94638 | 88.37420 |
| 732 | 535 824 | 27.05550 | 85.55700 | 782 | 611 524 | 27.96426 | 88.43076 |
| 733 | 537 289 | 27.07397 | 85.61542 | 783 | 613 089 | 27.98214 | 88.48729 |
| 734 | 538 756 | 27.09243 | 85.67380 | 784 | 614 656 | 28.00000 | 88.54377 |
| 735 | 540 225 | 27.11088 | 85.73214 | 785 | 616 225 | 28.01785 | 88.60023 |
| 736 | 541 696 | 27.12932 | 85.79044 | 786 | 617 796 | 28.03569 | 88.65664 |
| 737 | 543 169 | 27.14774 | 85.84870 | 787 | 619 369 | 28.05352 | 88.71302 |
| 738 | 544 644 | 27.16616 | 85.90693 | 788 | 620 944 | 28.07134 | 88.76936 |
| 739 | 546 121 | 27.18455 | 85.96511 | 789 | 622 521 | 28.08914 | 88.82567 |
| 740 | 547 600 | 27.20294 | 86.02325 | 790 | 624 100 | 28.10694 | 88.88194 |
| 741 | 549 081 | 27.22132 | 86.08136 | 791 | 625 681 | 28.12472 | 88.93818 |
| 742 | 550 564 | 27.23968 | 86.13942 | 792 | 627 264 | 28.14249 | 88.99438 |
| 743 | 552 049 | 27.25803 | 86.10745 | 793 | 628 849 | 28.16026 | 89.05055 |
| 744 | 553 536 | 27.27636 | 86.25543 | 794 | 630 436 | 28.17801 | 89.10668 |
| 745 | 555 025 | 27.29469 | 86.31338 | 795 | 632 025 | 28.19574 | 89.16277 |
| 746 | 556 516 | 27.31300 | 86.37129 | 796 | 633 616 | 28.21347 | 89.21883 |
| 747 | 558 009 | 27.33130 | 86.42916 | 797 | 635 209 | 28.23119 | 89.27486 |
| 748 | 559 504 | 27.34959 | 86.48609 | 798 | 636 804 | 28.24889 | 89.33085 |
| 749 | 561 001 | 27.36786 | 86.54479 | 799 | 638 401 | 28.26659 | 89.38680 |
| 750 | 562 500 | 27.38613 | 86.60254 | 800 | 640 000 | 28.28427 | 89.44272 |

| N | N² | √N | √10N | N | N² | √N | √10N |
|---|---|---|---|---|---|---|---|
| 800 | 640 000 | 28.28427 | 89.44272 | 850 | 722 500 | 29.15476 | 92.19544 |
| 801 | 641 601 | 28.30194 | 89.49860 | 851 | 724 201 | 29.17190 | 92.24966 |
| 802 | 643 204 | 28.31960 | 89.55445 | 852 | 725 904 | 29.18904 | 92.30385 |
| 803 | 644 809 | 28.33725 | 89.61027 | 853 | 727 609 | 29.20616 | 92.35800 |
| 804 | 646 416 | 28.35489 | 89.66605 | 854 | 729 316 | 29.22328 | 92.41212 |
| 805 | 648 025 | 28.37252 | 89.72179 | 855 | 731 025 | 29.24038 | 92.46621 |
| 806 | 649 636 | 28.39014 | 89.77750 | 856 | 732 736 | 29.25748 | 92.52027 |
| 807 | 651 249 | 28.40775 | 89.83318 | 857 | 734 449 | 29.27456 | 92.57429 |
| 808 | 652 864 | 28.42534 | 89.88882 | 858 | 736 164 | 29.29164 | 92.62829 |
| 809 | 654 481 | 28.44293 | 89.94443 | 859 | 737 881 | 29.30870 | 92.68225 |
| 810 | 656 100 | 28.46050 | 90.00000 | 860 | 739 600 | 29.32576 | 92.73618 |
| 811 | 657 721 | 28.47806 | 90.05554 | 861 | 741 321 | 29.34280 | 92.79009 |
| 812 | 659 344 | 28.49561 | 90.11104 | 862 | 743 044 | 29.35984 | 92.84396 |
| 813 | 660 969 | 28.51315 | 90.16651 | 863 | 744 769 | 29.37686 | 92.89779 |
| 814 | 662 596 | 28.53069 | 90.22195 | 864 | 746 496 | 29.39388 | 92.95160 |
| 815 | 664 225 | 28.54820 | 90.27735 | 865 | 748 225 | 29.41088 | 93.00538 |
| 816 | 665 856 | 28.56571 | 90.33272 | 866 | 749 956 | 29.42788 | 93.05912 |
| 817 | 667 489 | 28.58321 | 90.38805 | 867 | 751 689 | 29.44486 | 93.11283 |
| 818 | 669 124 | 28.60070 | 90.44335 | 868 | 753 424 | 29.46184 | 93.16652 |
| 819 | 670 761 | 28.61818 | 90.49862 | 869 | 755 161 | 29.47881 | 93.22017 |
| 820 | 672 400 | 28.63564 | 90.55385 | 870 | 756 900 | 29.49576 | 93.27379 |
| 821 | 674 041 | 28.65310 | 90.60905 | 871 | 758 641 | 29.51271 | 93.32738 |
| 822 | 675 684 | 28.67054 | 90.66422 | 872 | 760 384 | 29.52965 | 93.38094 |
| 823 | 677 329 | 28.68798 | 90.71935 | 873 | 762 129 | 29.54657 | 93.43447 |
| 824 | 678 976 | 28.70540 | 90.77445 | 874 | 763 876 | 29.56349 | 93.48797 |
| 825 | 680 625 | 28.72281 | 90.82951 | 875 | 765 625 | 29.58040 | 93.54143 |
| 826 | 682 276 | 28.74022 | 90.88454 | 876 | 767 376 | 29.59730 | 93.59487 |
| 827 | 683 929 | 28.75761 | 90.93954 | 877 | 769 129 | 29.61419 | 93.64828 |
| 828 | 685 584 | 28.77499 | 90.99451 | 878 | 770 884 | 29.63106 | 93.70165 |
| 829 | 687 241 | 28.79236 | 91.04944 | 879 | 772 641 | 29.64793 | 93.75500 |
| 830 | 688 900 | 28.80972 | 91.10434 | 880 | 774 400 | 29.66479 | 93.80832 |
| 831 | 690 561 | 28.82707 | 91.15920 | 881 | 776 161 | 29.68164 | 93.86160 |
| 832 | 692 224 | 28.84441 | 91.21403 | 882 | 777 924 | 29.69848 | 93.91486 |
| 833 | 693 889 | 28.86174 | 91.26883 | 883 | 779 689 | 29.71532 | 93.96808 |
| 834 | 695 556 | 28.87906 | 91.32360 | 884 | 781 456 | 29.73214 | 94.02027 |
| 835 | 697 225 | 28.89637 | 91.37833 | 885 | 783 225 | 29.74895 | 94.07444 |
| 836 | 698 896 | 28.91366 | 91.43304 | 886 | 784 996 | 29.76575 | 94.12757 |
| 837 | 700 569 | 28.93095 | 91.48770 | 887 | 786 769 | 29.78255 | 94.18068 |
| 838 | 702 244 | 28.94823 | 91.54234 | 888 | 788 544 | 29.79933 | 94.23375 |
| 839 | 703 921 | 28.96550 | 91.59694 | 889 | 790 321 | 29.81610 | 94.28680 |
| 840 | 705 600 | 28.98275 | 91.65151 | 890 | 792 100 | 29.83287 | 94.33981 |
| 841 | 707 281 | 29.00000 | 91.70605 | 891 | 793 881 | 29.84962 | 94.39280 |
| 842 | 708 964 | 29.01724 | 91.76056 | 892 | 795 664 | 29.86637 | 94.44575 |
| 843 | 710 649 | 29.03446 | 91.81503 | 893 | 797 449 | 29.88311 | 94.49868 |
| 844 | 712 336 | 29.05168 | 91.86947 | 894 | 799 236 | 29.89983 | 94.55157 |
| 845 | 714 025 | 29.06888 | 91.92388 | 895 | 801 025 | 29.91655 | 94.60444 |
| 846 | 715 716 | 29.08608 | 91.97826 | 896 | 802 816 | 29.93326 | 94.65728 |
| 847 | 717 409 | 29.10326 | 92.03260 | 897 | 804 609 | 29.94996 | 94.71008 |
| 848 | 719 104 | 29.12044 | 92.08692 | 898 | 806 404 | 29.96665 | 94.76286 |
| 849 | 720 801 | 29.13760 | 92.14120 | 899 | 808 201 | 29.98333 | 94.81561 |
| 850 | 722 500 | 29.15476 | 92.19544 | 900 | 810 000 | 30.00000 | 94.86833 |

| N | N² | $\sqrt{N}$ | $\sqrt{10N}$ | N | N² | $\sqrt{N}$ | $\sqrt{10N}$ |
|---|---|---|---|---|---|---|---|
| 900 | 810 000 | 30.00000 | 94.86833 | 950 | 902 500 | 30.82207 | 97.46794 |
| 901 | 811 801 | 30.01666 | 94.92102 | 951 | 904 401 | 30.83829 | 97.51923 |
| 902 | 813 604 | 30.03331 | 94.97368 | 952 | 906 304 | 30.85450 | 97.57049 |
| 903 | 815 409 | 30.04996 | 95.02631 | 953 | 908 209 | 30.87070 | 97.62172 |
| 904 | 817 216 | 30.06659 | 95.07891 | 954 | 910.116 | 30.88689 | 97.67292 |
| 905 | 819 025 | 30.08322 | 95.13149 | 955 | 912 025 | 30.90307 | 97.72410 |
| 906 | 820 836 | 30.09983 | 95.18403 | 956 | 913 936 | 30.91925 | 97.77525 |
| 907 | 822 649 | 30.11644 | 95.23655 | 957 | 915 849 | 30.93542 | 97.82638 |
| 908 | 824 464 | 30.13304 | 95.28903 | 958 | 917 764 | 30.95158 | 97.87747 |
| 909 | 826 281 | 30.14963 | 95.34149 | 959 | 919 681 | 30.96773 | 97.92855 |
| 910 | 828 100 | 30.16621 | 95.39392 | 960 | 921 600 | 30.98387 | 97.97959 |
| 911 | 829 921 | 30.18278 | 95.44632 | 961 | 928 521 | 31.00000 | 98.03061 |
| 912 | 831 744 | 30.19934 | 95.49869 | 962 | 925 444 | 31.01612 | 98.08160 |
| 913 | 833 569 | 30.21589 | 95.55103 | 963 | 927 369 | 31.03224 | 98.13256 |
| 914 | 835 396 | 30.23243 | 95.60335 | 964 | 929 296 | 31.04835 | 98.18350 |
| 915 | 837 225 | 30.24897 | 95.65563 | 965 | 931 225 | 31.06445 | 98.23441 |
| 916 | 839 056 | 30.26549 | 95.70789 | 966 | 933 156 | 31.08054 | 98.28530 |
| 917 | 840 889 | 30.28201 | 95.76012 | 967 | 935 089 | 31.09662 | 98.33616 |
| 918 | 842 724 | 30.29851 | 95.81232 | 968 | 937 024 | 31.11270 | 98.38699 |
| 919 | 844 561 | 30.31501 | 95.86449 | 969 | 938 961 | 31.12876 | 98.43780 |
| 920 | 846 400 | 30.33150 | 95.91663 | 970 | 940 900 | 31.14482 | 98.48858 |
| 921 | 848 241 | 30.34798 | 95.96874 | 971 | 942 841 | 31.16087 | 98.53933 |
| 922 | 850 084 | 30.36445 | 96.02083 | 972 | 944 784 | 31.17691 | 98.59006 |
| 923 | 851 929 | 30.38092 | 96.07289 | 973 | 946 729 | 31.19295 | 98.64076 |
| 924 | 853 776 | 30.39735 | 96.12492 | 974 | 948 676 | 31.20897 | 98.69144 |
| 925 | 855 625 | 30.41381 | 96.17692 | 975 | 950 625 | 31.22499 | 98.74209 |
| 926 | 857 476 | 30.43025 | 96.22889 | 976 | 952 576 | 31.24100 | 98.79271 |
| 927 | 859 329 | 30.44667 | 96.28084 | 977 | 954 529 | 31.25700 | 98.84331 |
| 928 | 861 184 | 30.46309 | 96.33276 | 978 | 956 484 | 31.27299 | 98.89388 |
| 929 | 863 041 | 30.47950 | 96.28465 | 979 | 958 441 | 31.28898 | 98.94443 |
| 930 | 864 900 | 30.49590 | 96.43651 | 980 | 960 400 | 31.30495 | 98.99495 |
| 931 | 866 761 | 30.51229 | 96.48834 | 981 | 962 361 | 31.32092 | 99.04544 |
| 932 | 868 624 | 30.52868 | 96.54015 | 982 | 964 324 | 31.33688 | 99.09591 |
| 933 | 870 489 | 30.54505 | 96.59193 | 983 | 966 144 | 31.43247 | 99.44848 |
| 934 | 872 356 | 30.56141 | 96.64368 | 984 | 968 256 | 31.36877 | 99.19677 |
| 935 | 874 225 | 30.57777 | 96.69540 | 985 | 970 225 | 31.38471 | 99.24717 |
| 936 | 876 096 | 30.59412 | 96.74709 | 986 | 972 196 | 31.40064 | 99.29753 |
| 937 | 877 969 | 30.61046 | 96.79876 | 987 | 974 169 | 31.41656 | 99.34787 |
| 938 | 879 844 | 30.62679 | 96.85040 | 988 | 976 144 | 31.43247 | 99.39819 |
| 939 | 881 721 | 30.64311 | 96.90201 | 989 | 978 121 | 31.44837 | 99.44848 |
| 940 | 883 600 | 30.65942 | 96.95360 | 990 | 980 100 | 31.46427 | 99.49874 |
| 941 | 885 481 | 30.67572 | 97.00515 | 991 | 982 081 | 31.48015 | 99.54898 |
| 942 | 887 364 | 30.69202 | 97.05668 | 992 | 984 064 | 31.49603 | 99.59920 |
| 943 | 889 249 | 30.70831 | 97.10819 | 993 | 986 049 | 31.51190 | 99.64939 |
| 944 | 891 136 | 30.72458 | 97.15966 | 994 | 988 036 | 31.52777 | 99.69955 |
| 945 | 893 025 | 30.74085 | 97.21111 | 995 | 990 025 | 31.54362 | 99.74969 |
| 946 | 894 916 | 30.75711 | 97.26253 | 996 | 992 016 | 31.55947 | 99.79980 |
| 947 | 896 809 | 30.77337 | 97.31393 | 997 | 994 009 | 31.57531 | 99.84989 |
| 948 | 898 704 | 30.78961 | 97.36529 | 998 | 996 004 | 31.59114 | 99.89995 |
| 949 | 900 601 | 30.80584 | 97.41663 | 999 | 998 001 | 31.60696 | 99.94999 |
| 950 | 902 500 | 30.82207 | 97.46794 | 1000 | 1 000 000 | 31.62278 | 100.00000 |

# APPENDIX 6
# F DISTRIBUTIONS TABLES

The following tables provide critical values of $F$ at the .05 and .01 levels of significance. The number of degrees of freedom for the *numerator* is indicated at the top of each *column*, and the number of degrees of freedom for the *denominator* determines the *row* to use.

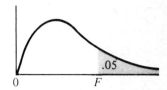

## Critical Values of $F_{v_1, v_2}$ for $\alpha = .05$

| $v_2$ | $v_1 = 1$ | 2 | 3 | 4 | 5 | 6 | 7 | 8 | 9 | 10 | 12 | 15 | 20 | 24 | 30 | 40 | 60 | 120 | ∞ |
|---|---|---|---|---|---|---|---|---|---|---|---|---|---|---|---|---|---|---|---|
| 1 | 161 | 200 | 216 | 225 | 230 | 234 | 237 | 239 | 241 | 242 | 244 | 246 | 248 | 249 | 250 | 251 | 252 | 253 | 254 |
| 2 | 18.5 | 19.0 | 19.2 | 19.2 | 19.3 | 19.3 | 19.4 | 19.4 | 19.4 | 19.4 | 19.4 | 19.4 | 19.4 | 19.5 | 19.5 | 19.5 | 19.5 | 19.5 | 19.5 |
| 3 | 10.1 | 9.55 | 9.28 | 9.12 | 9.01 | 8.94 | 8.89 | 8.85 | 8.81 | 8.79 | 8.74 | 8.70 | 8.66 | 8.64 | 8.62 | 8.59 | 8.57 | 8.55 | 8.53 |
| 4 | 7.71 | 6.94 | 6.59 | 6.39 | 6.26 | 6.16 | 6.09 | 6.04 | 6.00 | 5.96 | 5.91 | 5.86 | 5.80 | 5.77 | 5.75 | 5.72 | 5.69 | 5.66 | 5.63 |
| 5 | 6.61 | 5.79 | 5.41 | 5.19 | 5.05 | 4.95 | 4.88 | 4.82 | 4.77 | 4.74 | 4.68 | 4.62 | 4.56 | 4.53 | 4.50 | 4.46 | 4.43 | 4.40 | 4.37 |
| 6 | 5.99 | 5.14 | 4.76 | 4.53 | 4.39 | 4.28 | 4.21 | 4.15 | 4.10 | 4.06 | 4.00 | 3.94 | 3.87 | 3.84 | 3.81 | 3.77 | 3.74 | 3.70 | 3.67 |
| 7 | 5.59 | 4.74 | 4.35 | 4.12 | 3.97 | 3.87 | 3.79 | 3.73 | 3.68 | 3.64 | 3.57 | 3.51 | 3.44 | 3.41 | 3.38 | 3.34 | 3.30 | 3.27 | 3.23 |
| 8 | 5.32 | 4.46 | 4.07 | 3.84 | 3.69 | 3.58 | 3.50 | 3.44 | 3.39 | 3.35 | 3.28 | 3.22 | 3.15 | 3.12 | 3.08 | 3.04 | 3.01 | 2.97 | 2.93 |
| 9 | 5.12 | 4.26 | 3.86 | 3.63 | 3.48 | 3.37 | 3.29 | 3.23 | 3.18 | 3.14 | 3.07 | 3.01 | 2.94 | 2.90 | 2.86 | 2.83 | 2.79 | 2.75 | 2.71 |
| 10 | 4.96 | 4.10 | 3.71 | 3.48 | 3.33 | 3.22 | 3.14 | 3.07 | 3.02 | 2.98 | 2.91 | 2.85 | 2.77 | 2.74 | 2.70 | 2.66 | 2.62 | 2.58 | 2.54 |
| 11 | 4.84 | 3.98 | 3.59 | 3.36 | 3.20 | 3.09 | 3.01 | 2.95 | 2.90 | 2.85 | 2.79 | 2.72 | 2.65 | 2.61 | 2.57 | 2.53 | 2.49 | 2.45 | 2.40 |
| 12 | 4.75 | 3.89 | 3.49 | 3.26 | 3.11 | 3.00 | 2.91 | 2.85 | 2.80 | 2.75 | 2.69 | 2.62 | 2.54 | 2.51 | 2.47 | 2.43 | 2.38 | 2.34 | 2.30 |
| 13 | 4.67 | 3.81 | 3.41 | 3.18 | 3.03 | 2.92 | 2.83 | 2.77 | 2.71 | 2.67 | 2.60 | 2.53 | 2.46 | 2.42 | 2.38 | 2.34 | 2.30 | 2.25 | 2.21 |
| 14 | 4.60 | 3.74 | 3.34 | 3.11 | 2.96 | 2.85 | 2.76 | 2.70 | 2.65 | 2.60 | 2.53 | 2.46 | 2.39 | 2.35 | 2.31 | 2.27 | 2.22 | 2.18 | 2.13 |
| 15 | 4.54 | 3.68 | 3.29 | 3.06 | 2.90 | 2.79 | 2.71 | 2.64 | 2.59 | 2.54 | 2.48 | 2.40 | 2.33 | 2.29 | 2.25 | 2.20 | 2.16 | 2.11 | 2.07 |
| 16 | 4.49 | 3.63 | 3.24 | 3.01 | 2.85 | 2.74 | 2.66 | 2.59 | 2.54 | 2.49 | 2.42 | 2.35 | 2.28 | 2.24 | 2.19 | 2.15 | 2.11 | 2.06 | 2.01 |
| 17 | 4.45 | 3.59 | 3.20 | 2.96 | 2.81 | 2.70 | 2.61 | 2.55 | 2.49 | 2.45 | 2.38 | 2.31 | 2.23 | 2.19 | 2.15 | 2.10 | 2.06 | 2.01 | 1.96 |
| 18 | 4.41 | 3.55 | 3.16 | 2.93 | 2.77 | 2.66 | 2.58 | 2.51 | 2.46 | 2.41 | 2.34 | 2.27 | 2.19 | 2.15 | 2.11 | 2.06 | 2.02 | 1.97 | 1.92 |
| 19 | 4.38 | 3.52 | 3.13 | 2.90 | 2.74 | 2.63 | 2.54 | 2.48 | 2.42 | 2.38 | 2.31 | 2.23 | 2.16 | 2.11 | 2.07 | 2.03 | 1.98 | 1.93 | 1.88 |
| 20 | 4.35 | 3.49 | 3.10 | 2.87 | 2.71 | 2.60 | 2.51 | 2.45 | 2.39 | 2.35 | 2.28 | 2.20 | 2.12 | 2.08 | 2.04 | 1.99 | 1.95 | 1.90 | 1.84 |
| 21 | 4.32 | 3.47 | 3.07 | 2.84 | 2.68 | 2.57 | 2.49 | 2.42 | 2.37 | 2.32 | 2.25 | 2.18 | 2.10 | 2.05 | 2.01 | 1.96 | 1.92 | 1.87 | 1.81 |
| 22 | 4.30 | 3.44 | 3.05 | 2.82 | 2.66 | 2.55 | 2.46 | 2.40 | 2.34 | 2.30 | 2.23 | 2.15 | 2.07 | 2.03 | 1.98 | 1.94 | 1.89 | 1.84 | 1.78 |
| 23 | 4.28 | 3.42 | 3.03 | 2.80 | 2.64 | 2.53 | 2.44 | 2.37 | 2.32 | 2.27 | 2.20 | 2.13 | 2.05 | 2.01 | 1.96 | 1.91 | 1.86 | 1.81 | 1.76 |
| 24 | 4.26 | 3.40 | 3.01 | 2.78 | 2.62 | 2.51 | 2.42 | 2.36 | 2.30 | 2.25 | 2.18 | 2.11 | 2.03 | 1.98 | 1.94 | 1.89 | 1.84 | 1.79 | 1.73 |
| 25 | 4.24 | 3.39 | 2.99 | 2.76 | 2.60 | 2.49 | 2.40 | 2.34 | 2.28 | 2.24 | 2.16 | 2.09 | 2.01 | 1.96 | 1.92 | 1.87 | 1.82 | 1.77 | 1.71 |
| 30 | 4.17 | 3.32 | 2.92 | 2.69 | 2.53 | 2.42 | 2.33 | 2.27 | 2.21 | 2.16 | 2.09 | 2.01 | 1.93 | 1.89 | 1.84 | 1.79 | 1.74 | 1.68 | 1.62 |
| 40 | 4.08 | 3.23 | 2.84 | 2.61 | 2.45 | 2.34 | 2.25 | 2.18 | 2.12 | 2.08 | 2.00 | 1.92 | 1.84 | 1.79 | 1.74 | 1.69 | 1.64 | 1.58 | 1.51 |
| 60 | 4.00 | 3.15 | 2.76 | 2.53 | 2.37 | 2.25 | 2.17 | 2.10 | 2.04 | 1.99 | 1.92 | 1.84 | 1.75 | 1.70 | 1.65 | 1.59 | 1.53 | 1.47 | 1.39 |
| 120 | 3.92 | 3.07 | 2.68 | 2.45 | 2.29 | 2.18 | 2.09 | 2.02 | 1.96 | 1.91 | 1.83 | 1.75 | 1.66 | 1.61 | 1.55 | 1.50 | 1.43 | 1.35 | 1.25 |
| ∞ | 3.84 | 3.00 | 2.60 | 2.37 | 2.21 | 2.10 | 2.01 | 1.94 | 1.88 | 1.83 | 1.75 | 1.67 | 1.57 | 1.52 | 1.46 | 1.39 | 1.32 | 1.22 | 1.00 |

$v_1 = $ Degrees of freedom for numerator

$v_2 = $ Degrees of freedom for denominator

## Critical Values of $F_{v_1, v_2}$ for $\alpha = .01$

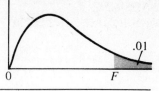

| $v_2$ = Degrees of freedom for denominator | $v_1$ = Degrees of freedom for numerator | | | | | | | | | | | | | | | | | | |
|---|---|---|---|---|---|---|---|---|---|---|---|---|---|---|---|---|---|---|---|
| | 1 | 2 | 3 | 4 | 5 | 6 | 7 | 8 | 9 | 10 | 12 | 15 | 20 | 24 | 30 | 40 | 60 | 120 | ∞ |
| 1 | 4,052 | 5,000 | 5,403 | 5,625 | 5,764 | 5,859 | 5,928 | 5,982 | 6,023 | 6,056 | 6,106 | 6,157 | 6,209 | 6,235 | 6,261 | 6,287 | 6,313 | 6,339 | 6,366 |
| 2 | 98.5 | 99.0 | 99.2 | 99.2 | 99.3 | 99.3 | 99.4 | 99.4 | 99.4 | 99.4 | 99.4 | 99.4 | 99.4 | 99.5 | 99.5 | 99.5 | 99.5 | 99.5 | 99.5 |
| 3 | 34.1 | 30.8 | 29.5 | 28.7 | 28.2 | 27.9 | 27.7 | 27.5 | 27.3 | 27.2 | 27.1 | 26.9 | 26.7 | 26.6 | 26.5 | 26.4 | 26.3 | 26.2 | 26.1 |
| 4 | 21.2 | 18.0 | 16.7 | 16.0 | 15.5 | 15.2 | 15.0 | 14.8 | 14.7 | 14.5 | 14.4 | 14.2 | 14.0 | 13.9 | 13.8 | 13.7 | 13.7 | 13.6 | 13.5 |
| 5 | 16.3 | 13.3 | 12.1 | 11.4 | 11.0 | 10.7 | 10.5 | 10.3 | 10.2 | 10.1 | 9.89 | 9.72 | 9.55 | 9.47 | 9.38 | 9.29 | 9.20 | 9.11 | 9.02 |
| 6 | 13.7 | 10.9 | 9.78 | 9.15 | 8.75 | 8.47 | 8.26 | 8.10 | 7.98 | 7.87 | 7.72 | 7.56 | 7.40 | 7.31 | 7.23 | 7.14 | 7.06 | 6.97 | 6.88 |
| 7 | 12.2 | 9.55 | 8.45 | 7.85 | 7.46 | 7.19 | 6.99 | 6.84 | 6.72 | 6.62 | 6.47 | 6.31 | 6.16 | 6.07 | 5.99 | 5.91 | 5.82 | 5.74 | 5.65 |
| 8 | 11.3 | 8.65 | 7.59 | 7.01 | 6.63 | 6.37 | 6.18 | 6.03 | 5.91 | 5.81 | 5.67 | 5.52 | 5.36 | 5.28 | 5.20 | 5.12 | 5.03 | 4.95 | 4.86 |
| 9 | 10.6 | 8.02 | 6.99 | 6.42 | 6.06 | 5.80 | 5.61 | 5.47 | 5.35 | 5.26 | 5.11 | 4.96 | 4.81 | 4.73 | 4.65 | 4.57 | 4.48 | 4.40 | 4.31 |
| 10 | 10.0 | 7.56 | 6.55 | 5.99 | 5.64 | 5.39 | 5.20 | 5.06 | 4.94 | 4.85 | 4.71 | 4.56 | 4.41 | 4.33 | 4.25 | 4.17 | 4.08 | 4.00 | 3.91 |
| 11 | 9.65 | 7.21 | 6.22 | 5.67 | 5.32 | 5.07 | 4.89 | 4.74 | 4.63 | 4.54 | 4.40 | 4.25 | 4.10 | 4.02 | 3.94 | 3.86 | 3.78 | 3.69 | 3.60 |
| 12 | 9.33 | 6.93 | 5.95 | 5.41 | 5.06 | 4.82 | 4.64 | 4.50 | 4.39 | 4.30 | 4.16 | 4.01 | 3.86 | 3.78 | 3.70 | 3.62 | 3.54 | 3.45 | 3.36 |
| 13 | 9.07 | 6.70 | 5.74 | 5.21 | 4.86 | 4.62 | 4.44 | 4.30 | 4.19 | 4.10 | 3.96 | 3.82 | 3.66 | 3.59 | 3.51 | 3.43 | 3.34 | 3.25 | 3.17 |
| 14 | 8.86 | 6.51 | 5.56 | 5.04 | 4.70 | 4.46 | 4.28 | 4.14 | 4.03 | 3.94 | 3.80 | 3.66 | 3.51 | 3.43 | 3.35 | 3.27 | 3.18 | 3.09 | 3.00 |
| 15 | 8.68 | 6.36 | 5.42 | 4.89 | 4.56 | 4.32 | 4.14 | 4.00 | 3.89 | 3.80 | 3.67 | 3.52 | 3.37 | 3.29 | 3.21 | 3.13 | 3.05 | 2.96 | 2.87 |
| 16 | 8.53 | 6.23 | 5.29 | 4.77 | 4.44 | 4.20 | 4.03 | 3.89 | 3.78 | 3.69 | 3.55 | 3.41 | 3.26 | 3.18 | 3.10 | 3.02 | 2.93 | 2.84 | 2.75 |
| 17 | 8.40 | 6.11 | 5.19 | 4.67 | 4.34 | 4.10 | 3.93 | 3.79 | 3.68 | 3.59 | 3.46 | 3.31 | 3.16 | 3.08 | 3.00 | 2.92 | 2.83 | 2.75 | 2.65 |
| 18 | 8.29 | 6.01 | 5.09 | 4.58 | 4.25 | 4.01 | 3.84 | 3.71 | 3.60 | 3.51 | 3.37 | 3.23 | 3.08 | 3.00 | 2.92 | 2.84 | 2.75 | 2.66 | 2.57 |
| 19 | 8.19 | 5.93 | 5.01 | 4.50 | 4.17 | 3.94 | 3.77 | 3.63 | 3.52 | 3.43 | 3.30 | 3.15 | 3.00 | 2.92 | 2.84 | 2.76 | 2.67 | 2.58 | 2.49 |
| 20 | 8.10 | 5.85 | 4.94 | 4.43 | 4.10 | 3.87 | 3.70 | 3.56 | 3.46 | 3.37 | 3.23 | 3.09 | 2.94 | 2.86 | 2.78 | 2.69 | 2.61 | 2.52 | 2.42 |
| 21 | 8.02 | 5.78 | 4.87 | 4.37 | 4.04 | 3.81 | 3.64 | 3.51 | 3.40 | 3.31 | 3.17 | 3.03 | 2.88 | 2.80 | 2.72 | 2.64 | 2.55 | 2.46 | 2.36 |
| 22 | 7.95 | 5.72 | 4.82 | 4.31 | 3.99 | 3.76 | 3.59 | 3.45 | 3.35 | 3.26 | 3.12 | 2.98 | 2.83 | 2.75 | 2.67 | 2.58 | 2.50 | 2.40 | 2.31 |
| 23 | 7.88 | 5.66 | 4.76 | 4.26 | 3.94 | 3.71 | 3.54 | 3.41 | 3.30 | 3.21 | 3.07 | 2.93 | 2.78 | 2.70 | 2.62 | 2.54 | 2.45 | 2.35 | 2.26 |
| 24 | 7.82 | 5.61 | 4.72 | 4.22 | 3.90 | 3.67 | 3.50 | 3.36 | 3.26 | 3.17 | 3.03 | 2.89 | 2.74 | 2.66 | 2.58 | 2.49 | 2.40 | 2.31 | 2.21 |
| 25 | 7.77 | 5.57 | 4.68 | 4.18 | 3.86 | 3.63 | 3.46 | 3.32 | 3.22 | 3.13 | 2.99 | 2.85 | 2.70 | 2.62 | 2.53 | 2.45 | 2.36 | 2.27 | 2.17 |
| 30 | 7.56 | 5.39 | 4.51 | 4.02 | 3.70 | 3.47 | 3.30 | 3.17 | 3.07 | 2.98 | 2.84 | 2.70 | 2.55 | 2.47 | 2.39 | 2.30 | 2.21 | 2.11 | 2.01 |
| 40 | 7.31 | 5.18 | 4.31 | 3.83 | 3.51 | 3.29 | 3.12 | 2.99 | 2.89 | 2.80 | 2.66 | 2.52 | 2.37 | 2.29 | 2.20 | 2.11 | 2.02 | 1.92 | 1.80 |
| 60 | 7.08 | 4.98 | 4.13 | 3.65 | 3.34 | 3.12 | 2.95 | 2.82 | 2.72 | 2.63 | 2.50 | 2.35 | 2.20 | 2.12 | 2.03 | 1.94 | 1.84 | 1.73 | 1.60 |
| 120 | 6.85 | 4.79 | 3.95 | 3.48 | 3.17 | 2.96 | 2.79 | 2.66 | 2.56 | 2.47 | 2.34 | 2.19 | 2.03 | 1.95 | 1.86 | 1.76 | 1.66 | 1.53 | 1.38 |
| ∞ | 6.63 | 4.61 | 3.78 | 3.32 | 3.02 | 2.80 | 2.64 | 2.51 | 2.41 | 2.32 | 2.18 | 2.04 | 1.88 | 1.79 | 1.70 | 1.59 | 1.47 | 1.32 | 1.00 |

SOURCE: From Maxine Merrington and Catherine M. Thompson, "Tables of the Percentage Points of the Inverted *F*-Distribution," *Biometrika*, vol. 33, pp. 73–88, 1943. Reprinted with the permission of the *Biometrika* Trustees.

# APPENDIX 7
# CHI-SQUARE DISTRIBUTION

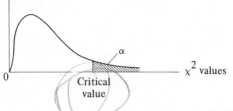

$x^2$ values

0

Critical value

$\alpha$

*Example of how to use this table:* In a chi-square distribution with 6 degrees of freedom (*df*), the area to the right of a critical value of 12.592—i.e., the $\alpha$ area—is .05.

| Degrees of freedom (*df*) | Area in shaded right tail ($\alpha$) | | |
|---|---|---|---|
| | .10 | .05 | .01 |
| 1 | 2.706 | 3.841 | 6.635 |
| 2 | 4.605 | 5.991 | 9.210 |
| 3 | 6.251 | 7.815 | 11.345 |
| 4 | 7.779 | 9.488 | 13.277 |
| 5 | 9.236 | 11.070 | 15.086 |
| 6 | 10.645 | 12.592 | 16.812 |
| 7 | 12.017 | 14.067 | 18.475 |
| 8 | 13.362 | 15.507 | 20.090 |
| 9 | 14.684 | 16.919 | 21.666 |
| 10 | 15.987 | 18.307 | 23.209 |
| 11 | 17.275 | 19.675 | 24.725 |
| 12 | 18.549 | 21.026 | 26.217 |
| 13 | 19.812 | 22.362 | 27.688 |
| 14 | 21.064 | 23.685 | 29.141 |
| 15 | 22.307 | 24.996 | 30.578 |
| 16 | 23.542 | 26.296 | 32.000 |
| 17 | 24.769 | 27.587 | 33.409 |
| 18 | 25.989 | 28.869 | 34.805 |
| 19 | 27.204 | 30.144 | 36.191 |
| 20 | 28.412 | 31.410 | 37.566 |
| 21 | 29.615 | 32.671 | 38.932 |
| 22 | 30.813 | 33.924 | 40.289 |
| 23 | 32.007 | 35.172 | 41.638 |
| 24 | 33.196 | 36.415 | 42.980 |
| 25 | 34.382 | 37.652 | 44.314 |
| 26 | 35.563 | 38.885 | 45.642 |
| 27 | 36.741 | 40.113 | 46.963 |
| 28 | 37.916 | 41.337 | 48.278 |
| 29 | 39.087 | 42.557 | 49.588 |
| 30 | 40.256 | 43.773 | 50.892 |

SOURCE: This table is abridged from Table IV of Fisher and Yates, *Statistical Tables for Biological, Agricultural and Medical Research,* published by Longman Group, Ltd., London (previously published by Oliver & Boyd, Ltd., Edinburgh). Reproduced with the permission of the authors and publishers.

# APPENDIX 8
## CRITICAL VALUES OF *T* FOR $\alpha = .05$ AND $\alpha = .01$ IN THE WILCOXON SIGNED RANK TEST

| | Two-tailed test | | One-tailed test | |
|---|---|---|---|---|
| *n* | .05 | .01 | .05 | .01 |
| 4 | | | | |
| 5 | | | 0 | |
| 6 | 0 | | 2 | |
| 7 | 2 | | 3 | 0 |
| 8 | 3 | 0 | 5 | 1 |
| 9 | 5 | 1 | 8 | 3 |
| 10 | 8 | 3 | 10 | 5 |
| 11 | 10 | 5 | 13 | 7 |
| 12 | 13 | 7 | 17 | 9 |
| 13 | 17 | 9 | 21 | 12 |
| 14 | 21 | 12 | 25 | 15 |
| 15 | 25 | 15 | 30 | 19 |
| 16 | 29 | 19 | 35 | 23 |
| 17 | 34 | 23 | 41 | 27 |
| 18 | 40 | 27 | 47 | 32 |
| 19 | 46 | 32 | 53 | 37 |
| 20 | 52 | 37 | 60 | 43 |
| 21 | 58 | 42 | 67 | 49 |
| 22 | 65 | 48 | 75 | 55 |
| 23 | 73 | 54 | 83 | 62 |
| 24 | 81 | 61 | 91 | 69 |
| 25 | 89 | 68 | 100 | 76 |
| 26 | 98 | 75 | 110 | 84 |
| 27 | 107 | 83 | 119 | 92 |
| 28 | 116 | 91 | 130 | 101 |
| 29 | 126 | 100 | 140 | 110 |
| 30 | 137 | 109 | 151 | 120 |
| 31 | 147 | 118 | 163 | 130 |
| 32 | 159 | 128 | 175 | 140 |
| 33 | 170 | 138 | 187 | 151 |
| 34 | 182 | 148 | 200 | 162 |
| 35 | 195 | 159 | 213 | 173 |
| 40 | 264 | 220 | 286 | 238 |
| 50 | 434 | 373 | 466 | 397 |
| 60 | 648 | 567 | 690 | 600 |
| 70 | 907 | 805 | 960 | 846 |
| 80 | 1211 | 1086 | 1276 | 1136 |
| 90 | 1560 | 1410 | 1638 | 1471 |
| 100 | 1955 | 1779 | 2045 | 1850 |

SOURCE: Abridged from Robert L. McCormack, "Extended Tables of the Wilcoxon Matched Pair Signed Rank Statistic," *Journal of the American Statistical Association*, September 1965, pp. 866–867.

# APPENDIX 9
# DISTRIBUTION OF U IN THE MANN-WHITNEY TEST

## One-tailed test tables

Critical $U$ values: $\alpha = .05$ for a one-tailed test
(and $\alpha = .10$ for a two-tailed test)

| $n_1$ \ $n_2$ | 1 | 2 | 3 | 4 | 5 | 6 | 7 | 8 | 9 | 10 | 11 | 12 | 13 | 14 | 15 | 16 | 17 | 18 | 19 | 20 |
|---|---|---|---|---|---|---|---|---|---|---|---|---|---|---|---|---|---|---|---|---|
| 1 | | | | | | | | | | | | | | | | | | | 0 | 0 |
| 2 | | | | 0 | 0 | 0 | 1 | 1 | 1 | 1 | 2 | 2 | 2 | 2 | 3 | 3 | 3 | 4 | 4 | 4 |
| 3 | | | 0 | 0 | 1 | 2 | 2 | 3 | 3 | 4 | 5 | 5 | 6 | 7 | 7 | 8 | 9 | 9 | 10 | 11 |
| 4 | | | 0 | 1 | 2 | 3 | 4 | 5 | 6 | 7 | 8 | 9 | 10 | 11 | 12 | 14 | 15 | 16 | 17 | 18 |
| 5 | | 0 | 1 | 2 | 4 | 5 | 6 | 8 | 9 | 11 | 12 | 13 | 15 | 16 | 18 | 19 | 20 | 22 | 23 | 25 |
| 6 | | 0 | 2 | 3 | 5 | 7 | 8 | 10 | 12 | 14 | 16 | 17 | 19 | 21 | 23 | 25 | 26 | 28 | 30 | 32 |
| 7 | | 0 | 2 | 4 | 6 | 8 | 11 | 13 | 15 | 17 | 19 | 21 | 24 | 26 | 28 | 30 | 33 | 35 | 37 | 39 |
| 8 | | 1 | 3 | 5 | 8 | 10 | 13 | 15 | 18 | 20 | 23 | 26 | 28 | 31 | 33 | 36 | 39 | 41 | 44 | 47 |
| 9 | | 1 | 3 | 6 | 9 | 12 | 15 | 18 | 21 | 24 | 27 | 30 | 33 | 36 | 39 | 42 | 45 | 48 | 51 | 54 |
| 10 | | 1 | 4 | 7 | 11 | 14 | 17 | 20 | 24 | 27 | 31 | 34 | 37 | 41 | 44 | 48 | 51 | 55 | 58 | 62 |
| 11 | | 1 | 5 | 8 | 12 | 16 | 19 | 23 | 27 | 31 | 34 | 38 | 42 | 46 | 50 | 54 | 57 | 61 | 65 | 69 |
| 12 | | 2 | 5 | 9 | 13 | 17 | 21 | 26 | 30 | 34 | 38 | 42 | 47 | 51 | 55 | 60 | 64 | 68 | 72 | 77 |
| 13 | | 2 | 6 | 10 | 15 | 19 | 24 | 28 | 33 | 37 | 42 | 47 | 51 | 56 | 61 | 65 | 70 | 75 | 80 | 84 |
| 14 | | 2 | 7 | 11 | 16 | 21 | 26 | 31 | 36 | 41 | 46 | 51 | 56 | 61 | 66 | 71 | 77 | 82 | 87 | 92 |
| 15 | | 3 | 7 | 12 | 18 | 23 | 28 | 33 | 39 | 44 | 50 | 55 | 61 | 66 | 72 | 77 | 83 | 88 | 94 | 100 |
| 16 | | 3 | 8 | 14 | 19 | 25 | 30 | 36 | 42 | 48 | 54 | 60 | 65 | 71 | 77 | 83 | 89 | 95 | 101 | 107 |
| 17 | | 3 | 9 | 15 | 20 | 26 | 33 | 39 | 45 | 51 | 57 | 64 | 70 | 77 | 83 | 89 | 96 | 102 | 109 | 115 |
| 18 | | 4 | 9 | 16 | 22 | 28 | 35 | 41 | 48 | 55 | 61 | 68 | 75 | 82 | 88 | 95 | 102 | 109 | 116 | 123 |
| 19 | 0 | 4 | 10 | 17 | 23 | 30 | 37 | 44 | 51 | 58 | 65 | 72 | 80 | 87 | 94 | 101 | 109 | 116 | 123 | 130 |
| 20 | 0 | 4 | 11 | 18 | 25 | 32 | 39 | 47 | 54 | 62 | 69 | 77 | 84 | 92 | 100 | 107 | 115 | 123 | 130 | 138 |

Critical $U$ values: $\alpha = .01$ for a one-tailed test
(and $\alpha = .02$ for a two-tailed test)

| $n_1$ \ $n_2$ | 1 | 2 | 3 | 4 | 5 | 6 | 7 | 8 | 9 | 10 | 11 | 12 | 13 | 14 | 15 | 16 | 17 | 18 | 19 | 20 |
|---|---|---|---|---|---|---|---|---|---|---|---|---|---|---|---|---|---|---|---|---|
| 1 | | | | | | | | | | | | | | | | | | | | |
| 2 | | | | | | | | | | | | | 0 | 0 | 0 | 0 | 0 | 0 | 1 | 1 |
| 3 | | | | | | | 0 | 0 | 1 | 1 | 1 | 2 | 2 | 2 | 3 | 3 | 4 | 4 | 4 | 5 |
| 4 | | | | | 0 | 1 | 1 | 2 | 3 | 3 | 4 | 5 | 5 | 6 | 7 | 7 | 8 | 9 | 9 | 10 |
| 5 | | | | 0 | 1 | 2 | 3 | 4 | 5 | 6 | 7 | 8 | 9 | 10 | 11 | 12 | 13 | 14 | 15 | 16 |
| 6 | | | | 1 | 2 | 3 | 4 | 6 | 7 | 8 | 9 | 11 | 12 | 13 | 15 | 16 | 18 | 19 | 20 | 22 |
| 7 | | | 0 | 1 | 3 | 4 | 6 | 7 | 9 | 11 | 12 | 14 | 16 | 17 | 19 | 21 | 23 | 24 | 26 | 28 |
| 8 | | | 0 | 2 | 4 | 6 | 7 | 9 | 11 | 13 | 15 | 17 | 20 | 22 | 24 | 26 | 28 | 30 | 32 | 34 |
| 9 | | | 1 | 3 | 5 | 7 | 9 | 11 | 14 | 16 | 18 | 21 | 23 | 26 | 28 | 31 | 33 | 36 | 38 | 40 |
| 10 | | | 1 | 3 | 6 | 8 | 11 | 13 | 16 | 19 | 22 | 24 | 27 | 30 | 33 | 36 | 38 | 41 | 44 | 47 |
| 11 | | | 1 | 4 | 7 | 9 | 12 | 15 | 18 | 22 | 25 | 28 | 31 | 34 | 37 | 41 | 44 | 47 | 50 | 53 |
| 12 | | | 2 | 5 | 8 | 11 | 14 | 17 | 21 | 24 | 28 | 31 | 35 | 38 | 42 | 46 | 49 | 53 | 56 | 60 |
| 13 | | 0 | 2 | 5 | 9 | 12 | 16 | 20 | 23 | 27 | 31 | 35 | 39 | 43 | 47 | 51 | 55 | 59 | 63 | 67 |
| 14 | | 0 | 2 | 6 | 10 | 13 | 17 | 22 | 26 | 30 | 34 | 38 | 43 | 47 | 51 | 56 | 60 | 65 | 69 | 73 |
| 15 | | 0 | 3 | 7 | 11 | 15 | 19 | 24 | 28 | 33 | 37 | 42 | 47 | 51 | 56 | 61 | 66 | 70 | 75 | 80 |
| 16 | | 0 | 3 | 7 | 12 | 16 | 21 | 26 | 31 | 36 | 41 | 46 | 51 | 56 | 61 | 66 | 71 | 76 | 82 | 87 |
| 17 | | 0 | 4 | 8 | 13 | 18 | 23 | 28 | 33 | 38 | 44 | 49 | 55 | 60 | 66 | 71 | 77 | 82 | 88 | 93 |
| 18 | | 0 | 4 | 9 | 14 | 19 | 24 | 30 | 36 | 41 | 47 | 53 | 59 | 65 | 70 | 76 | 82 | 88 | 94 | 100 |
| 19 | | 1 | 4 | 9 | 15 | 20 | 26 | 32 | 38 | 44 | 50 | 56 | 63 | 69 | 75 | 82 | 88 | 94 | 101 | 107 |
| 20 | | 1 | 5 | 10 | 16 | 22 | 28 | 34 | 40 | 47 | 53 | 60 | 67 | 73 | 80 | 87 | 93 | 100 | 107 | 114 |

## Two-tailed test tables

Critical $U$ values: $\alpha = .05$ for a two-tailed test
(and $\alpha = .025$ for a one-tailed test)

| $n_1$ \ $n_2$ | 1 | 2 | 3 | 4 | 5 | 6 | 7 | 8 | 9 | 10 | 11 | 12 | 13 | 14 | 15 | 16 | 17 | 18 | 19 | 20 |
|---|---|---|---|---|---|---|---|---|---|---|---|---|---|---|---|---|---|---|---|---|
| 1 |  |  |  |  |  |  |  |  |  |  |  |  |  |  |  |  |  |  |  |  |
| 2 |  |  |  |  |  |  |  | 0 | 0 | 0 | 0 | 1 | 1 | 1 | 1 | 1 | 2 | 2 | 2 | 2 |
| 3 |  |  |  |  | 0 | 1 | 1 | 2 | 2 | 3 | 3 | 4 | 4 | 5 | 5 | 6 | 6 | 7 | 7 | 8 |
| 4 |  |  |  | 0 | 1 | 2 | 3 | 4 | 4 | 5 | 6 | 7 | 8 | 9 | 10 | 11 | 11 | 12 | 13 | 13 |
| 5 |  |  | 0 | 1 | 2 | 3 | 5 | 6 | 7 | 8 | 9 | 11 | 12 | 13 | 14 | 15 | 17 | 18 | 19 | 20 |
| 6 |  |  | 1 | 2 | 3 | 5 | 6 | 8 | 10 | 11 | 13 | 14 | 16 | 17 | 19 | 21 | 22 | 24 | 25 | 27 |
| 7 |  |  | 1 | 3 | 5 | 6 | 8 | 10 | 12 | 14 | 16 | 18 | 20 | 22 | 24 | 26 | 28 | 30 | 32 | 34 |
| 8 |  | 0 | 2 | 4 | 6 | 8 | 10 | 13 | 15 | 17 | 19 | 22 | 24 | 26 | 29 | 31 | 34 | 36 | 38 | 41 |
| 9 |  | 0 | 2 | 4 | 7 | 10 | 12 | 15 | 17 | 20 | 23 | 26 | 28 | 31 | 34 | 37 | 39 | 42 | 45 | 48 |
| 10 |  | 0 | 3 | 5 | 8 | 11 | 14 | 17 | 20 | 23 | 26 | 29 | 33 | 36 | 39 | 42 | 45 | 48 | 52 | 55 |
| 11 |  | 0 | 3 | 6 | 9 | 13 | 16 | 19 | 23 | 26 | 30 | 33 | 37 | 40 | 44 | 47 | 51 | 55 | 58 | 62 |
| 12 |  | 1 | 4 | 7 | 11 | 14 | 18 | 22 | 26 | 29 | 33 | 37 | 41 | 45 | 49 | 53 | 57 | 61 | 65 | 69 |
| 13 |  | 1 | 4 | 8 | 12 | 16 | 20 | 24 | 28 | 33 | 37 | 41 | 45 | 50 | 54 | 59 | 63 | 67 | 72 | 76 |
| 14 |  | 1 | 5 | 9 | 13 | 17 | 22 | 26 | 31 | 36 | 40 | 45 | 50 | 55 | 59 | 64 | 67 | 74 | 78 | 83 |
| 15 |  | 1 | 5 | 10 | 14 | 19 | 24 | 29 | 34 | 39 | 44 | 49 | 54 | 59 | 64 | 70 | 75 | 80 | 85 | 90 |
| 16 |  | 1 | 6 | 11 | 15 | 21 | 26 | 31 | 37 | 42 | 47 | 53 | 59 | 64 | 70 | 75 | 81 | 86 | 92 | 98 |
| 17 |  | 2 | 6 | 11 | 17 | 22 | 28 | 34 | 39 | 45 | 51 | 57 | 63 | 67 | 75 | 81 | 87 | 93 | 99 | 105 |
| 18 |  | 2 | 7 | 12 | 18 | 24 | 30 | 36 | 42 | 48 | 55 | 61 | 67 | 74 | 80 | 86 | 93 | 99 | 106 | 112 |
| 19 |  | 2 | 7 | 13 | 19 | 25 | 32 | 38 | 45 | 52 | 58 | 65 | 72 | 78 | 85 | 92 | 99 | 106 | 113 | 119 |
| 20 |  | 2 | 8 | 13 | 20 | 27 | 34 | 41 | 48 | 55 | 62 | 69 | 76 | 83 | 90 | 98 | 105 | 112 | 119 | 127 |

Critical $U$ values: $\alpha = .01$ for a two-tailed test
(and $\alpha = .005$ for a one-tailed test)

| $n_1$ \ $n_2$ | 1 | 2 | 3 | 4 | 5 | 6 | 7 | 8 | 9 | 10 | 11 | 12 | 13 | 14 | 15 | 16 | 17 | 18 | 19 | 20 |
|---|---|---|---|---|---|---|---|---|---|---|---|---|---|---|---|---|---|---|---|---|
| 1 |  |  |  |  |  |  |  |  |  |  |  |  |  |  |  |  |  |  |  |  |
| 2 |  |  |  |  |  |  |  |  |  |  |  |  |  |  |  |  |  |  | 0 | 0 |
| 3 |  |  |  |  |  |  |  |  | 0 | 0 | 0 | 1 | 1 | 1 | 2 | 2 | 2 | 2 | 3 | 3 |
| 4 |  |  |  |  |  | 0 | 0 | 1 | 1 | 2 | 2 | 3 | 3 | 4 | 5 | 5 | 6 | 6 | 7 | 8 |
| 5 |  |  |  |  | 0 | 1 | 1 | 2 | 3 | 4 | 5 | 6 | 7 | 7 | 8 | 9 | 10 | 11 | 12 | 13 |
| 6 |  |  |  | 0 | 1 | 2 | 3 | 4 | 5 | 6 | 7 | 9 | 10 | 11 | 12 | 13 | 15 | 16 | 17 | 18 |
| 7 |  |  |  | 0 | 1 | 3 | 4 | 6 | 7 | 9 | 10 | 12 | 13 | 15 | 16 | 18 | 19 | 21 | 22 | 24 |
| 8 |  |  |  | 1 | 2 | 4 | 6 | 7 | 9 | 11 | 13 | 15 | 17 | 18 | 20 | 22 | 24 | 26 | 28 | 30 |
| 9 |  |  | 0 | 1 | 3 | 5 | 7 | 9 | 11 | 13 | 16 | 18 | 20 | 22 | 24 | 27 | 29 | 31 | 33 | 36 |
| 10 |  |  | 0 | 2 | 4 | 6 | 9 | 11 | 13 | 16 | 18 | 21 | 24 | 26 | 29 | 31 | 34 | 37 | 39 | 42 |
| 11 |  |  | 0 | 2 | 5 | 7 | 10 | 13 | 16 | 18 | 21 | 24 | 27 | 30 | 33 | 36 | 39 | 42 | 45 | 48 |
| 12 |  |  | 1 | 3 | 6 | 9 | 12 | 15 | 18 | 21 | 24 | 27 | 31 | 34 | 37 | 41 | 44 | 47 | 51 | 54 |
| 13 |  |  | 1 | 3 | 7 | 10 | 13 | 17 | 20 | 24 | 27 | 31 | 34 | 38 | 42 | 45 | 49 | 53 | 56 | 60 |
| 14 |  |  | 1 | 4 | 7 | 11 | 15 | 18 | 22 | 26 | 30 | 34 | 38 | 42 | 46 | 50 | 54 | 58 | 63 | 67 |
| 15 |  |  | 2 | 5 | 8 | 12 | 16 | 20 | 24 | 29 | 33 | 37 | 42 | 46 | 51 | 55 | 60 | 64 | 69 | 73 |
| 16 |  |  | 2 | 5 | 9 | 13 | 18 | 22 | 27 | 31 | 36 | 41 | 45 | 50 | 55 | 60 | 65 | 70 | 74 | 79 |
| 17 |  |  | 2 | 6 | 10 | 15 | 19 | 24 | 29 | 34 | 39 | 44 | 49 | 54 | 60 | 65 | 70 | 75 | 81 | 86 |
| 18 |  |  | 2 | 6 | 11 | 16 | 21 | 26 | 31 | 37 | 42 | 47 | 53 | 58 | 64 | 70 | 75 | 81 | 87 | 92 |
| 19 |  | 0 | 3 | 7 | 12 | 17 | 22 | 28 | 33 | 39 | 45 | 51 | 56 | 63 | 69 | 74 | 81 | 87 | 93 | 99 |
| 20 |  | 0 | 3 | 8 | 13 | 18 | 24 | 30 | 36 | 42 | 48 | 54 | 60 | 67 | 73 | 79 | 86 | 92 | 99 | 105 |

SOURCE: Reprinted with permission from William H. Beyer (ed.), *Handbook of Tables for Probability and Statistics,* 2d ed., 1968, Copyright The Chemical Rubber Co., CRC Press, Inc.

# APPENDIX 10
## CRITICAL VALUES FOR $r$ IN THE RUNS TEST FOR RANDOMNESS

Any sample value of $r$ which is equal to or less than that shown in table ($a$) or which is equal to or greater than that shown in table ($b$) is cause for rejection of $H_0$ at the .05 level of significance.

| $n_1$ \ $n_2$ | 2 | 3 | 4 | 5 | 6 | 7 | 8 | 9 | 10 | 11 | 12 | 13 | 14 | 15 | 16 | 17 | 18 | 19 | 20 |
|---|---|---|---|---|---|---|---|---|---|---|---|---|---|---|---|---|---|---|---|
| 2 | | | | | | | | | | | 2 | 2 | 2 | 2 | 2 | 2 | 2 | 2 | 2 |
| 3 | | | 2 | 2 | 2 | 2 | 2 | 2 | 2 | 2 | 2 | 2 | 2 | 3 | 3 | 3 | 3 | 3 | 3 |
| 4 | | 2 | 2 | 2 | 3 | 3 | 3 | 3 | 3 | 3 | 3 | 3 | 3 | 3 | 4 | 4 | 4 | 4 | 4 |
| 5 | | 2 | 2 | 3 | 3 | 3 | 3 | 3 | 4 | 4 | 4 | 4 | 4 | 4 | 4 | 4 | 5 | 5 | 5 |
| 6 | 2 | 2 | 3 | 3 | 3 | 3 | 4 | 4 | 4 | 4 | 5 | 5 | 5 | 5 | 5 | 5 | 6 | 6 | 6 |
| 7 | 2 | 2 | 3 | 3 | 3 | 4 | 4 | 5 | 5 | 5 | 5 | 5 | 6 | 6 | 6 | 6 | 6 | 6 | 6 |
| 8 | 2 | 3 | 3 | 3 | 4 | 4 | 5 | 5 | 5 | 6 | 6 | 6 | 6 | 6 | 7 | 7 | 7 | 7 | 7 |
| 9 | 2 | 3 | 3 | 4 | 4 | 5 | 5 | 5 | 6 | 6 | 6 | 7 | 7 | 7 | 7 | 8 | 8 | 8 | 8 |
| 10 | 2 | 3 | 3 | 4 | 5 | 5 | 5 | 6 | 6 | 7 | 7 | 7 | 7 | 8 | 8 | 8 | 8 | 9 | 9 |
| 11 | 2 | 3 | 4 | 4 | 5 | 5 | 6 | 6 | 7 | 7 | 7 | 8 | 8 | 8 | 9 | 9 | 9 | 9 | 9 |
| 12 | 2 | 2 | 3 | 4 | 4 | 5 | 6 | 6 | 7 | 7 | 7 | 8 | 8 | 8 | 9 | 9 | 9 | 10 | 10 |
| 13 | 2 | 2 | 3 | 4 | 5 | 5 | 6 | 6 | 7 | 7 | 8 | 8 | 9 | 9 | 9 | 10 | 10 | 10 | 10 |
| 14 | 2 | 2 | 3 | 4 | 5 | 5 | 6 | 7 | 7 | 8 | 8 | 9 | 9 | 9 | 10 | 10 | 10 | 11 | 11 |
| 15 | 2 | 3 | 3 | 4 | 5 | 6 | 6 | 7 | 7 | 8 | 8 | 9 | 9 | 10 | 10 | 11 | 11 | 11 | 12 |
| 16 | 2 | 3 | 4 | 4 | 5 | 6 | 6 | 7 | 8 | 8 | 9 | 9 | 10 | 10 | 11 | 11 | 11 | 12 | 12 |
| 17 | 2 | 3 | 4 | 4 | 5 | 6 | 7 | 7 | 8 | 9 | 9 | 10 | 10 | 11 | 11 | 11 | 12 | 12 | 13 |
| 18 | 2 | 3 | 4 | 5 | 5 | 6 | 7 | 8 | 8 | 9 | 9 | 10 | 10 | 11 | 11 | 12 | 12 | 13 | 13 |
| 19 | 2 | 3 | 4 | 5 | 6 | 6 | 7 | 8 | 8 | 9 | 10 | 10 | 11 | 11 | 12 | 12 | 13 | 13 | 13 |
| 20 | 2 | 3 | 4 | 5 | 6 | 6 | 7 | 8 | 9 | 9 | 10 | 10 | 11 | 12 | 12 | 13 | 13 | 13 | 14 |

($a$)

| $n_1$ \ $n_2$ | 2 | 3 | 4 | 5 | 6 | 7 | 8 | 9 | 10 | 11 | 12 | 13 | 14 | 15 | 16 | 17 | 18 | 19 | 20 |
|---|---|---|---|---|---|---|---|---|---|---|---|---|---|---|---|---|---|---|---|
| 2 | | | | | | | | | | | 6 | 6 | 6 | 6 | 6 | 6 | 6 | 6 | 6 |
| 3 | | | | 8 | 8 | 8 | 8 | 8 | 8 | 8 | 8 | 8 | 8 | 8 | 8 | 8 | 8 | 8 | 8 |
| 4 | | | 9 | 9 | 10 | 10 | 10 | 10 | 10 | 10 | 10 | 10 | 10 | 10 | 10 | 10 | 10 | 10 | 10 |
| 5 | | | 9 | 10 | 10 | 11 | 11 | 12 | 12 | 12 | 12 | 12 | 12 | 12 | 12 | 12 | 12 | 12 | 12 |
| 6 | | 8 | 9 | 10 | 11 | 12 | 12 | 13 | 13 | 13 | 13 | 14 | 14 | 14 | 14 | 14 | 14 | 14 | 14 |
| 7 | | 8 | 10 | 11 | 12 | 13 | 13 | 14 | 14 | 14 | 14 | 15 | 15 | 15 | 16 | 16 | 16 | 16 | 16 |
| 8 | | 8 | 10 | 11 | 12 | 13 | 14 | 14 | 15 | 15 | 16 | 16 | 16 | 16 | 17 | 17 | 17 | 17 | 17 |
| 9 | | 8 | 10 | 12 | 13 | 14 | 14 | 15 | 16 | 16 | 16 | 17 | 17 | 18 | 18 | 18 | 18 | 18 | 18 |
| 10 | | 8 | 10 | 12 | 13 | 14 | 15 | 16 | 16 | 17 | 17 | 18 | 18 | 18 | 19 | 19 | 19 | 20 | 20 |
| 11 | | 8 | 10 | 12 | 13 | 14 | 15 | 16 | 17 | 17 | 18 | 19 | 19 | 19 | 20 | 20 | 20 | 21 | 21 |
| 12 | 6 | 8 | 10 | 12 | 13 | 14 | 16 | 16 | 17 | 18 | 19 | 19 | 20 | 20 | 21 | 21 | 21 | 22 | 22 |
| 13 | 6 | 8 | 10 | 12 | 14 | 15 | 16 | 17 | 18 | 19 | 19 | 20 | 20 | 21 | 21 | 22 | 22 | 23 | 23 |
| 14 | 6 | 8 | 10 | 12 | 14 | 15 | 16 | 17 | 18 | 19 | 20 | 20 | 21 | 22 | 22 | 23 | 23 | 23 | 24 |
| 15 | 6 | 8 | 10 | 12 | 14 | 15 | 16 | 18 | 18 | 19 | 20 | 21 | 22 | 22 | 23 | 23 | 24 | 24 | 25 |
| 16 | 6 | 8 | 10 | 12 | 14 | 16 | 17 | 18 | 19 | 20 | 21 | 21 | 22 | 23 | 23 | 24 | 25 | 25 | 25 |
| 17 | 6 | 8 | 10 | 12 | 14 | 16 | 17 | 18 | 19 | 20 | 21 | 22 | 23 | 23 | 24 | 25 | 25 | 26 | 26 |
| 18 | 6 | 8 | 10 | 12 | 14 | 16 | 17 | 18 | 19 | 20 | 21 | 22 | 23 | 24 | 25 | 25 | 26 | 26 | 27 |
| 19 | 6 | 8 | 10 | 12 | 14 | 16 | 17 | 18 | 20 | 21 | 22 | 23 | 23 | 24 | 25 | 26 | 26 | 27 | 27 |
| 20 | 6 | 8 | 10 | 12 | 14 | 16 | 17 | 18 | 20 | 21 | 22 | 23 | 24 | 25 | 25 | 26 | 27 | 27 | 28 |

($b$)

# INDEX

Secular trend:
misuse of, 31–32
for odd number of years, 301–304
projection of, 306–307
properties of trend line, 300–301
reasons for measuring, 299
straight-line equation for, 299
Sign test, 353–358
procedure: with large samples, 356–358
with small samples, 353–356
use of, 353
Simple random sample, 119–120
Simple regression (*see* Regression analysis)
Simulation, 391–392
Skewness, 49–50, 60–61, 85–86
coefficient of, 85–86
meaning of, 49–50
Spearman rank correlation coefficient, 368–370
computational procedure for, 368–370
use of, 368
Specific seasonals, 311–315
Spurious accuracy, 33–34
Spurious correlation, 329
Standard deviation, 73–80, 83
characteristics of, 83
computation: for grouped data, 76–79
for ungrouped data, 73–76
direct method of computation, 76–78
interpretation of, 79–80
shortcut method of computation, 78–80
Standard error:
of difference: between means, 207
between percentages, 213
of estimate, 335–339
computation of, 336–337
meaning of, 336, 338–339
use of, 336
of mean, 125–130, 224
computation of, 126–127
defined, 126
estimator of, 151–152
sample size and, 128
of percentage, 132–133, 158
computation of, 132–133
estimator of, 158
Standard unit, 174
Statistic, 7*n*., 115–116
Statistical inference, 6–10
defined, 6–7
procedures in, 7–8
Statistical problem-solving methodology, 11–15
Statistics, 6–7, 8–16, 19–35
computer role in, 15–16

Statistics:
defined, 6–7
descriptive procedures in, 6–7
misuse of, 19–36
need for, 8–11
problem-solving methodology in, 11–15
Stratified sampling, 121, 386
Stratton, Sam, 20
Strong, Lydia, 328
Stuart, Alan, 5*n*.
Sukoff, Albert, 34*n*.
Summarizing, 13–14

*t* distribution, 153–157, 178
defined, 153
degrees of freedom and, 155
effects of sample size on, 153–154
use of, 153–157
Tables, 23–24
Tennyson, Alfred Lord, 114
Thompson, Catherine, 416
Time series, 293–298
components of, 295–298
defined, 293
reasons for studying, 294–295
Time series analysis, 11, 31–32, 293–325
approach to, 298
cyclical fluctuations in, 298, 317–319
defined, 293
in forecasting, 319
irregular movements in, 298, 317–319
misuse of, 31–32
problems in use of, 319–320
reasons for, 294–295
seasonal variations in, 297, 307–317
secular trend in, 295–296, 299–307
use of, 7, 11, 293–294
variables in, 295–298
Twain, Mark, 11, 19, 32
Two-sample hypothesis tests, 204–219
Two-tailed tests, 182–185, 190–191, 194–195, 208–211, 213–214
of differences: between means, 208–211
between percentages, 213–214
of means, 182–185, 190–191
when $\sigma$ is known, 182–185
when $\sigma$ is unknown, 190–191
of percentages, 194–195
Type I error, 177*n*.
Type II error, 177*n*.

Uniform distribution, 255–256
Universe (*see* Population)
Urban, W. E., 33–34